OMICS-Based Approaches in Plant Biotechnology

Scrivener Publishing
100 Cummings Center, Suite 541J
Beverly, MA 01915-6106

Publishers at Scrivener
Martin Scrivener (martin@scrivenerpublishing.com)
Phillip Carmical (pcarmical@scrivenerpublishing.com)

OMICS-Based Approaches in Plant Biotechnology

Edited by

Rintu Banerjee, Garlapati Vijay Kumar and S.P. Jeevan Kumar

Scrivener Publishing

WILEY

This edition first published 2019 by John Wiley & Sons, Inc., 111 River Street, Hoboken, NJ 07030, USA and Scrivener Publishing LLC, 100 Cummings Center, Suite 541J, Beverly, MA 01915, USA

For more information about Scrivener publications please visit www.scrivenerpublishing.com.

Wiley Global Headquarters
111 River Street, Hoboken, NJ 07030, USA

For details of our global editorial offices, customer services, and more information about Wiley products visit us at www.wiley.com.

Library of Congress Cataloging-in-Publication Data

ISBN 978-1-119-50993-6

Cover image: The Editors
Cover design by Russell Richardson

Set in size of 11pt and Minion Pro by Manila Typesetting Company, Makati, Philippines

Contents

Part 2: Metabolomics 105

Part 3: Bioinformatics 183

Introduction

Climate change challenges could be tackled with the advent of techniques in plant biotechnology, which is a key component to usher sustainable food production and productivity. Plant biotechnology has been started with the culturing of plant cells in various media, which depend on the totipotency of the plant cell. Further, with the advancement of genetic engineering, introducing foreign genes into cell and tissue has become an important tool to develop genetically modified (GM) transgenic crops with enhanced/improved characteristics and traits. In recent studies, the plant biotechnology domain has been tremendously changed/shifted from GM crops and gene manipulation to "OMICS"-based approaches to decipher the underlying mechanisms for abiotic and biotic stress tolerance. Advances in instrumentation and technologies revealed that the genomics, proteomics, metabolomics, methylome (epigenetic regulation), bioinformatics, and phenomics have great potential for identifying and characterizing novel traits in plants to meet environmental challenges. To understand the underlying tolerant mechanisms for climate change conditions, an attempt has been made to conglomerate all interdisciplinary branches under one umbrella to emphasize the essentiality of inter-allied sciences for tackling the problem.

To meet the nation's food demand, fusion of improved varieties with superior genetics to seed chain at appropriate time intervals is inevitable. The primary objective of developing new varieties and hybrids of various crop species will only be achieved through embracing new strategies/technologies and practical implementation for increased productivity. This book is a reflection of the role played by new OMICS technologies in improving the food and nutritional security. Moreover, OMICS potential to use resources effectively for sustainable production has been illustrated vividly to understand the roles of newer technologies.

Agricultural scientists are striving toward the development of appropriate technologies in the form of improved varieties with higher yielding capacity, a wide range of adaptability and resistance to multiple pests, apt for complex and diverse agro-ecological situations. Continued innovations in the field of plant breeding along with the availability of modern tools and techniques provided dividends, enabling varietal development at a much higher pace. Accordingly, in Chapter 1, legume resources such as *velvet bean*, *winged bean*, *rice bean* and *lablab bean*, which are rich in protein, have been studied using a genomics approach to facilitate toward molecular breeding and gene discovery programs in the near future. Chapter 2 mainly emphasizes the dissection of insecticidal genes and their application for crop improvement. In addition, this chapter throws light over genetically modified crops in controlling pests such as BT technology, and expression of enzymes like chitinase has been explored; it is concluded that transgenic technology coupled with integrated pest management could alleviate the pest problem and enhance the crop productivity. In Chapter 3, miRNA (noncoding small endogenous regulatory RNAs) technologies for crop improvement have been placed for appraisal of latest developments in the domain. miRNAs mediate gene silencing (fully or nearly complementary targets) either through cleavage of target mRNA or translational repression in plants. In this domain, right from first identification of miRNA genes, *let-7* and *lin-4* from *Caenorhabditis elegans*, thousands of miRNAs have been identified in plants, and the current MiRBase entries for plants (viridiplantae) have 10,504 mature sequences, which indicates that these molecules could play a prominent role in crop improvement.

Forward genetics approaches are very appealing than conventional genetics; as a result, few chapters are dedicated to forward genetic approaches and their utility in identifying a gene function, investigating the causal locus/gene, and developing thermotolerant plants (see Chapters 4 and 5). Besides genomics and forward genetics approaches, metabolomics is a new branch, which is emerging and helpful to understand the stress-tolerant mechanisms in the plants. As this field is new, Chapter 6 has been focused on metabolomics methodology using single cell type such as the stomatal guard cells that have been used for analyses of the stress-responsive metabolomes when challenged with stressors. Using bioinformatics and statistical approaches, the dataset of explored guard cell metabolome response to a given treatment (bicarbonate) could decipher the cellular mechanisms pertaining to biological events. In Chapter 7, regular procedures for metabolite profiling and metabolomics analysis in plant systems using proton nuclear magnetic resonance (^{1}H NMR) spectroscopy and gas chromatography–mass spectroscopy (GC-MS) have been elucidated. Further, explanation on general experimental workflow, metabolite data acquisition, metabolite profiling, and statistical analysis for metabolomics using MetaboAnalyst have been dealt with.

OMICS techniques have been dynamic and possess massive potential for crop improvement, which is yet to be reaped particularly in medicinal plants. Application of OMICS in the identification of alkaloids and other secondary metabolites could aid in developing novel insecticides. Chapters 8, 9, and 11 particularly deal with OMICS and poly-OMICS approaches not only on medicinal plants but also on systems biology and biofuel production. Moreover, bioinformatics amalgamating with recent techniques such as CRISPR-Cas9, methylome (epigenetic regulation), and phenomics could be more promising and have been duly accorded in Chapters 10, 12, 13, and 14. It is noteworthy to observe that genome editing (GE), a new breeding technology (NBT), has shown potential to transform not only in fundamental research of plant biology but more importantly also for addressing growing challenges of food security. Hence, due accord has been given to introduce the basic concepts of genome editing with CRISPR-Cas9 in Chapter 12. Regulation of copy number of genes has been maintained by methylation pattern and is considered as epigenetic regulation. Recent reports reveal that nucleic acid methylation in plants like *Arabidopsis*, maize, and rice has shown that H3K9me2-dependent pathway, ribonucleic acid directed nucleic acid methylation pathway, and mobile siRNAs are the key pathways in the regulation of gene copy number, which is explained in Chapter 13. In addition to these domains, phenomics field is emerging, which has great potential to determine the physiological changes occurring in the plants in response to metabolites and physical factors, and also helps in the development of microfluidic devices. To appraise the advancements taking place in the field, Chapter 14 deals about the know-how, interpretation, and applications in plant biology and crop improvement.

This book aims to keep abreast with the advances taking place in OMICS studies that ultimately aid in confronting climatic challenges. Unprecedentedly, this book is diverse, encompassing several chapters with the latest information, emphasizing new aspects. Indeed, this book would be helpful to plant biotechnologists, plant breeders, agricultural biotechnologists, policymakers, and plant physiologists. Students could refer to this book for competitive exams. It is hoped that this book will be an enriching reference material.

Part 1
GENOMICS

1

Exploring Genomics Research in the Context of Some Underutilized Legumes—A Review

Patrush Lepcha, Pittala Ranjith Kumar and N. Sathyanarayana*

Department of Botany, Sikkim University, Gangtok, East Sikkim, India

Abstract
Broadening legume resource base is imperative to meet the ever-increasing demand for protein-rich diet in the developing world. Many legumes species considered to be minor on a global scale have now been investigated and found to possess excellent nutritional value. Some of them are even a storehouse of rare drug molecules. Till date, their large-scale adoption for cultivation has remained unmet owing to poor research investments in these crops. Many of them have skipped genomics revolution and lack targeted genetic improvement programs. Recently, there has been renewed interest in these crops, and progress in genetic and genomics resources development is catching up, fueling greater promise toward molecular breeding and gene discovery programs in the near future. This review focuses on providing nutritional potential and prospects of genomic research in four lesser-known legume species: velvet bean, winged bean, rice bean, and lablab bean, which are grown as minor crops across the Indian subcontinent.

Keywords: Genomics, legumes, genomic resources, transcriptome, nutritional potential, segregant population, genetic map

1.1 Introduction

Trends in human population growth and pattern of consumption imply that the global demand for food will continue to grow for the next 40 years. This, along with depleting land and water resources in addition to climate change,

**Corresponding author*: nsathyanarayana@cus.ac.in

Rintu Banerjee, Garlapati Vijay Kumar, and S.P. Jeevan Kumar (eds.) OMICS-Based Approaches in Plant Biotechnology, (3–18) © 2019 Scrivener Publishing LLC

poses serious threats to food security, particularly in the developing countries [1]. The burgeoning problem may attain serious dimensions in future years as the current yield-increase trends in major food crops may not be adequate in dealing with the growing demand [2, 3]. The grain legumes provide humans with important sources of food, fodder, oil, and fodder products [4]. They are also the vital source of dietary protein, vitamins, minerals, as well as omega-3 fatty acids [5] and can supply rare pharmaceuticals [6]. Even though quite a few proteinaceous edible legumes are available on the market, their production rate vis-à-vis consumption in most instances has remained unachieved and an ever-rising demand has been witnessed [7]. Also, a rising penchant for protein-rich vegetarian-based diet in world population has created unusual scarcity to plant resources [8]. There are several minor food legumes whose potential is untapped and underexploited. Bambara groundnut (*Vigna subterranean* L.), adzuki bean [*Vigna angularis*], faba bean (*Vicia faba* L.), velvet bean (*Mucuna* spp.), grass pea (*Lathyrus sativus* L.), horse gram [*Macrotyloma uniflorum*], hyacinth bean (*Lablab purpureus* L.), moth bean [*Vigna aconitifolia*], rice bean [*Vigna umbellata*], and winged bean [*Psophocarpus tetragonolobus* (L.) DC.] are important members of this grouping [6]. They possess excellent nutritional value and can offer a vital source of protein, vitamins, and minerals in LIFDC (low-income-food-deficit) countries. Since many of them are well adapted to marginal conditions, they may also be a warehouse of important genes associated with biotic and abiotic stress tolerance. However, to varying extents, almost all these crops have suffered from scantily developed resources for genetic and genomic research, thus limiting use of enabling biotechnologies for their improvement. In this review, we have focused on the nutritional potential and the accessibility and deployment of advanced genetic and genomic tools for diversity assessment, trait mapping, and molecular breeding in four underutilized legume species cultivated in and around the Indian subcontinent (Table 1.1). Further, an insight based on newly emerging biological approaches for early deployment of molecular breeding and development of improved cultivars has been provided, though many of these methods are yet to be tested for improving quality, nutritional abundance, and productivity in these legume species.

1.2 Velvet Bean [*Mucuna pruriens* (L.) DC. var. *utilis* (Wall. ex Wight)] Baker ex Burck

Common Names: Velvet bean, Bengal velvet bean, Florida velvet bean.

Description: Self-pollinated tropical species [9] belonging to phaseoloid clade of leguminosae. Chromosome number of 2n = 2x = 22 [10] and genome

Table 1.1 Comparison of genomic resources in four lesser-known legume species.

	Mucuna pruriens	*Psophocarpus tetragonolobus*	*Vigna umbellata*	*Lablab purpureus*
Chromosome number	2n = 2x = 22 [10]	2n = 2x = 18 [27]	2n = 2x = 22 [44, 45]	2n = 2x = 22 or 24 [58]
Genome size	~1,361 Mbp [6]	1.22 Gbp [28]		367 Mb [59]
RAPD	Padmesh *et al.*, [21], Patil *et al.*, [23]	Mohanty *et al.*, [39]	Kaga *et al.*, [49], Shafiqul *et al.*, [50], Muthusamy *et al.*, [52]	Rai *et al.*, [69], Biswas *et al.*, [70]
AFLP	Capo-chichi *et al.*, [19, 20], Sathyanarayana *et al.*, [22]	–	–	Maass *et al.*, [71], Venkatesha *et al.*, [72]
ISSR	Patil *et al.*, [23]	Mohanty *et al.*, [39], Chen *et al.*, [40]	Muthusamy *et al.*, [51]	
SSR	Sathyanarayana *et al.*, [24]		Chen *et al.*, [52], Wang *et al.*, [53], Ingrai *et al.*, [54], Thakur *et al.*, [55]	Wang *et al.*, [73], Yao *et al.*, [74], Shivakumar *et al.*, [75], Guwen *et al.*, [76]
Transcriptome	Sathyanarayana *et al.*, [24]	Vatanparast *et al.*, [28], Wong *et al.*, [41, 43],	Chen *et al.*, [52]	Chapman [42]
Genetic mapping	Capo-chichi *et al.*, [25], Mahesh *et al.*, [26]	–	Kaga *et al.*, [49], Ismura *et al.*, [57]	Kondur *et al.*, [77], Humphry *et al.*, [78], Yuan [79, 80]

size is ~1,361 Mbp [6]. The plant is a climber bearing large white or purple flowers; seeds (3–6/pod) are black/mottled/white in color and highly variable in color, size, and texture (Figure 1.1). Pods are non-itching. Matured seeds, immature pods, and leaves are consumed as food and used as supplement for ruminant livestock feed in several parts of the Asia and Africa [11–13]. Different plant parts are used in traditional Ayurvedic system of medicine for the treatment of diabetes, gout, tuberculosis, and nervous disorders and also as an aphrodisiac [14]. Most importantly, *Mucuna* spp. are a chief source of 3,4-dihydroxy-L-phenylalanine (L-Dopa, 1–9%)—a precursor of the dopamine widely used in the treatment of Parkinson's disease [15].

Nutritional potential: Good source of protein (28%), carbohydrates (33%), lipids (7%), fibers (8%), moisture (8%), ash (6%), and minerals such as sodium, potassium, calcium, magnesium, phosphorus, manganese, iron, zinc, and amino acids [16, 17]. Presence of antinutritional factors such as saponins, phytic acids, phenolic compounds, tannins, hemagglutinins, as well as protease inhibitors such as trypsin inhibitors and chymotrypsin inhibitors are also reported [17, 18].

Genetic and genomic resources: The first-ever marker study on velvet bean was reported by Capo-chichi *et al.*, [19] who studied genetic diversity among 40 US landraces using amplified fragment length polymorphism (AFLP) markers, which revealed narrow genetic base (3–13%). An extended study by the same authors on 64 accessions [20] revealed enhanced genetic diversity (0–0.32%). In India, Padmesh *et al.*, [21] carried out the first diversity study using six accessions of *M. pruriens* comprising both wild (var. *pruriens*) and cultivated (var. *utilis*) varieties from Kerala using 15 randomly amplified polymorphic DNA (RAPD) primers. The results found overall good diversity

Figure 1.1 Variability for seed characters in *M. pruriens* germplasm.

(10–61%) with var. *pruriens* genetically more diverse vis-à-vis var. *utilis*. Later, similar results were reported in other germplasm collections using AFLP and inter simple sequence repeat (ISSR) markers [22, 23]. Recently, *de novo* transcriptome assembly comprising 67,561 assembled transcripts with N50 length of 987 bp and a mean transcript length of 641 bp has been reported [24]. From a total of 7,493 SSR motifs accounted from this work, 134 SSRs have been validated, offering an important resource for genetic studies and ongoing breeding programs. Linkage map based on AFLP markers has been developed, in addition to segregation analysis of pod color and pod pubescence in F2 population [25]. Recently, another genetic map has been reported from Indian *M. pruriens* [26] defining quantitative trait loci (QTL) positions for floral, pod, and seed traits using F2 intraspecific population. Beyond this, there are no reports in the direction of trait-based mapping, QTL studies, or any other works related to genomic resource development in this species.

1.3 *Psophocarpus tetragonolobus* (L.) DC.

Common Names: Winged bean, asparagus bean, asparagus pea, Goa bean

Description: Winged bean is another promising tropical legume with high protein content. It has diploid genome with chromosome number 2n = 2x = 18 [27] and an estimated genome size of 1.22 Gbp [28]. Winged bean is a perennial twining herb (Figure 1.2a, b), but is mostly grown as an

Figure 1.2 (a) Flowering and (b) fruiting in *P. tetragonolobus* (L.) DC.

annual plant. It bears flowers with colors ranging from blue, bluish-white to purple and a pod with 5–21 seeds. Some varieties produce starchy underground tubers [29]. Seeds, young pods, flower, leaf, and tuberous root of winged bean are fit for human consumption due to which it has earned the distinction as "super market on stalk" [30, 31]. Winged bean grows naturally in Indonesia, Malaysia, Thailand, The Philippines, Burma, Sri Lanka, and Bangladesh [32]. It is also introduced to several African and tropical American countries [33].

Nutritional potential: Winged bean is analogous to soybean in nutritional content [34]. High amounts of protein (33%), carbohydrates (22%), moisture (9%), ash (4%), fibers (12%), minerals, vitamins including vitamin A, B1, B2, B3, B6, B9, C, and E, and amino acids are reported [35–37]. Anti-nutritional factors such as free phenolics, phytic acid, tannins, saponins, flatulence factors, and hydrogen cyanide are some of the concerns [38].

Diversity, genetic and genomic resources: Only a few studies related to genetic analyses employing DNA markers have been reported, so far. Mohanty *et al.*, [39] established the superiority of ISSR over RAPD markers as part of genetic diversity analysis of 24 winged bean collections. Genetic diversity among 45 winged bean accessions revealed narrow genetic base [40]. On genomics front, the first transcriptome assembly has been published using 198,554 contigs derived from leaf, root, and reproductive tissues. The work identified 24,598 SSRs of which 84 have been validated [41]. Subsequently, 1,800 conserved orthologous set (COS) loci and 1,900 microsatellite markers have been developed from seedling transcriptome of winged bean genotype *Ibadan Local-1*, which produced 52,083 transcripts with an N50 of 1420 bp [42]. Of late, in an effort that can offer greater impetus to the genomics-assisted programs, Vatanparast *et al.*, [28] sequenced transcriptomes of multiple tissues from two Sri Lankan winged bean genotypes and reported large-scale marker development. This work generated a combined assembly with 97,241 transcripts and identified 12,956 SSRs and 5,190 high-confidence SNPs. Most recently, by transcriptome sequencing of Malaysian accessions, 9,682 genic SSR markers have been developed from an assembly built on 198,554 contigs with an N50 of 1462 bp of which 138,958 (70%) has been annotated [43].

1.4 *Vigna umbellata* (Thunb.) Ohwiet. Ohashi

Common Names: Rice bean or ricebean, mambi bean.

Description: Rice bean is a short-day, warm-season annual vine legume with chromosome number 2n = 2x = 22 [44, 45]. The plant bears bright yellow flowers and produces large numbers of pods [46]. The seed coat color is highly variable including green, yellow, or shades of yellowish-maroon, brown, green, speckled, and mottled [47]. Dry seeds, young pods, and leaves are eaten as vegetables as well as used as fodder, cover crop, green manure, and hedge [48]. Rice bean is cultivated mainly in Nepal, Bhutan, and northeast India reaching up to Myanmar, Southern China, Northern Thailand, Laos, Vietnam, Indonesia, and East Timor [44].

Nutritional potential: Nutritional constituents such as protein (26%), lipids (3%), dietary fibers (4%), carbohydrates (52%), vitamins, minerals, amino acids, and fatty acids are comparable with other legumes of this genus [47]. It is also rich in vitamins as well as methionine and tryptophan. Anti-nutritional factors such as phenolics, tannins, saponin, phytic acid, trypsin inhibitors, and lipoxygenase activity are present [47].

Diversity, genetic and genomic resources: Reports on molecular markers-based diversity studies are scarce and scattered in rice bean. This includes restriction fragment length polymorphism (RFLP), RAPD, and ISSR analyses of local germplasm collections [49–51]. Recently, the first study on transcriptome sequencing reported assembly of 71,929 unigenes with an average length of 986 bp [52]. A total of 3,011 genic SSRs have been identified, which supplements 221 genomic microsatellites [53] and 28 specific SSR primer pairs [54] developed by other groups. These initiatives triggered studies on population genetics, etc. using microsatellite markers [55]. Besides, the first genome map has been developed with a total of 326 loci (103 AFLP loci, 7 common bean SSR loci, 44 cowpea SSR loci, and 172 adzuki bean SSR loci), which are assigned on 11 linkage groups covering a total of 796.1 cM of the rice bean genome at an average marker density of 2.5 cM [56]. Using this map, 31 domestication-related traits have been dissected into 69 QTLs.

1.5 *Lablab purpureus* (L.) Sweet

Common names: Lablab-bean, Egyptian kidney bean, Australian pea, hyacinth bean.

Description: Lablab bean is a climbing or bushy perennial plant with chromosome number $2n = 2x = 22$ or 24 [57] and genome size 367 Mbp, much smaller as compared to other closely related species [58]. It has flowers with colors ranging from white, pink, and violet to purple [59]. The pods are flat or inflated, straight or curved, and usually contain three to six seeds of variable colors and sizes [60]. Seeds are ovoid to oblong and the seed coat color varies from white, cream, brown, and black [61]. There are two crop types of lablab plants that are available such as the vine garden type and the erect, bushy field type. Seeds and pods of lablab bean are the most popular foods in South Asia, especially in India, China, West Africa, Japan, and the Caribbean Islands [62]. Different parts of the plant are used in the treatment of cholera and inflamed ear and throat [63]. It is believed to be native of Southeast Asia (particularly India) or Africa [64].

Nutritional potential: The lablab bean possesses a good amount of protein (28%), carbohydrates (34%), dietary fibers (9%), and ash (4%) [65–67]. Presence of minerals such as P, K, Ca, Fe, etc. as well as amino acids in adequate quantities has also been reported [68]. Anti-nutritional factors such as trypsin inhibitors, tannins, phytates, and hemagglutinins or lectins occur as in other legumes [65, 66].

Diversity, genetic and genomic resources: Genetic diversity and relationships among wild and cultivated lablab accessions have been studied extensively using molecular markers such as RAPD [69, 70], AFLP [71, 72], cross-species SSRs [73–75], as well as EST-SSRs from legume databases [76]. Transcriptome sequencing has been carried out from the seedling tissue, which generated an assembly of 52,019 transcripts with an N50 of 1570 bp. A total number of 2,567 SSRs has been discovered [42]. In addition, linkage map has been constructed using RFLP and RAPD markers on an F2 population, which revealed 17 linkage groups covering 1610 cM with an average distance of 7 cM between markers [77]. Comparative analysis of this map with mung bean genetic map revealed a high level of synteny between the genomic regions of these two legumes [78]. Further, QTL mapping found that traits of agronomic importance such as flowering time, podding time, pod length, pod diameter, pod fresh thickness, and harvest maturity period each has stable QTLs [79, 80].

1.6 Avenues for Future Research

Velvet bean: The medicinal and agronomic potential of *M. pruriens* has remained largely underexploited. Efforts are needed to breed improved varieties not only for high or low L-Dopa content, but also for developing self-supporting determinate cultivars, resistant to biotic and abiotic stresses with enhanced nutritional value. For initiating molecular breeding, characterizing worldwide germplasm and developing functional markers such as EST-SSRs, SNPs, intron spanning regions, etc. are immediate needs. Also, so far, only two genetic linkage maps are available, that too with dominant AFLP markers. More efforts are needed to develop high-density QTL maps based on codominant markers as well as association analysis of target traits. In addition, diminishing genotyping and sequencing costs are fueling new hopes of whole genome sequencing as well as sequence-based trait mapping such as genotyping by sequencing (GBS) and/or genome-wide association studies (GWAS). These efforts in combination with other "omics" techniques such as proteomics and metabolomics can foster discovery of candidate genes and pathways involved in important biochemical and agronomic traits.

Winged bean: In winged bean, varietal development focusing on traits such as early maturing, erect, dwarf, and non-shattering pod with reduced anti-nutritional factors is the major breeding objective [28]. With the recent efforts in genomics and transcriptomics, the genetic resource development is grabbing pace. As genome sequencing is becoming increasingly affordable, efforts such as deep sequencing of winged bean genome should be put forward to enable large-scale analyses of gene content, evolution of repetitive elements, linkage and association mapping. Prior to that, it is important to answer key questions related to trait-specific germplasm characterization, detection of genes, and molecular markers responsible for the key agronomic and chemical traits and contribute to the development of genetic as well as physical maps.

Rice bean: Rice bean has rich genetic diversity with promising agronomic potential in terms of its adaptability to grow well in less fertile soils of hot and humid climates and resistance to storage pests and many diseases [81]. Despite many useful traits, it has been subjected to little systematic breeding and thus the potential of this highest grain yielder among Ceratotropis species has remained underexploited. Recent works on development of EST resources and marker development have enabled genomic resources in rice bean. Future works thus should focus on wider germplasm characterization, discovery of functional markers from sequencing efforts, and development of high-resolution genetic maps, all of which will aid trait mapping of agronomically and nutritionally important traits. Newer opportunities arising from the whole genome sequencing of cowpea and availability of several transcriptome data should be used to advance candidate genes discovery and functional genomics studies in this crop.

Lablab bean: Lablab bean offers a multitude of uses as vegetable for human consumption, fodder for livestock, and fixing biological nitrogen. A drought-tolerant feature of lablab is typically greater than common bean and cowpea, and there is a great scope for cultivation in water-deficient semi-arid regions. Thus, speedy effort towards large-scale marker development, construction of saturated genetic linkage maps, physical maps, as well as association mapping will be needed to move forward in the direction of marker-assisted breeding for improvement of these key agronomic traits.

In addition, information generation through transcriptomics, proteomics, and metabolomics should be duly advanced to facilitate better understanding of the traits and crop to devise an effective strategy for achieving higher gains in a rapid and cost-effective manner.

1.7 Conclusions

As can be seen from the foregoing discussion, the four legume species discussed in this review, i.e., velvet bean, winged bean, rice bean, and lablab bean as well as several other lesser-known legume species, possess excellent nutritional value and offer unique opportunities for cultivation in the underdeveloped world where farming of popular legume species is not only difficult but also expensive. It is vital therefore, to raise investments in genetic improvement of these crops through molecular breeding programs. Reducing costs of next-generation sequencing and high-throughput genotyping platforms are rendering genomic tools affordable even for lesser-known species. In this context, the future research must focus on utilizing these resources toward (a) characterizing worldwide germplasm of these species to reflect biodiversity and potential genetic solutions hitherto unexplored, (b) accelerate trait discovery through linkage and association mapping, (c) organize physical maps and functional genomics tools to facilitate candidate genes discovery, and (d) target genes that underlie key agronomic traits rapidly and precisely. Such efforts will enable an era of genomics-enabled crop improvement and creation of market place for these crops in near future.

Acknowledgments

The authors thank Sikkim University, Gangtok, for the necessary resources toward completion of this manuscript.

References

1. Varshney, R.K., Glaszmann, J.C., Leung, H., Ribaut, J.M., More genomic resources for less-studied crops. *Trends Biotechnol.*, 28, 452, 2010.
2. Ray, D.K., Mueller, N.D., West, P.C., Foley, J.A., Yield trends are insufficient to double global crop production by 2050. *PLoS One*, 8, 6, e66428, 2013.

3. Fedoroff, N.V., Food in a future of 10 billion. *Agric. FoodSecur.*, 2015, https://doi.org/10.1186/s40066-015-0031-7.
4. Varshney, R.K., Nayak, S.N., May, G.D., Jackson, S.A., Next-generation sequencing technologies and their implications for crop genetics and breeding. *Trends Biotechnol.*, 27, 9, 522, 2009.
5. Pandey, M.K., Roorkiwal, M., Singh, V.K., Ramalingam, A., Kudapa, H., Thudi, M., Chitikineni, A., Rathore, A., Varshney, R.K., Emerging genomic tools for legume breeding: Current status and future prospects. *Front. Plant Sci.*, 7, 455, 2016.
6. Sathyanarayana, N., Mahesh, S., Leelambika, M., Jaheer, M., Chopra, R., Rashmi, K.V., Role of genetic resources and molecular markers in *Mucunapruriens* (L.) DC improvement. *Plant Genet. Resour.*, 14, 4, 70, 2016.
7. Ali, M. and Kumar, S., Problems and prospects of pulses research in India. *Indian Farm*, 50, 4, 2000.
8. Bhat, R. and Karim, A.A., Exploring the nutritional potential of wild and underutilized legumes. *Compr. Rev. Food Sci. Food Saft.*, 8, 305, 2009.
9. Duke, J.A., *Handbook of legumes of world economic importance*, pp. 199–265, Plenum Press, New York, 1981.
10. Sastrapradja, S., Sastrapradja, D., Aminah, S.H., Lubis, I., Idris, S., Morphological and cytological investigation on some species of *Mucuna* (Papilionaceae). *Ann. Bogor.*, 5, 173, 1974.
11. Ravindran, V. and Ravindran, G., Nutritional and anti-nutritional characteristics of *Mucuna* (*Mucunautilis*) bean seeds. *J. Sci. Food Agric.*, 46, 71, 1988.
12. Osei-Bonsu, P., Buckles, D., Soza, F.R., Asibuo, J.Y., Edible cover crops. *ILEIA Newsletter*, 12, 30, 1996.
13. Muinga, R.W., Saha, H.M., Mureithi, J.G., The effect of *Mucuna* (*Mucunapruriens*) forage on the performance of lactating cows. *Trop. Subtrop. Agroecosyst.*, 1, 87, 2003.
14. Sathiyanarayanan, L. and Arulmozhi, S., *Mucunapruriens*—A comprehensive review. *Pharmacognosy Rev.*, 1, 157, 2007.
15. Haq, N., New food legume crops for the tropics, in: *Better Crops for Food*, J. Nugent and M.O. Connor (Eds.), pp. 144–160, Pitman Books, London (Cuba Foundation symposium, 97), 1983.
16. Gurumoorthi, P., Pugalenthi, M., Janardhanan, K., Nutritional potential of five accessions of a south Indian tribal pulse *Mucuna pruriens* var. *utilis*; II Investigation on total free phenolics, tannins, trypsin and chymotrypsin inhibitors, phytohaemagglutinins, and *in vitro* protein digestibility. *Trop. Subtrop. Agroecosyst.*, 1, 153, 2003.
17. Kalidass, C. and Mohan, V.R., Nutritional and antinutritional composition of itching bean [*Mucuna pruriens* (L.) DC. var. *pruriens*]: An underutilized tribal pulse in Western Ghats, Tamil Nadu. *Trop. Subtrop. Agroecosyst.*, 14, 279, 2011.

18. Kalidass, C. and Mahapatra, A.K., Evaluation of the proximate and phytochemical compositions of an underexploited legume *Mucunapruriens* var. *utilis* (Wall ex Wight) Baker ex Burck. *Int. Food Res. J.*, 21, 1, 303, 2014.
19. Capo-chichi, L.J.A., Weaver, D.B., Morton, C.M., AFLP assessment of genetic variability among velvet bean (*Mucuna* sp.) accessions. *Theor. Appl. Genet.*, 103, 1180, 2001.
20. Capo-chichi, L.J.A., Weaver, D.B., Morton, C.M., The use of molecular markers to study genetic diversity in *Mucuna. Trop. Subtrop. Agroecosyst.*, 1, 309, 2003.
21. Padmesh, P., Reji, J.V., JinishDhar, M., Seeni, D., Estimation of genetic diversity in varieties of *M. pruriens* using RAPD. *Biologia Plantarum*, 50, 367, 2006.
22. Sathyanarayana, N., Leelambika, M., Mahesh, S., Jaheer, M., AFLP assessment of genetic diversity among Indian *Mucuna* accessions. *Physiol. Mol. Biol. Plants*, 17, 2, 171, 2011.
23. Patil, R.R., Pawar, K.D., Rane, M.R., Yadav, S.R., Bapat, V.A., Jadhav, J.P., Assessment of genetic diversity in *Mucuna* species of India using randomly amplified polymorphic DNA and inter simple sequence repeat markers. *Physiol. Mol. Biol. Plants*, 22, 207, 2016.
24. Sathyanarayana, N., Pittala, R.K., Tripathi, P.K., Chopra, R., Singh, H.K., Belamkar, V., Bhardwaj, P.K., Doyle, J.J., Egan, A.N., Transcriptomic resources for the medicinal legume *Mucunapruriens*: *De novo* transcriptome assembly, annotation, identification and validation of EST-SSR markers. *BMC Genomics*, 18, 409, 2017.
25. Capo-chichi, L.J.A., Morton, C.M., Weaver, D.B., An intraspecific genetic map of velvet bean (*Mucuna* sp.) based on AFLP markers. *Theor. Appl. Genet.*, 108, 814, 2004.
26. Mahesh, S., Leelambika, M., Jaheer, M., Anitha kumari, A., Sathyanarayana, N., Genetic mapping and QTL analysis of agronomic traits in Indian *Mucuna pruriens* using an intraspecific F2 population. *J. Genet.*, 95, 35, 2016.
27. Harder, D. and Smartt, J., Further evidence on the origin of cultivated winged bean, *Psophocarpus tetragonolobus* (L.) DC. (Fabaceae): Chromosome numbers and the presence of host specific fungus. *Econ. Bot.*, 46, 2, 187, 1992.
28. Vatanparast, M., Shetty, P., Chopra, R., Doyle, J.J., Sathyanarayana, N., Egan, A.N., Transcriptome sequencing and marker development in winged bean (*Psophocarpus tetragonolobus;* Leguminosae). *Sci. Rep.*, 6, 29070, 2016.
29. National Academy of Sciences, *The winged bean: A high protein crop for the tropics.* Washington, USA, 1975.
30. Garcia, V.V. and Palmer, J.N., Proximate analysis of five varieties of winged beans, *Psophocarpus tetragonolobus* (L.) DC. *J. Food Technol.*, 15, 469, 1980.
31. Sri Kantha, S. and Erdman, J.W., The winged bean as an oil and protein source: A review. *J. Am. Oil Chem. Soc.*, 61, 3, 515, 1984.
32. Hymowitz, T. and Boyd, J., Origin, ethnobotany and agricultural potential of the winged bean—*Psophocarpus tetragonolobus. Econ. Bot.*, 31, 2, 180, 1977.

33. Khan, T.N., Bohn, J.C., Stevenson, R.A., Winged bean: Cultivation in Papua New Guinea. *World Crops*, 29, 5, 208, 1977.
34. Lepcha, P., Egan, A.N., Doyle, J.J., Sathyanarayana, N., A review on current status and future prospects of winged bean (*Psophocarpus tetragonolobus*) in tropical agriculture. *Plant Foods Hum. Nutr.*, 72, 225, 2017.
35. Jaffe, W.G. and Korte, R., Nutritional characteristics of the winged bean in rats. *Nutr. Rep. Int.*, 4, 4, 449, 1976.
36. Amoo, I.A., Adebayo, O.T., Oyeleye, A.O., Chemical evaluation of winged beans (*Psophocarpus tetragonolobus*), Pitanga cherries (*Eugina uniflora*) and orchid fruit (orchid fruit myristica). *Afr. J. Food Agric. Nutr. Dev.*, 6, 1, 2006.
37. Mohanty, C.S., Verma, S., Singh, V., Khan, S., Gaur, P., Gupta, P.M., Nizar, A., Dikshit, N., Pattanayak, R., Shukla, A., Niranjan, A., Sahu, N., Behera, S.K., Rana, T.S., Characterization of winged bean (*Psophocarpustetragonolobus* (L.) DC.) based on molecular, chemical and physiological parameters. *AJMB.*, 3, 187, 2013.
38. Kortt, A.A., Purification and properties of the basic lectins from winged bean seed (*Psophocarpus tetragonolobus* (L.) DC.). *Eur. J. Biochem.*, 138, 519, 1984.
39. Mohanty, C.S., Verma, S., Singh, V., Characterization of winged bean (*Psophocarpus tetragonolobus* (L.) DC.) based on molecular, chemical and physiological parameters. *Am. J. Mol. Biol.*, 3, 187, 2013.
40. Chen, D., Yi, X., Yang, H., Zhou, H., Yu, Y., Tian, Y., Lu, X., Genetic diversity evaluation of winged bean (*Psophocarpus tetragonolobus* (L.) DC.) using inter-simple sequence repeat (ISSR). *Genet. Resour. Crop. Evol.*, 62, 6, 823, 2015.
41. Wong, Q.N., Massawe, F., Mayes, K., Blythe, M., Mayes, S., Molecular genetic tools to support genetic improvement of winged bean (*Psophocarpus tetragonolobus*) for food and nutrition security. *ISHS Acta Horticulturae 1110*, 2014, https://doi.org/10.17660/ActaHortic.2016.1110.1.
42. Chapman, M.A., Transcriptome sequencing and marker development for four underutilized legumes. *Appl. Plant Sci.*, 3, 2, 1400111, 2015.
43. Wong, Q.N., Tanzi, A.S., Ho, W.K., Malla, S., Blythe, M., Karunaratne, A., Massawe, F., Mayes, S., Development of gene-based SSR markers in winged bean (*Psophocarpustetragonolobus* (L.) DC.) for diversity assessment. *Genes*, 8, 3, 100, 2017.
44. Tian, J., Isemura, T., Kaga, A., Vaughan, D.A., Tomooka, N., Genetic diversity of the rice bean (*Vigna umbellata*) genepool as assessed by SSR markers. *Genome*, 56, 12, 717, 2013.
45. Froni-Martinus, E.R., A new chromosome number in the genus Vigna (Leguminosae-Papilionoidae). *Bull. deJardin Bot. Nationale de Belgique*, 56, 129, 1986.
46. Katoch, R., Morpho-physiological and nutritional characterization of rice bean (*Vignaumbellata*). *Acta Agronomica Hungarica*, 59, 2, 125, 2011.
47. Katoch, R., Nutritional potential of rice bean (*Vigna umbellata*): An underutilized legume. *J. Food Sci.*, 78, C8, 2013.

48. Facciola, S., *Cornucopia II: A source book of edible plants*, Kampong Publications, 1998.
49. Kaga, A., Ishii, T., Tsukimoto, K., Tokoro, E., Kamijima, O., Comparative molecular mapping in Ceratotropis species using an interspecific cross between azuki bean (*Vigna angularis*) and rice bean (*V. umbellata*). *Theor. Appl. Genet.*, 100, 2, 207, 2000.
50. Shafiqul, I., Dutta, M., Shachi, S., Kumar, P., Bhatt, K.V., Assessment of genetic diversity of rice bean [(*Vigna umbellata*) (Thunb.) Ohwiet & Ohashi)] varieties and their narrow leaf cross derivatives using RAPD markers. *IJAEB.*, 10, 4, 415, 2017.
51. Muthusamy, S., Kanagarajan, S., Ponnusamy, S., Efficiency of RAPD and ISSR markers system in accessing genetic variation of rice bean (*Vigna umbellata*) landraces. *Electron. J. Biotechn.*, 2008, http://www.ejbiotechnology.cl/ontent/vol11/issue3/full/8/index.html.
52. Chen, H., Chen, X., Tian, J., Yang, Y., Liu, Z., Hao, X., Wang, L., Wang, S., Liang, J., Zhang, L., Yin, F., Cheng, X., Development of gene-based SSR markers in rice bean (*Vigna umbellata* L.) based on transcriptome data. *PLoS One*, 2016, https://doi.org/10.1371/journal.pone.0151040.
53. Wang, L., Kim, K.D., Gao, D., Chen, H., Wang, S., Lee, S., Jackson, S.A., Cheng, X., Analysis of simple sequence repeats in rice bean (*Vigna umbellata*) using an SSR-enriched library. *Crop J.*, 4, 40, 2016.
54. Iangrai, B., Pattanayak, A., Evanoreen, D., Khongwir, A., Pale, G., Gatphoh, E.M., Das, A., Chrungoo, N.K., Development and characterization of a new set of genomic microsatellite markers in rice bean (*Vigna umbellata* (Thunb.) Ohwi and Ohashi) and their utilization in genetic diversity analysis of collections from North East India. *PLoS One*, 12, 7, e0179801, 2017.
55. Thakur, S., Bhardwaj, N., Chahota, R.K., Evaluation of genetic diversity in rice bean [*Vigna umbellata* (Thunb.) Ohwi and Ohashi] germplasm using SSR markers. *Electron. J. Plant Breed.*, 8, 2, 674, 2017.
56. Isemura, T., Kaga, A., Tomooka, N., Shimizu, T., Vaughan, D.A., The genetics of domestication of rice bean, *Vigna umbellata*. *Ann. Bot.*, 106, 6, 927, 2010.
57. Verdcourt, B., Lablab Adans, in: *Studies in the Leguminosae Papilionoideae for the 'Flora of Tropical East Africa': III*, vol. 24, p. 409, Kew Bull, 1970.
58. Iwata, A., Greenland, C.M., Jackson, S.A., Cytogenetics of legumes in the phaseoloid clade. *Plant Genome*, 6, 3, 2013.
59. NAS/NRC, *Report by an ad hoc advisory panel of the Advisory Committee on Technology Innovation, Board on Science and Technology for International Development, Commission on International Relations on Tropical legumes: Resources for the future*, pp. 59–67, National Academy of Sciences and the National Research Council, Washington DC, 1979.
60. Shivashankar, G. and Kulkarni, R.S., *Lablab purpureus* (L.) Sweet, in: *Plant Resources of South-East Asia No 1. Pulses*, L.J.G. van der Maesen and S. Somaatmadja (Eds.), pp. 48–50, Pudoc Scientific Publishers, Wageningen, the Netherlands, 1989.

61. Pengelly, B.C. and Maass, B.L., *Lablab purpureus* (L.) Sweet—Diversity, potential use and determination of a core collection of this multi-purpose tropical legume. *Genet. Resour. Crop. Evol.*, 48, 3, 261, 2001.
62. Valenzuela, H., Smith, J., *Lablab*, pp. 1–3, Sustainable agriculture green manure crops, SA-GM-7, 2002.
63. Morris, J.B., Wang, M.L., Anthocyanin content in seeds, leaves and flowers of *Lablab purpureus*. In ASA Meeting Abstracts [CD-ROM], 2008.
64. Shivashankar, G., Kulkarni, R.S., Shashidhar, H.E., Mahishi, D.M., Improvement of Field Bean, in: *Advances in Horticulture*, K.L. Chandha and G. Kallo (Eds.), pp. 277–286, Vegetable crops, Malhotra Publishing House, New Delhi, 1993.
65. Deka, R.K. and Sarkar, C.R., Nutrient composition and ant nutritional factors of *Dolichos lablab* L. seeds. *Food Chem.*, 38, 4, 239, 1990.
66. Shastry, M. and John, E., Biochemical changes and *in-vitro* protein digestibility of the endosperm of germinating *Dolichos lablab. J. Sci. Food Agric.*, 55, 4, 529, 1991.
67. Naeem, M., Masroor, M., Khan, A., Uddin, M., Hyacinth bean (*Lablab purpureus* (L.) Sweet): A hidden treasure of useful phytochemicals. *J. Int. Legum. Soc.*, 13, 30, 2016.
68. Habib, H.M., Theuri, S.W., Kheadr, E.E., Mohamed, F.E., Functional, bioactive, biochemical, and physicochemical properties of the *Dolichos lablab* bean. *Food Funct.*, 8, 2, 872, 2017.
69. Rai, N., Kumar, A., Singh, P.K., Singh, M., Datta, D., Rai, M., Genetic relationship among Indian bean (*Lablab purpureus*) genotypes cultivars from different races based on quantitative traits and random amplified polymorphic DNA marker. *Afr. J. Biotechnol.*, 9, 137, 2010.
70. Biswas, S.M., Mohammad, Z., Mizanur, R.M., Assessments of genetic diversity in country bean (*Lablab purpureus* L.) using RAPD marker against photo-insensitivity. *J. Plant Develop.*, 19, 65, 2012.
71. Maass, B.L., Jamnadass, R.H., Hanson, J., Pengelly, B.C., Determining sources of diversity in cultivated and wild *Lablab purpureus* related to provenance of germplasm by using amplified fragment length polymorphism. *Genet. Resour. Crop Evol.*, 52, 6, 683, 2005.
72. Venkatesha, S.C., Ramanjini Gowda, P.H., Ganapathy, K.N., Gowda, M.B., Ramachandra, R., Giris, G., Channamallikarjuna, V., Shantala, L., Gowda, T.K.S., Genetic fingerprinting in Dolichos bean using AFLP Markers and morphological traits. *Int. J. Biotech. Biochem.*, 6, 3, 395, 2010.
73. Wang, M.L., Gillaspie, A.G., Newman, M.L., Dean, R.E., Pittman, R.N., Morris, J.B., Pederson, G.A., Transfer of simple sequence repeat (SSR) markers across the legume family for germplasm characterization and evaluation. *Plant Genet. Resour.*, 2, 2, 107, 2004.

74. Yao, L.M., Zhang, L.D., Hu, Y.L., Wang, B., Wu, T.L., Characterization of novel soybean derived simple sequence repeat markers and their transferability in hyacinth bean [*Lablab purpureus* (L.) Sweet]. *Indian J. Genet.*, 72, 46, 2012.
75. Shivakumar, M.S., Ramesh, S., Mohan Rao, A., Uday kumar, H.R., Keerthi, C.M., Cross legume species/genera transferability of SSR markers and their utility in assessing polymorphism among advanced breeding lines in Dolichos bean (*Lablab purpureus* L.). *Int. J. Curr. Microbiol. App. Sci.*, 6, 8, 656, 2017.
76. Guwen, Z., Shengchun, X., Weihua, M., Yaming, G., Qizan, H., Development of EST-SSR markers to study genetic diversity in hyacinth bean (*Lablab purpureus* L.). *POJ*, 6, 4, 295, 2013.
77. Konduri, V., Godwin, I.D., Liu, C., Genetic mapping of the *Lablab purpureus* genome suggests the presence of 'cuckoo' gene(s) in this species. *Theor. Appl. Genet.*, 100, 6, 866, 2000.
78. Humphry, M., Konduri, V., Lambrides, C., Magner, T., McIntyre, C., Aitken, E., Liu, C., Development of a mung bean (*Vigna radiata*) RFLP linkage map and its comparison with lablab (*Lablab purpureus*) reveals a high level of collinearity between the two genomes. *Theor. Appl. Genet.*, 105, 1, 160, 2002.
79. Yuan, J., Yang, R., Wu, T., Bayesian mapping QTL for fruit and growth phenological traits in *Lablab purpureus* (L.) Sweet. *Afr. J. Biotechnol.*, 8, 2, 167, 2009.
80. Yuan, J., Wang, B., Wu, T.L., Quantitative trait loci (QTL) mapping for inflorescence length traits in *Lablab purpureus* (L.) sweet. *Afr. J. Biotechnol.*, 10, 18, 3558, 2011.
81. Chandel, K.P.S., Joshi, B.S., Arora, R.K., Pant, K.C., Rice bean-A new pulse with higher potential. *Indian Farming*, 28, 9, 19, 1978.

2

Overview of Insecticidal Genes Used in Crop Improvement Program

Neeraj Kumar Dubey[1*], Prashant Kumar Singh[1], Satyendra Kumar Yadav[2] and Kunwar Deelip Singh[2]

[1]*Botany Department, Rashtriya Snatkottar, Mahavidyalaya, Jamuhai, Jaunpur, UP, India*
[2]*Zoology Department, Rashtriya Snatkottar, Mahavidyalaya, Jamuhai, Jaunpur, UP, India*

Abstract
Crop productivity is largely influenced by different factors divided into broad category as biotic and abiotic factors. Insects and pest compose bigger constituent of biotic factors influencing the crop productivity. Phytophagus insects destroy the crop productivity by eating the plant body through masticating or by sucking mode. These insects damage directly the crop by physical interaction with plants or indirectly by transmitting several viral diseases. Many approaches have been used to manage the loss of insect-mediated crop productivity, but recently, transgenic-mediated genetically modified strategy is one of the attractive practices in the world. Time-saving beauty of incorporating new traits via transgenic techniques is proving to be one of the best techniques among other techniques like breeding. This chapter will summarize the current knowledge on the biology and utilization of toxic genes or gene components to cope with attacking insects in the improved crop productivity.

Keywords: Transgenic plants, herbivorous insects, resistant plants, crop productivity

2.1 Introduction

Crop-derived foods are one of the important sources of eating and nutritional material in developing and underdeveloped countries, since people

**Corresponding author*: neerajd19@gmail.com

Rintu Banerjee, Garlapati Vijay Kumar, and S.P. Jeevan Kumar (eds.) OMICS-Based Approaches in Plant Biotechnology, (19–54) © 2019 Scrivener Publishing LLC

have low income to buy animal-derived nutritious food. Price hike of crop-derived raw and manufactured food is due to increasing demand caused by increasing population. Another reason for the price hike of crop-derived foods is the increasing input cost of agricultural practices. These include seed purchasing, field maintenance, irrigation, fertigation, and spraying insecticides/pesticides. Among these agricultural practices, the use of pesticide is one that requires big financial budget after seed purchasing. The purchase of seeds and pesticide affects poor farmers in developing countries on one hand; on the other hand, spraying of pesticide leads to food safety, environmental damages, and development of resistant biotypes [1, 2]. Overuse of insecticide leads to intensification of agriculture practices, but it also has negative impact especially with regards to environmental and ecological biodiversity and development of secondary pests [3]. Biological strategies to control insects attract attention, but rearing of parasitoids, their controlled release, and the huge increase in the insect population compared to parasitoids are challenging tasks with these methods [4]. Insects affect cultivation process by chewing or sap-sucking methods. They generally attack food and textile fiber crops during flowering or seed setting [1]. Loss of mechanical tissue, damage of root system, and nutritional deprivation in plants lead to suffering growth of desired crops, which ultimately leads to low crop production. Sometimes these insects transmit other insect-mediated disease, which contributes to additional yield loss [1, 5, 6]. Different types of herbivorous insects showed different types of eating patterns when attacking host plants. Some insects are chewing (e.g., *Helicoverpa armigera*) and some are sucking (e.g., aphids, whitefly), with some monophagous (e.g., Jowar shoot fly), oligophagous (e.g., groundnut leaf miner), and polyphagous in nature (e.g., *Helicoverpa armigera*). Less than 10% of crop productivity is reduced by different insect pests [7]. Genetically modified crops are used as one of the best tools for insect pest control [8]. The transgenic crop-mediated agriculture led to reduction in dependency on insecticide and pesticides.

Commercially released transgenic crops change the agriculture productivity due to low input to high output ratio. About 28 countries growing more than 179.7 million hectares of genetically modified crops have been reported since 2015 [9, 10]. Most GM crops contain genes of insecticidal activity and that are herbicide resistant, improving nutrient quality and quantity [11]. Several genes producing Bt (*Bacillus thuringiensis*) toxin, enzyme, enzyme inhibitors, translational inhibitors, carbohydrate binding lectins, etc. have been used to generate insect-resistant transgenic crops.

Bacillus thuringiensis derived delta-endotoxin shows toxicity to many insect species and thus is used to make transgenic plant for insect resistance [12].

However, a narrow range of d-endotoxin for some pest requires an urgent need of several other insecticidal genes [12]. Different strategies such as pyramiding of DvSnf7 dsRNA and *Bacillus thuringiensis* (Bt) Cry3Bb1 to control transgenic maize against western corn rootworm, *Diabrotica virgifera* (LeConte) (WCR) [13, 14], have been used. Similarly, pest movement between two different plants of the same crop having single gene (Cry1Ab) and stacked trait (Cry1A.105 and Cry2Ab2) were studied to revise the disease-resistant development [8]. Different combinations of a novel cry2AX1 gene consisting of a sequence of cry2Aa and cry2Ac genes have also been used to develop transgenic plants [15] against insects. Similarly, more than 10,000 arthropod species are currently targeted by different insecticides derived from viruses, fusion proteins, peptides, spider-venom peptides, etc. through transgenic plant preparation [16]. Further, spider insecticidal gene [17], cry1C gene [18], lectins [19], nematode resistance gene [20], neurotoxin AaIT [21], cathepsin L-like cysteine protease [22], AOS-like gene of soybean [23], and cowpea trypsin inhibitor gene (CpTI) [24] are also few examples used for insect-resistant transgenic plant preparation. Some plant-derived toxins such as AtSerpin1, a serpin from *Arabidopsis thaliana* (inhibited proteases from all pest), inactive 2S seed storage protein of Indian mustard *Brassica juncea*, and AHA gene of *Amaranthus hypochondriacus* [25–27] also have been used sources of toxins during transgenic plant preparation against attacking insects. In this chapter, different transgenic plants generated for insect resistance have been discussed and categorized into models—dicots and monocots transgenic plants.

2.2 Insect-Resistant Transgenic Model Plant

Several model plants like *Arabidopsis*, tobacco, *Torenia fournieri* (torenia), *Medicago truncatula*, *Populus*, *Boechera*, etc. are mostly used to explain and prove genetic concept for transgenic strategies [28]. Among them, due to easy transgenic preparation with respect to time and consumption, *Arabidopsis* and tobacco are mostly used to generate transgenic plants against insects. Several toxic genes like *Photorhabdus* toxic protein, cytosolic farnesyldiphosphate synthase (FPS2), pollen specific gene SKS13, and terpene synthase (TPS10) are used to generate transgenic *Arabidopsis* plants against attacking insects like tobacco hornworm (*Manduca sexta*), aphids (*Myzus persicae*), and *Spodoptera littoralis*, respectively (Table 2.1) [29–31]. Similarly, a complete pathway expressing tyrosine-derived cyanogenic glucoside dhurrin biosynthesis in *Arabidopsis thaliana* to generate resistance against flea beetle (*Phyllotreta nemorum*) has been made [32]. *Arabidopsis* plants expressing aphid alarm pheromone also showed resistance level against aphids (*Myzus persicae*) [33]. Different fusion proteins like scorpion venom neurotoxin (*Androctonus australis* toxin, AaIT) with snowdrop lectin (*Galanthus nivalis* agglutinin, GNA) expressed in *Arabidopsis*, rice, and tobacco showed resistance against whitefly, *Bemisia tabaci*, and the rice brown planthopper, *Nilaparvata lugens* [34]. Similarly, Hv1a/GNA, containing the spider venom toxin ω-ACTX-Hv1a linked to snowdrop lectin (GNA), showed development of resistance against peach-potato aphid *Myzus persicae* in *Arabidopsis* (Table 2.1) [35].

Transgenic plants of tobacco expressing RNAi against MsCYP6B46 and proteinase inhibitors I and II are used to generate resistance against *Manduca sexta* [36, 37]. Several other genes expressing caffeine biosynthetic pathway, potato proteinase inhibitor 2, Cry1Ec, CpBV-CST1 of polydnavirus *Cotesia plutellae* Bracovirus (CpBV), and *Macrothele gigas* (Magi 6) spider venom peptide were also used to generate transgenic tobacco against species of *Spodoptera* like *S. litura*, *S. exigua*, and *S. frugiperda* (Table 2.1) [12, 38–41]. Similarly, other genes like ribosome-inactivating proteins, cry1Ah, synthesis of cry1Ac and cry2Ab, dsRNA of EcR, maize ribosome-inactivating protein, and *Solanum americanum* proteinase inhibitor (SaPIN2a) are used to generate transgenic tobacco against species of *Helicoverpa* like *Helicoverpa zea* and *Helicoverpa armigera* [42–46] (Table 2.1). Some lectins like PTA (*Pinellia ternata* agglutinin) (Mannose), *Allium sativum* leaf lectin (ASAL), and Si-RNAs against v-ATPase A gene are used to generate transgenic plants against sap-sucking insects like peach potato aphid (*Myzus persicae* Sulzer), *Myzus nicotianae*, and white fly (*Bemisia tabaci*) [47–50]. Similarly, cholesterol oxidase targeting cotton boll weevil (*Anthonomus grandis* Boheman) [51], Cry1Ab Cry1AC, and Cry3A targeting tobacco hornworm (THW), Colorado potato beetle [52, 53], *Hadronyche versuta* (Blue Mountains funnel-web spider), neurotoxin (Hvt) and onion leaf lectin targeting *Phenacoccus solenopsis* (cotton mealybug), *Myzus persicae* (green peach aphids), and *Bemisia tabaci* (silver leaf whitefly) [54], spider toxin (Hvt) gene targeting American bollworm (*Heliothis armigera*) [55], and biotin-binding proteins (BBPs) targeting potato tuber moth, *Phthorimaea operculella* (Zeller) [56] are used to make insect-resistant transgenic tobacco (Table 2.1). Similarly, several transgenic plants of model trees like poplar were also developed. Poplar plants expressing Bt-Cry3A and Oryzacystatin I, chitinase-BmkIT, fusion gene of spider, *Atrax robustus* Simon ω-ACTX-Ar1 toxin and Bt-toxin C-peptide, and scorpion neurotoxin AaIT showed resistance against beetle (*Chrysomela tremulae*), *Plagiodera versicolora* larvae, Asian gypsy moth (*Lymantria dispar* L.), and *Apocheimia cinerarius* [57–61] (Table 2.1).

Table 2.1 Insect-resistant transgenic model plant.

Gene used for transgenic preparation	Promoter used	Plant species	Targeted insects	Reference
Pathway for tyrosine-derived cyanogenic glucoside dhurrin biosynthesis	–	*Arabidopsis thaliana*	Flea beetle (*Phyllotreta nemorum*)	[32]
Photorhabdus toxin	–	*Arabidopsis thaliana*	Tobacco hornworm (*Manduca sexta*)	[29]
Aphid alarm pheromone	CaMV35S	*Arabidopsis thaliana*	Aphid (*Myzus persicae*)	[33]
Terpene synthase (TPS10)	CaMV35S	*Arabidopsis thaliana*	*Spodoptera littoralis*	[31]
Fusion protein Hv1a/GNA, containing the spider venom toxin ω-ACTX-Hv1a linked to snowdrop lectin (GNA)	CaMV35S	*Arabidopsis thaliana*	Aphid	[35]
Cytosolic farnesyldiphosphate synthase (FPS2)	CaMV35S	*Arabidopsis thaliana*	Aphid	[30]
Scorpion venom neurotoxin (*Androctonus australis* toxin, AaIT) is fused to snowdrop lectin (*Galanthus nivalis* agglutinin, GNA)	–	Tobacco, *Arabidopsis* and rice	Whitefly (*Bemisia tabaci*), and rice brown planthopper, (*Nilaparvata lugens*)	[36]
RNAi against MsCYP6B46	CaMV35S	Tobacco	*Manduca sexta*	[36]

(*Continued*)

Table 2.1 Insect-resistant transgenic model plant. (*Continued*)

Gene used for transgenic preparation	Promoter used	Plant species	Targeted insects	Reference
Caffeine biosynthetic pathway	–	Tobacco	*Spodoptera litura*	[38]
Ribosome-inactivating proteins	CaMV35S	Tobacco	*Helicoverpa zea*	[42]
PTA: *Pinellia ternata* agglutinin (mannose)	CaMV35S	Tobacco	Aphid	[47]
Allium sativum leaf lectin (ASAL)	CaMV35S	Tobacco	Aphid	[48]
,,	ASus1	Tobacco	*Myzus nicotianae*	[49]
Proteinase inhibitors I and II	CaMV35S	Tobacco	*Manduca sexta*	[37]
Si-RNAs against v-ATPase A gene	CaMV35S	Tobacco	White fly	[50]
Cholesterol oxidase	Figwort mosaic viruses	Tobacco	Cotton boll weevil (*Anthonomus grandis grandis*)	[51]
Cry1Ab Cry1AC, Cry3A	CaMV35S	Tobacco and tomato	Tobacco hornworm (THW), Colorado potato beetle	[52, 53]
Potato proteinase inhibitor 2	CaMV35S	Tobacco and rice	*Spodoptera exigua*	[39]

(*Continued*)

Table 2.1 Insect-resistant transgenic model plant. (*Continued*)

Gene used for transgenic preparation	Promoter used	Plant species	Targeted insects	Reference
Cry1Ec	CaMV35S	Tobacco and cotton	*Spodoptera litura*	[12]
CpBV-CST1 of polydnavirus *Cotesia plutellae* bracovirus (CpBV)	–	Tobacco	*Spodoptera exigua*, *Helicoverpa assulta*, and aphid	[40]
Phadronyche versuta (Blue Mountains funnel-web spider) neurotoxin (Hvt) and onion leaf lectin	Phloem-specific	Tobacco	Cotton mealybug (*Phenacoccus solenopsis*), aphid, and whitefly	[54]
Cry1Ah	Chloroplast	Tobacco	*Helicoverpa armigera*	[43]
Synthesis of cry1Ac and cry2Ab	–	Tobacco	*Helicoverpa armigera* and *Spodoptera exigua*	[44]
dsRNA of EcR	CaMV35S	Tobacco	*Helicoverpa armigera*	[45]
Spider toxin (Hvt) gene	RSs1 and RolC	Tobacco	*American bollworm* (*Heliothis armigera*)	[55]
Biotin-binding proteins (BBPs)	–	Tobacco	Potato tuber moth (*Phthorimaea operculella*)	[56]

(*Continued*)

Table 2.1 Insect-resistant transgenic model plant. (*Continued*)

Gene used for transgenic preparation	Promoter used	Plant species	Targeted insects	Reference
Solanum americanum proteinase inhibitor (SaPIN2a)	CaMV35S	Tobacco	*Helicoverpa armigera* and *Spodoptera litura*	[46]
Macrothele gigas spider, Magi 6 spider venom peptide	CaMV35S	Tobacco	*Spodoptera frugiperda*	[41]
Bt-Cry3A and oryzacystatin I	CaMV35S	Poplar	*Plagiodera versicolora*	[60]
Fusion gene consisting of the spider, *Atrax robustus* (ω-ACTX-Ar1) sequence and Bt-toxin C-peptide	–	Poplar	Asian gypsy moth (*Lymantria dispar* (L.))	[59]
B.t. toxin	CaMV35S	Poplar	Asian gypsy moth and *Apocheimia cinerarius*	[57]
Scorpion neurotoxin AaIT	–	Poplar	Asian gypsy moth	[58]

2.3 Insect-Resistant Transgenic Dicot Plants

Among dicots, plants of castor, potato, *Brassica*, lettuce, soybean, different peas, tomato, cabbage, okra, cotton, etc. are used to generate transgenic plants. Castor (*Ricinus communis* L.) plants expressing cry1EC were generated. These plants showed resistance against tobacco caterpillar (*Spodoptera litura* Fabr) and castor semilooper (*Achoea janata* L.) [62] (Table 2.2). Peanut (*Arachis hypogaea* L.) expressing Cry1EC and Cr1Ac and rice chitinase showed resistance against *Spodoptera litura* and cornstalk borer [63–66] (Table 2.2). Potato plants expressing Concanavalin A (ConA), *Galanthus nivalis* agglutinin (GNA), Cry1Ab, Oryzacystatin I (OCI), and sea anemone equistatin showed resistance against green peach aphid (*Myzus persicae*), tomato moth (*Lacanobia oleracea*), tuber moth (*Phthorimaea operculella*, Zeller), and Colorado potato beetle (*Leptinotarsa decemlineata* Say) [67–71] (Table 2.2). *Brassica juncea* expressing *Allium sativum* leaf agglutinin (ASAL), wheat germ agglutinin (WGA), Cry proteins–cry IA, chitinase gene, and a scorpion insect toxin showed resistance against sap-sucking hemipteran pest-mustard aphid (*Lipaphis erysimi*), cabbage caterpillar (*Pieris rapae*), *Plutella xylostella*, and diamondback moth (*Plutella maculipenis*) [1, 72–75] (Table 2.2). Lettuce (*Lactuca sativa*) plants expressing dsRNA against v-ATPase of whitefly (*Bemisia tabaci*) showed resistance against whitefly (*Bemisia tabaci*) [4] (Table 2.2). Soybean expressing double-stranded RNA against ribosomal protein P0 gene and Cry1Ac showed resistance against soybean pod borer (*Leguminivora glycinivorella*) and *Spodoptera litura* [76, 77] (Table 2.2). Pigeon pea expressing Cry1Ac and Cry2Aa showed resistance against *Helicoverpa armigera* [2]. Pea (*Pisum sativum*) expressing alpha-amylase inhibitor showed resistance against cowpea weevil and Bruchus beetles [78, 79] (Table 2.2). Chickpea (*Cicer arietinum*) and cowpea (*Vigna unguiculata*) expressing ASA: *Allium sativum* agglutinin (Mannose), fusion of cry1Ab/Ac and αAI-1, an α-amylase inhibitor from common bean (*Phaseolus vulgaris*), showed resistance against chickpea aphid (*Aphis craccivora*), *H. armigera*, and bruchid beetles [80–82] (Table 2.2). Tomato expressing RNAi against chitinase gene of *Helicoverpa armigera* and tobacco anionic peroxidase showed resistance against *Helicoverpa armigera*, *Helicoverpa zea*, and *Manduca sexta* [83, 84] (Table 2.2). Cabbage plants expressing Cry1Ia8 showed resistance against *Plutella xylostella* (Linnaeus) and *Pieris rapae* (Linnaeus) [85]. Okra (*Abelmoschus esculentus* (L.) Moench) plants expressing cry1Ac gene showed resistance against fruit and shoot borer (*Earias vittella*) [86] (Table 2.2). Medicinal plant *Isatis indigotica* expressing Cry1Ac gene and *Inellia ternata* agglutinin gene (Pta) showed resistance

Table 2.2 Insect-resistant transgenic dicot plant.

Gene used for transgenic preparation	Tissue of expression	Plant species	Targeted insects	Reference
Cry1EC	CaMV35S	Castor (*Ricinus communis* L.)	*Spodoptera litura* and castor semilooper (*Achoea janata* L.)	[62]
Cry1EC	CaMV35S	Peanut (*Arachis hypogaea* L.)	*Spodoptera litura*	[64]
Cry1EC and rice chitinase	CaMV35S	Peanut (*Arachis hypogaea* L.)	*Spodoptera litura*	[65]
Cry1EC	Pathogenesis responsive (PR-1a)	Peanut (*Arachis hypogaea* L.)	*Spodoptera litura*	[66]
CryIAc	–	Peanut (*Arachis hypogaea* L.)	Cornstalk borer	[63]
Concanavalin A (ConA)	CaMV35S	Potato	Aphid and tomato moth (*Lacanobia oleracea*)	[68]
Galanthus nivalis agglutinin (GNA)	CaMV35S	Potato	Aphid	[69]
Cry1Ab	C(4)-PEPC-	Potato	Potato tuber moth	[70]
Oryzacystatin I (OCI)	CaMV35S	Potato	Colorado potato beetle (*Leptinotarsa decemlineata* Say)	[67]

(*Continued*)

Table 2.2 Insect-resistant transgenic dicot plant. (*Continued*)

Gene used for transgenic preparation	**Tissue of expression**	**Plant species**	**Targeted insects**	**Reference**
Sea anemone equistatin	CaMV35S, Lhca.3.St1 of potato, and the RbcS1 of chrysanthemum	Potato	Colorado potato beetle (*Leptinotarsa decemlineata* Say)	[71]
Allium sativum leaf agglutinin (*ASAL*)	CaMV35S	*Brassica juncea*	Mustard aphid (*Lipaphis erysimi*)	[1]
Wheat germ agglutinin (WGA)	,,	*Brassica juncea*	Mustard aphid	[73]
Cry IA	–	Rutabaga (*Brassica napobrassica*)	Cabbage caterpillar (*Pieris rapae*)	[72]
Cry proteins	–	Canola (*Brassica napus* L.)	*Plutella xylostella*	[74]
Chitinase gene and a scorpion insect toxin	–	Canola (*Brassica napus* L.)	Diamondback moth (*Plutella maculipenis*)	[75]
Novel v-ATPase of whitefly	CaMV35S	Lettuce (*Lactuca sativa*)	Whitefly	[4]
Double-stranded RNA against ribosomal protein P0 gene	–	Soybean	Soybean pod borer (*Leguminivora glycinivorella*)	[76]

(*Continued*)

Table 2.2 Insect-resistant transgenic dicot plant. (*Continued*)

Gene used for transgenic preparation	Tissue of expression	Plant species	Targeted insects	Reference
Cry1Ac	–	Soybean	*Spodoptera litura*	[77]
Cry1Ac and Cry2Aa	CaMV35S	Pigeon pea	*Helicoverpa armigera*	[2]
Alpha-amylase inhibitor	CaMV35S	Peas (*Pisum sativum*)	Cowpea weevil	[78]
Alpha-amylase inhibitor	–	Peas (*Pisum sativum*)	Bruchus beetles	[79]
ASA: *Allium sativum* agglutinin (mannose)	CaMV35S	Chickpea	Chickpea aphid (*Aphis craccivora*)	[80]
Fused cry1Ab/Ac	Pod-specific (msg)	Chickpea	*Helicoverpa armigera*	[82]
αAI-1, an α-amylase inhibitor from common bean (*Phaseolus vulgaris*)	Cotyledon-specific	Chickpea (*Cicer arietinum*) and cowpea (*Vigna unguiculata*)	Bruchid beetles	[81]
RNAi against chitinase gene of *Helicoverpa armigera*	–	Tobacco and tomato	*Helicoverpa armigera*	[84]
Tobacco anionic peroxidase	–	Tomato	*Helicoverpa zea* and *Manduca sexta*	[83]

(*Continued*)

Table 2.2 Insect-resistant transgenic dicot plant. (*Continued*)

Gene used for transgenic preparation	Tissue of expression	Plant species	Targeted insects	Reference
Cry1Ia8	–	Cabbage	*Plutella xylostella* and *Pieris rapae*	[85]
cry1Ac gene	–	Okra (*Abelmoschus esculentus* (L.))	Fruit and shoot borer (*Earias vittella*)	[86]
Cry1Ac gene and *Iinellia ternata* agglutinin gene (Pta)	CaMV35S	*Isatis indigotica*	Diamondback moths (*Plutella xylostella* L.) and aphids	[87]
BT toxin	–	Broccoli	*Plutella xylostella*	[88]
Cry1C	–	Broccoli	Diamondback moth	[89]
Avidin or streptavidin	–	Apple (*Malus domestica*) and tobacco	Light brown apple moth (LBAM) (*Epiphyas postvittana*) and larval potato tuber moth	[90]
CrylAc	–	Pinus	*Teia anartoides*	[91]
Tma12 protein from *Tectaria macrodontal*	CaMV35S	Cotton	Whitefly	[97]

(*Continued*)

Table 2.2 Insect-resistant transgenic dicot plant. (*Continued*)

Gene used for transgenic preparation	Tissue of expression	Plant species	Targeted insects	Reference
dsRNA-HaHR3 a cotton	CaMV35S	Cotton	Cotton bollworm (*Helicoverpa armigera*)	[98]
dsRNA targeting to cytochrome P450 gene (CYP6AE14)	CaMV35S	Cotton	Cotton bollworm (*Helicoverpa armigera*)	[105]
Down-regulation of AsFAR (fatty acyl-Coa reductase)	CaMV35S	Cotton	Bugs (*Adelphocoris suturalis*)	[10]
Transgenic cotton pyramids (Bt + RNAi)		Cotton	Cotton bollworm	[14]
Allium sativum leaf agglutinin encoding gene (ASAL) cotton	CaMV35S	Cotton	Jassid and whitefly	[96]

(*Continued*)

Table 2.2 Insect-resistant transgenic dicot plant. (*Continued*)

Gene used for transgenic preparation	Tissue of expression	Plant species	Targeted insects	Reference
Chimeric TVip3A	Chloroplasts	Cotton	Armyworm (*Spodoptera frugiperda*) and beet armyworm (*Spodoptera exigua*)	[139]
Potato type I and II serine protease inhibitors	CaMV35S	Cotton	*Helicoverpa punctigera*	[95]
Cry2Ab	–	Cotton	Pink bollworm (*Pectinophora gossypiella*)	[92]
Cry1EC	Insect bite and wound inducible promoter PR-1a	Cotton	*Spodoptera litura*	[94]

against diamondback moths (*Plutella xylostella* L.) and peach potato aphids (*Myzus persicae* Sulzer) [87] (Table 2.2). Broccoli expressing BT toxin such as Cry1C showed resistance against *Plutella xylostella* [88, 89]. Trees like apple (*Malus domestica*) expressing avidin or streptavidin showed resistance against light brown apple moth (LBAM) (*Epiphyas postvittana* (Walker)) and larval potato tuber moth [90] (Table 2.2). Gymnosperm plants like *Pinus radiate* expressing crylAc showed resistance against *Teia anartoides* [91].

Among the fiber crops, most of the transgenic plants generated belong to cotton plants (Table 2.2). Cotton plants expressing Tma12 protein from *Tectaria macrodontal*, dsRNA-HaHR3, dsRNA-P450 gene (CYP6AE14), ds-AsFAR (FATTY ACYL-COA REDUCTASE), *Allium sativum* leaf agglutinin encoding gene (ASAL), chimeric TVip3A, potato type I and II serine protease inhibitors, and Cry2Ab and Cry1EC genes showed resistance against whitefly (*Bemisia tabaci*), cotton bollworm (*Helicoverpa armigera*), bugs (*Adelphocoris suturalis*), jassid, armyworm (*Spodoptera frugiperda*) and beet armyworm (*Spodoptera exigua*), *Helicoverpa punctigera*, pink bollworm (*Pectinophora gossypiella*), and *Spodoptera litura* [10, 92–98] (Table 2.2). Different strategies like transgenic cotton pyramiding Bt toxin and RNAi have also been used to generate resistance against *Helicoverpa armigera* [14].

2.4 Insect-Resistant Transgenic Monocot Plants

Many monocot plants like wheat, rice, and maize are utilized to generate transgenic plants against insect pests. Wheat plants expressing cowpea trypsin inhibitor gene (CpTI) were shown resistance to stored grain insect of wheat, viz., the grain moth (*Sitotroga cerealella* Olivier) [99] (Table 2.3). Sugarcane plant expressing δ-endotoxin like Cry1Ab, modified cry1Ac, and bovine pancreatic trypsin inhibitor (aprotinin) gene showed resistance to borers (*Diatraea saccharalis* F.), stem borers (*Scirpophaga nivela*), top borers (*Scirpophaga excerptalis*), and western corn rootworm (WCR) (*Diabrotica virgifera* LeConte) [11, 100–102] (Table 2.3). The corn plant expressing dsRNA against vacuolar ATPase (V-ATPase), insecticidal protein (PIP-47Aa) of *Pseudomonas mosselii*, avidin, insecticidal protein from *Pseudomonas* IPD072Aa, chitinase of cotton leaf worm (*Spodoptera littoralis*), maize ribosome-inactivating protein (MRIP), wheat germ agglutinin (WGA), and snowdrop lectin (*Galanthus nivalis* L. agglutinin; GNA) showed resistance against western corn rootworm (WCR) (*Diabrotica virgifera* LeConte), corn borers (*Sesamia cretica*), fall armyworms (*Spodoptera*

Table 2.3 Insect-resistant transgenic monocot plant.

Gene used for transgenic preparation	Promoter used	Plant species	Targeted insects	Reference
Cowpea trypsin inhibitor gene (CpTI)		Wheat	Grain moth (*Sitotroga cerealella*)	[99]
Cry1Ab	Ubi-1	Sugarcane	Borer (*Diatraea saccharalis*)	[11]
Modified cry1Ac	–	Sugarcane	Stem borers	[101]
δ-endotoxin	Phosphoenolpyruvate (PEP) carboxylase	Sugarcane	*Scirpophaga nivela*	[102]
Bovine pancreatic trypsin inhibitor (aprotinin) gene	Maize ubiquitin	Sugarcane	Top borer (*Scirpophaga excerptalis*)	[100]
RNAi against V-ATPase	–	Maize	Western corn rootworm (WCR) (*Diabrotica virgifera virgifera*)	[106]
Insecticidal protein (PIP-47Aa) of *Pseudomonas mosselii*	Sorghum Sb-RCc3	Maize	Western corn rootworm (WCR) (*Diabrotica virgifera virgifera*)	[110]
Avidin	–	Maize	Stored-product insect pests	[103]
Insecticidal protein from *Pseudomonas* IPD072Aa	–	Maize	Western corn rootworm	[109]

(*Continued*)

Table 2.3 Insect-resistant transgenic monocot plant. (*Continued*)

Gene used for transgenic preparation	Promoter used	Plant species	Targeted insects	Reference
Chitinase cDNA from cotton leaf worm (*Spodoptera littoralis*)	–	Western corn rootworm (WCR) (*Diabrotica virgifera virgifera*)	Corn borer (*Sesamia cretica*)	[108]
Maize ribosome-inactivating protein (MRIP) and wheat germ agglutinin (WGA)	–	Western corn rootworm (WCR) (*Diabrotica virgifera virgifera*)	Fall armyworms (*Spodoptera frugiperda*) and corn earworms (*Helicoverpa zea*)	[107]
Wheat oxalate oxidase	–	Western corn rootworm (WCR) (*Diabrotica virgifera virgifera*)	European corn borer (*Ostrinia nubilalis*)	[93]
Snowdrop lectin (*Galanthus nivalis* L. agglutinin; GNA)	Phloem-specific	Western corn rootworm (WCR) (*Diabrotica virgifera virgifera*)	Leaf aphid (*Rhopalosiphum maidis*)	[104]
Cry1Ab/Cry1Ac	Actin1	Rice (*Oryza sativa* L.)	Leaf folder and yellow stem borer (*Scirpophaga incertulas*)	[111]
Cry1Ab–Cry1Ac hybrid	CaMV35S	Rice (*Oryza sativa* L.)		[112]

(*Continued*)

Table 2.3 Insect-resistant transgenic monocot plant. (*Continued*)

Gene used for transgenic preparation	Promoter used	Plant species	Targeted insects	Reference
Cry2A* gene driven by maize ubiquitin promoter	–	Rice (*Oryza sativa* L.)	Striped stem borer (*Chilo suppressalis*), yellow stem borer (*Tryporyza incertulas*), and rice leaf folder (*Cnaphalocrocis medinalis*)	[115]
A novel cry2AX1 gene consisting of a sequence of cry2Aa and cry2Ac genes	rbcS	Rice (*Oryza sativa* L.)	Rice leaf folder	[15]
Fusion protein of Cry1Ac and Cry1I	Green tissue-specific promoter	Rice (*Oryza sativa* L.)	Rice leaf folder and the striped stem borer	[116]
Cry1Ia5	CaMV35S	Rice (*Oryza sativa* L.)	Stem borer (*Chilo agamemnon*)	[114]
Cry1Ca1	–	Rice (*Oryza sativa* L.)	*Chilo suppressalis* and *Spodoptera litura*	[113]
Snowdrop lectin (*Galanthus nivalis* agglutinin; GNA)	RSs1 and maize ubiquitin	Rice (*Oryza sativa* L.)	Rice brown hopper	[117]
Allium sativum leaf agglutinin (ASAL) gene	RSS1 and rolC	Rice (*Oryza sativa* L.)	*Aphis craccivora, Myzus persicae*	[119]

(*Continued*)

Table 2.3 Insect-resistant transgenic monocot plant. (*Continued*)

Gene used for transgenic preparation	Promoter used	Plant species	Targeted insects	Reference
GNA snowdrop lectin (*Galanthus nivalis* agglutinin; GNA)	RSS	Rice (*Oryza sativa* L.)	Green leaf hopper (*Nephotettix virescens*)	[118]
Garlic lectin gene (ASAL)	Phloem specific	Rice (*Oryza sativa* L.)	Rice brown hopper	[121]
Rice expressing *Allium sativum* leaf agglutinin (ASAL)	CaMV35S	Rice (*Oryza sativa* L.)	Brown plant hopper, green leafhopper, and white backed plant hopper	[120]
Avidin	Endosperm-specific (GluB-1)	Rice (*Oryza sativa* L.)	*Tribolium confusum* and *Sitotroga cerealella*	[123]
NlHT1, Nlcar, Nltry	Ubiquitin1	Rice (*Oryza sativa* L.)	Brown plant hopper	[125]
BPH18 gene from the BPH-resistant wild rice species *Oryza australiensis*	CaMV35S	Rice (*Oryza sativa* L.)	Brown plant hopper	[127]
Vegetative insecticidal protein (Syn vip3BR) of *Bacillus thuringiensis*	Rubisco	Rice (*Oryza sativa* L.)	Yellow stem borer (*Scirpophaga incertulas*), rice leaf folder, and rice horn caterpillar (*Melanitis leda*)	[126]
Brassica rapa Defensin 1 (BrD1)	–	Rice (*Oryza sativa* L.)	Brown plant hopper	[124]
Potato proteinase inhibitor II	Wound-inducible	Rice (*Oryza sativa* L.)	Pink stem borer (*Sesamia inferens*)	[122]

frugiperda), corn earworms (*Helicoverpa zea*), leaf aphid (*Rhopalosiphum maidis* Fitch), and European corn borers (*Ostrinia nubilalis*) [103–110] (Table 2.3).

Among monocots, rice is one of the mostly used crops to generate transgenic because of its easy regeneration processes. Transgenic rice plants were generated expressing several Cry proteins like Cry1Ab/Cry1Ac, Cry1Ab–Cry1Ac hybrid under a different promoter, cry2A* gene driven by maize ubiquitin promoter, a novel cry2AX1 gene consisting a sequence of cry2Aa and cry2Ac genes, fusion protein of Cry1Ac and Cry1I, cry1Ia5, and Cry1Ca1 and these plants showed resistance against leaf folder and yellow stem borer (*Scirpophaga incertulas*), striped stem borer (*Chilo suppressalis*), yellow stem borer (*Tryporyza incertulas*) and rice leaf folder (*Cnaphalocrocis medinalis*), and *Spodoptera litura* [15, 111–116] (Table 2.3). Similarly, different lectins like snowdrop lectin (*Galanthus nivalis* agglutinin; GNA), *Allium sativum* leaf agglutinin (ASAL) gene, and garlic lectin gene (ASAL) under a different promoter were generated, and these plants showed resistance against brown planthopper (*Nilaparvata lugens* Stål), *Aphis craccivora*, *Myzus persicae*, leafhopper (*Nephotettix virescens*; GLH) and brown planthopper (BPH), green leafhopper (GLH), and white backed planthopper (WBPH) [117–121] (Table 2.3). Another strategy like expression of *Brassica rapa* Defensin 1 (BrD1), potato proteinase inhibitor II, vegetative insecticidal protein (Syn vip3BR) of *Bacillus thuringiensis*, BPH18 gene from the BPH-resistant wild rice species *Oryza australiensis*, avidin, and knockdown of NlHT1/Nlcar/Nltry were also used, and these plants showed resistance against planthopper (*Nilaparvata lugens*), pink stem borer (*Sesamia inferens*), yellow stem borer (*Scirpophaga incertulas*), rice leaf folder (*Cnaphalocrocis medinalis*), rice horn caterpillar (*Melanitis leda ismene*), the striped stem borer *Chilo suppressalis* (Walker), *Tribolium confusum*, and *Sitotroga cerealella* [122–127] (Table 2.3).

2.5 Working Principle of Insecticidal Genes Used in Transgenic Plant Preparation

Most transgenic plants express δ-endotoxin (Cry protein) obtained from *Bacillus thuringiensis* bacteria. Several Cry genes like Cry1Ab, Cry1AC, Cry1EC, Cry 1Ie, Cry1Ah, Cry3A, synthesized Cry1Ac and Cry2Ab, Cry3A, fused Cry1Ab/Ac, Cry1Ia8, Cry1C, Cry1Ac and Cry1I, cry1Ia5, and Cry1Ca1 are used to generate transgenic plants (Tables 2.1–2.3). Scientists reported that these proteins destroy the digestive system of attacking insects and thus insects die after parasitizing on transgenic plants [11, 12, 62, 64, 65, 101, 102]. Several other bacterial and viral derived toxin genes such as *Photorhabdus* toxin, CpBV-CST1 of polydnavirus *Cotesia plutellae* bracovirus (CpBV), insecticidal protein (PIP-47Aa) of *Pseudomonas mosselii*, insecticidal protein of *Pseudomonas* IPD072Aa, and vegetative insecticidal protein (Syn vip3BR) of *Bacillus thuringiensis* are also used to generate transgenic plants. Being both parasitoid and predators of attacking insects, scorpions and spiders are one of the best enemies of herbivorous insects. Several genes expressing toxin of spider and scorpion such as venom toxin ω-ACTX-Hv1a of spider, venom neurotoxin (AaIT) of *Androctonus australis* scorpion, toxin of Blue Mountains funnel-web spider, other spider toxin (Hvt) gene, Magi 6 of *Macrothele gigas* spider toxin, scorpion neurotoxin AaIT, and their different fusions are also used to generate transgenic plants (Tables 2.1–2.3). Some authors suggested that these toxins paralyze attacking insects and reduce their performance on host plants, thus improving crop productivity compared to nontransgenic plants.

Several lectins such as snowdrop lectin (*Galanthus nivalis* agglutinin; GNA), *Allium sativum* leaf agglutinin (ASAL), wheat germ agglutinin (WGA), and PTA (*Pinellia ternata* agglutinin) are also used to generate transgenic plants. Some authors suggested that these lectins bind to the carbohydrate moiety of insect gut and thus reduce the herbivorous performance and improve the resistance against attacking insects [117–119]. Several other inhibitors such as *Solanum americanum* proteinase inhibitor (SaPIN2a), cowpea trypsin inhibitor gene (CpTI), bovine pancreatic trypsin inhibitor (aprotinin), proteinase inhibitors I and II, alpha-amylase inhibitor, and potato type I and II serine protease inhibitors are also used to generate the transgenic plants (Tables 2.1–2.3). Some authors suggested that these enzyme inhibitors reduce the digestive power of insects and thus improve the resistant nature of transgenic plants [24, 37, 39, 122]. Further, several enzymes such as chitinase, cholesterol oxidase, tobacco anionic peroxidase, and wheat oxalate oxidase are also used to generate transgenic plants, and these enzymes destroy the cementing and building blocks of attacking insects [51, 75, 93]. RNAi is a new developing technology in which transgenic plants express small RNA specific to attacking insects and thus the desired protein formation is reduced and plants get resistant power against attacking insects [10, 84, 98, 105, 106]. Several new strategies from exploring biodiversity such as sea anemone equistatin [71] and pteridophyte (Tma12 protein from *Tectaria macrodontal*) [97] are also used as toxic genes, and their mode of function is yet to be discovered. The fusion of different insecticides targeting different insects would be a great approach in pest management.

2.6 Discussion

Although transgenic strategy is one of the better and fastest strategies among other crop improvement programs, regulated expression of insecticidal genes is very important to cope with their effect on transgenic plants themselves as well as on the environment. Generally, plant virus-derived constitutive promoters have been used to make transgenic plants, and it leads to constitutive expression of transgene that has an abnormal effect on plants. Viral promoters like CaMV35S (Tables 2.1–2.3) are generally used to make transgenic plants to express several insecticidal genes, e.g., δ-endotoxin Cry1EC, terpene synthase (TPS10), lectins, etc. [12, 31, 64, 128]. The need for a precise tissue- or site-specific expressive promoter that can facilitate regulated expression of insecticides at the targeted time and site is the current demand [129]. For example, no detection of any δ-endotoxin in cane juice and Cry1C free endosperm of rice [130] of genetically modified plants against the respective pathogen [102] showed that the fine-tuning of transgene expression would lead to less load of transgenes on plants themselves as well as the environment. The transgenic strategy has many advantages like acceleration on gene incorporation related to food productivity, quality, and pharmaceutical farming, pest resistance due to lack of resistant trait for pest in wild crossing partner in classical breeding, super crops having best nutritional quality, etc., [131–135]; it has some major disadvantages like safety of transgenic material [131], horizontal/vertical herbicide-resistant gene transfer to other weeds, development of antibiotic resistance pathogens, development of allergens, killing of beneficial insects, development of resistance against insecticide, time consumption in getting clearance from government, and ethical issues [1, 136–138]. Similarly, development of resistance from attacking insects is also one potential future problem [135]. Several strategies like fusion of different genes are also used to generate transgenic plants like chimeric TVip3A accumulating in chloroplasts and cotton [139], and to avoid development of resistance from insects and pests. But in the war between farmers and pests, the best strategy of using insect inducible promoters expressing transgene at the site can reduce the potential risk of insecticidal molecules. In this reference, several strategies like expression of insecticide under tissue-specific expression (Tables 2.1 and 2.3) [54, 55, 123], pathogen responsive expression (Table 2.2) [64, 94], or organ-specific expression (Table 2.2) [81, 82], have been proposed and proved. Another strategy like Cre/lox mediated recombination strategy also has been used to remove these genes [1] to avoid the suggested drawback. Further, the antibiotic-resistant gene can be removed and a marker-free plant can be generated either by site-specific recombination strategy or by co-transformation method [140]. No adverse effect of BT technology has been reported on studied animals like rats or insects like honey bee and silkworm [141–143]. Similarly, transgenic poplar leaf expressing chitinase-BmkIT also did not show any adverse effect on animals like rabbits [61]. The above studies prove that few technologies would be safe for humans and the environment and also provide hope to develop new strategies similar to them. Further, in the future, combinations of different approaches like transgenic, classical pest management, breeding programs, and use of greenhouses can reduce the risk of transgenic plants [144]. Here we conclude that insect and pest control along with transgenic strategy and use of integrated pest management would be effective and secure for different environmental issues.

References

1. Bala, A., Roy, A., Das, A., Chakraborti, D., Das, S., Development of selectable marker free, insect resistant, transgenic mustard (*Brassica juncea*) plants using Cre/lox mediated recombination. *BMC Biotechnol.*, 13, 88, 2013.
2. Ghosh, G., Ganguly, S., Purohit, A., Chaudhuri, R.K., Das, S., Chakraborti, D., Transgenic pigeonpea events expressing Cry1Ac and Cry2Aa exhibit resistance to *Helicoverpa armigera*. *Plant Cell Rep.*, 36, 1037–1051, 2017.
3. Catarino, R., Ceddia, G., Areal, F.J., Park, J., The impact of secondary pests on *Bacillus thuringiensis* (Bt) crops. *Plant Biotechnol. J.*, 13, 601–612, 2015.
4. Ibrahim, A.B., Monteiro, T.R., Cabral, G.B., Aragão, F.J., RNAi-mediated resistance to whitefly (*Bemisia tabaci*) in genetically engineered lettuce (*Lactuca sativa*). *Transgenic Res.*, 26, 613–624, 2017.
5. Sharma, S. and Gill, C.K., Comparative efficiency of *Myzus persicae* (Sulzer) and *Lipaphis erysimi* (Kaltenbach) in transmitting radish mosaic virus. *J. Res.*, 41, 239–245, 2004.
6. Dombrovsky, A., Huet, H., Chejanovsky, N., Raccah, B., Aphid transmission of a potyvirus depends on suitability of the helper component and the N terminus of the coat protein. *Arch. Virol.*, 150, 287–298, 2005.
7. Chen, H., Lin, Y., Zhang, Q., Review and prospect of transgenic rice research. *Chin. Sci. Bull.*, 54, 4049, 2009.
8. Erasmus, A., Marais, J., Van den Berg, J., Movement and survival of *Busseola fusca* (Lepidoptera: Noctuidae) larvae within maize plantings with different ratios of non-Bt and Bt seed. *Pest Manag. Sci.*, 72, 2287–2294, 2016.
9. James, C., 20th Anniversary (1996–2015) of the global commercialization of biotech crops and biotech crop highlights in 2015. ISAAA Brief no. 51.
10. Luo, J., Liang, S., Li, J., Xu, Z., Li, L., Zhu, B., Li, Z., Lei, C., Lindsey, K., Chen, L., Jin, S., A transgenic strategy for controlling plant bugs (*Adelphocoris*

suturalis) through expression of double-stranded RNA homologous to fatty acyl-coenzyme A reductase in cotton. *New Phytol.*, 215, 1173–1185, 2017.
11. Wang, W.Z., Yang, B.P., Feng, X.Y., Cao, Z.Y., Feng, C.L., Wang, J.G., Xiong, G.R., Shen, L.B., Zeng, J., Zhao, T.T., Zhang, S.Z., Development and characterization of transgenic sugarcane with insect resistance and herbicide tolerance. *Front. Plant Sci.*, 8, 1535, 2017.
12. Singh, P., Kumar, M., Chaturvedi, C., Yadav, D., Tuli, R., Development of a hybrid delta-endotoxin and its expression in tobacco and cotton for control of a polyphagous pest *Spodoptera litura*. *Transgenic Res.*, 13, 397–410, 2004.
13. Moar, W., Khajuria, C., Pleau, M., Ilagan, O., Chen, M., Jiang, C., Price, P., McNulty, B., Clark, T., Head, G., Cry3Bb1-resistant western corn rootworm, *Diabrotica virgifera virgifera* (LeConte) does not exhibit cross-resistance to DvSnf7 dsRNA. *PloS One*, 12, e0169175, 2017.
14. Ni, M., Ma, W., Wang, X., Gao, M., Dai, Y., Wei, X., Zhang, L., Peng, Y., Chen, S., Ding, L., Tian, Y., Next-generation transgenic cotton: Pyramiding RNAi and Bt counters insect resistance. *Plant Biotechnol. J.*, 15, 1204–1213, 2017.
15. Manikandan, R., Balakrishnan, N., Sudhakar, D., Udayasuriyan, V., Development of leaf folder resistant transgenic rice expressing cry2AX1 gene driven by green tissue-specific rbcS promoter. *World J. Microbiol. Biotechnol.*, 32, 37, 2016.
16. Windley, M.J., Herzig, V., Dziemborowicz, S.A., Hardy, M.C., King, G.F., Nicholson, G.M., Spider-venom peptides as bioinsecticides. *Toxins*, 4, 191–227, 2012.
17. Huang, J.Q., Wei, Z.M., An, H.L., Zhu, Y.X., *Agrobacterium tumefaciens*-mediated transformation of rice with the spider insecticidal gene conferring resistance to leaffolder and striped stem borer. *Cell Res.*, 11, 149–155, 2001.
18. Christov, N.K., Imaishi, H., Ohkawa, H., Green-tissue-specific expression of a reconstructed cry1C gene encoding the active fragment of *Bacillus thuringiensis* δ-endotoxin in haploid tobacco plants conferring resistance to *Spodoptera litura*. *Biosci. Biotechnol. Biochem.*, 63, 1433–1444, 1999.
19. Zhou, Y., Tian, Y., Wu, B., Mang, K., Inhibition effect of transgenic tobacco plants expressing snowdrop lectin on the population development of *Myzus persicae*. *Chin. J. Biotechnol.*, 14, 9–16, 1998.
20. Rossi, M., Goggin, F.L., Milligan, S.B., Kaloshian, I., Ullman, D.E., Williamson, V.M., The nematode resistance gene Mi of tomato confers resistance against the potato aphid. *Proc. Nat. Acad. Sci.*, 95, 9750–9754, 1998.
21. Yao, B., Fan, Y., Zeng, Q., Zhao, R., Insect-resistant tobacco plants expressing insect-specific neurotoxin AaIT. *Chin. J. Biotechnol.*, 12, 67–72, 1996.
22. Zhang, Y., Ma, F., Wang, Y., Yang, B., Chen, S., Expression of v-cath gene from HearNPV in tobacco confers an antifeedant effect against *Helicoverpa armigera*. *J. Biotechnol.*, 138, 52–55, 2008.
23. Wu, J., Wu, Q., Wu, Q., Gai, J., Yu, D., Constitutive overexpression of AOS-like gene from soybean enhanced tolerance to insect attack in transgenic tobacco. *Biotechnol. Lett.*, 30, 1693–1698, 2008.

24. Bi, R.M., Jia, H.Y., Feng, D.S., Wang, H.G., Transgenic wheat (*Triticum aestivum* L.) with increased resistance to the storage pest obtained by *Agrobacterium tumefaciens*–mediated. Sheng wu gong cheng xue bao. *Chin. J. Biotechnol.*, 22, 431–437, 2006.
25. Zhou, Y.G., Tian, Y.C., Mang, K.Q., Cloning of AHA gene from *Amaranthus hypochondriacus* and its aphid inhibitory effect in transgenic tabacco plants. Sheng wu gong cheng xue bao. *Chin. J. Biotechnol.*, 17, 34–39, 2001.
26. Alvarez-Alfageme, F., Maharramov, J., Carrillo, L., Vandenabeele, S., Vercammen, D., Van Breusegem, F., Smagghe, G., Potential use of a serpin from Arabidopsis for pest control. *PLoS One*, 6, e20278, 2011.
27. Mandal, S., Kundu, P., Roy, B., Mandal, R.K., Precursor of the inactive 2S seed storage protein from the Indian mustard *Brassica juncea* is a novel trypsin inhibitor characterization, post-translational processing studies, and transgenic expression to develop insect-resistant plants. *J. Biol. Chem.*, 277, 37161–37168, 2002.
28. Aida, R., *Torenia fournieri* (torenia) as a model plant for transgenic studies. *Plant Biotechnol.*, 25, 541–545, 2008.
29. Liu, D., Burton, S., Glancy, T., Li, Z.S., Hampton, R., Meade, T., Merlo, D.J., Insect resistance conferred by 283-kDa *Photorhabdus luminescens* protein TcdA in *Arabidopsis thaliana. Nat. Biotechnol.*, 21, 1222, 2003.
30. Bhatia, V., Maisnam, J., Jain, A., Sharma, K.K., Bhattacharya, R., Aphid-repellent pheromone E-β-farnesene is generated in transgenic *Arabidopsis thaliana* over-expressing farnesyl diphosphate synthase2. *Ann. Bot.*, 115, 581–591, 2014.
31. Schnee, C., Köllner, T.G., Held, M., Turlings, T.C., Gershenzon, J., Degenhardt, J., The products of a single maize sesquiterpene synthase form a volatile defense signal that attracts natural enemies of maize herbivores. *Proc. Nat. Acad. Sci. USA*, 103, 1129–1134, 2006.
32. Tattersall, D.B., Bak, S., Jones, P.R., Olsen, C.E., Nielsen, J.K., Hansen, M.L., Høj, P.B., Møller, B.L., Resistance to an herbivore through engineered cyanogenic glucoside synthesis. *Science*, 293, 1826–1828, 2001.
33. Beale, M.H., Birkett, M.A., Bruce, T.J., Chamberlain, K., Field, L.M., Huttly, A.K., Martin, J.L., Parker, R., Phillips, A.L., Pickett, J.A., Prosser, I.M., Aphid alarm pheromone produced by transgenic plants affects aphid and parasitoid behavior. *Proc. Nat. Acad. Sci.*, 103, 10509–10513, 2006.
34. Liu, S.M., Li, J., Zhu, J.Q., Wang, X.W., Wang, C.S., Liu, S.S., Chen, X.X., Li, S., Transgenic plants expressing the AaIT/GNA fusion protein show increased resistance and toxicity to both chewing and sucking pests. *Insect Sci.*, 23, 265–276, 2016.
35. Nakasu, E.Y., Edwards, M.G., Fitches, E., Gatehouse, J.A., Gatehouse, A.M., Transgenic plants expressing ω-ACTX-Hv1a and snowdrop lectin (GNA) fusion protein show enhanced resistance to aphids. *Front. Plant Sci.*, 5, 673, 2014.
36. Kumar, P., Pandit, S.S., Baldwin, I.T., Tobacco rattle virus vector: A rapid and transient means of silencing *Manduca sexta* genes by plant mediated RNA interference. *PLoS One*, 7, e31347, 2012.

37. Johnson, R., Narvaez, J., An, G., Ryan, C., Expression of proteinase inhibitors I and II in transgenic tobacco plants: Effects on natural defense against Manduca sexta larvae. *Proc. Nat. Acad. Sci.*, 86, 9871–9875, 1989.
38. Kim, Y.S., Uefuji, H., Ogita, S., Sano, H., Transgenic tobacco plants producing caffeine: A potential new strategy for insect pest control. *Transgenic Res.*, 15, 667–672, 2006.
39. Jongsma, M.A. and Bolter, C., The adaptation of insects to plant protease inhibitors. *J. Insect Physiol.*, 43, 885–895, 1997.
40. Kim, E., Kim, Y., Yeam, I., Kim, Y., Transgenic expression of a viral cystatin gene CpBV-CST1 in tobacco confers insect resistance. *Environ. Entomol.*, 45, 1322–1331, 2016.
41. Hernández-Campuzano, B., Suárez, R., Lina, L., Hernández, V., Villegas, E., Corzo, G., Iturriaga, G., Expression of a spider venom peptide in transgenic tobacco confers insect resistance. *Toxicon*, 53, 122–128, 2009.
42. Dowd, P.F., Zuo, W.N., Gillikin, J.W., Johnson, E.T., Boston, R.S., Enhanced resistance to *Helicoverpa zea* in tobacco expressing an activated form of maize ribosome-inactivating protein. *J. Agricult. Food Chem.*, 51, 3568–3574, 2003.
43. Li, X., Li, S., Lang, Z., Zhang, J., Zhu, L., Huang, D., Chloroplast-targeted expression of the codon-optimized truncated cry1Ah gene in transgenic tobacco confers a high level of protection against insects. *Plant Cell Rep.*, 32, 1299–1308, 2013.
44. Sohail, M.N., Karimi, S.M., Asad, S., Mansoor, S., Zafar, Y., Mukhtar, Z., Development of broad-spectrum insect-resistant tobacco by expression of synthetic cry1Ac and cry2Ab genes. *Biotechnol. Lett.*, 34, 1553–1560, 2012.
45. Zhu, J.Q., Liu, S., Ma, Y., Zhang, J.Q., Qi, H.S., Wei, Z.J., Yao, Q., Zhang, W.Q., Li, S., Improvement of pest resistance in transgenic tobacco plants expressing dsRNA of an insect-associated gene EcR. *PloS One*, 7, e38572, 2012.
46. Luo, M., Wang, Z., Li, H., Xia, K.F., Cai, Y., Xu, Z.F., Overexpression of a weed (*Solanum americanum*) proteinase inhibitor in transgenic tobacco results in increased glandular trichome density and enhanced resistance to *Helicoverpa armigera* and *Spodoptera litura*. *Int. J. Mol. Sci.*, 10, 1896–1910, 2009.
47. Yao, J., Pang, Y., Qi, H., Wan, B., Zhao, X., Kong, W., Sun, X., Tang, K., Transgenic tobacco expressing *Pinellia ternata* agglutinin confers enhanced resistance to aphids. *Transgenic Res.*, 12, 715–722, 2003.
48. Dutta, I., Saha, P., Majumder, P., Sarkar, A., Chakraborti, D., Banerjee, S., Das, S., The efficacy of a novel insecticidal protein, *Allium sativum* leaf lectin (ASAL), against homopteran insects monitored in transgenic tobacco. *Plant Biotechnol. J.*, 3, 601–611, 2005.
49. Sadeghi, A., Broeders, S., De Greve, H., Hernalsteens, J.P., Peumans, W.J., Van Damme, E.J., Smagghe, G., Expression of garlic leaf lectin under the control of the phloem-specific promoter Asus1 from *Arabidopsis thaliana* protects tobacco plants against the tobacco aphid (*Myzus nicotianae*). *Pest Manag. Sci.*, 63, 1215–1223, 2007.

50. Thakur, N., Upadhyay, S.K., Verma, P.C., Chandrashekar, K., Tuli, R., Singh, P.K., Enhanced whitefly resistance in transgenic tobacco plants expressing double stranded RNA of v-ATPase A gene. *PloS One*, 9, e87235, 2014.
51. Corbin, D.R., Grebenok, R.J., Ohnmeiss, T.E., Greenplate, J.T., Purcell, J.P., Expression and chloroplast targeting of cholesterol oxidase in transgenic tobacco plants. *Plant Physiol.*, 126, 1116–1128, 2001.
52. Perlak, F.J., Fuchs, R.L., Dean, D.A., McPherson, S.L., Fischhoff, D.A., Modification of the coding sequence enhances plant expression of insect control protein genes. *Proc. Nat. Acad. Sci.*, 88, 8, 3324–3328, 1991.
53. Perlak, F.J., Stone, T.B., Muskopf, Y.M., Petersen, L.J., Parker, G.B., McPherson, S.A., Wyman, J., Love, S., Reed, G., Biever, D., Fischhoff, D.A., Genetically improved potatoes: Protection from damage by Colorado potato beetles. *Plant Mol. Biol.*, 22, 313–321, 1993.
54. Javaid, S., Amin, I., Jander, G., Mukhtar, Z., Saeed, N.A., Mansoor, S., A transgenic approach to control hemipteran insects by expressing insecticidal genes under phloem-specific promoters. *Scientific Rep.*, 6, 34706, 2016.
55. Shah, A.D., Ahmed, M., Mukhtar, Z., Khan, S.A., Habib, I., Malik, Z.A., Mansoor, S., Saeed, N.A., Spider toxin (Hvt) gene cloned under phloem specific RSs1 and RolC promoters provides resistance against American bollworm (*Heliothis armigera*). *Biotechnol. Lett.*, 33, 1457–1463, 2011.
56. Murray, C., Markwick, N.P., Kaji, R., Poulton, J., Martin, H., Christeller, J.T., Expression of various biotin-binding proteins in transgenic tobacco confers resistance to potato tuber moth, *Phthorimaea operculella* (Zeller) (fam. Gelechiidae). *Transgenic Res.*, 19, 1041–1051, 2010.
57. Tian, Y., Li, T., Mang, K., Han, Y., Li, L., Wang, X., Lu, M., Dai, L., Yan, J., Insect tolerance of transgenic *Populus nigra* plants transformed with *Bacillus thuringiensis* toxin gene. *Chin. J. Biotechnol.*, 9, 219–227, 1993.
58. Wu, N.F., Sun, Q., Yao, B., Fan, Y.L., Rao, H.Y., Huang, M.R., Wang, M.X., Insect-resistant transgenic poplar expressing AaIT gene. Sheng wu gong cheng xue bao. *Chin. J. Biotechnol.*, 16, 129–133, 2000.
59. Cao, C.W., Liu, G.F., Wang, Z.Y., Yan, S.C., Ma, L., Yang, C.P., Response of the gypsy moth, Lymantria dispar to transgenic poplar, *Populus simonii*× *P. nigra*, expressing fusion protein gene of the spider insecticidal peptide and Bt-toxin C-peptide. *J. Insect Sci.*, 10, 200, 2010.
60. Zhang, B., Chen, M., Zhang, X., Luan, H., Diao, S., Tian, Y., Su, X., Laboratory and field evaluation of the transgenic *Populus alba*× *Populus glandulosa* expressing double coleopteran-resistance genes. *Tree Physiol.*, 31, 567–573, 2011.
61. Yang, L., Sun, Y., Wang, Y., Hao, Y., Effects of dietary transgenic poplar leaf pellets on performance and tissues in rabbits. *J. Sci. Food Agricult.*, 94, 1163–1167, 2014.
62. Sujatha, M., Lakshminarayana, M., Tarakeswari, M., Singh, P.K., Tuli, R., Expression of the cry1EC gene in castor (*Ricinus communis* L.) confers field resistance to tobacco caterpillar (*Spodoptera litura* Fabr) and castor semilooper (*Achoea janata* L.). *Plant Cell Rep.*, 935–946, 2009.

63. Singsit, C., Adang, M.J., Lynch, R.E., Anderson, W.F., Wang, A., Cardineau, G., Ozias-Akins, P., Expression of a *Bacillus thuringiensis* cryIA (c) gene in transgenic peanut plants and its efficacy against lesser cornstalk borer. *Transgenic Res.*, 6, 169–176, 1997.
64. Tiwari, S., Mishra, D.K., Singh, A., Singh, P.K., Tuli, R., Expression of a synthetic cry1EC gene for resistance against *Spodoptera litura* in transgenic peanut (*Arachis hypogaea* L.). *Plant Cell Rep.*, 27, 1017–1025, 2008.
65. Beena, M.R., Tuli, R., Gupta, A.D., Kirti, P.B., Transgenic peanut (*Arachis hypogaea* L.) plants expressing cry1EC and rice chitinase cDNA (Chi11) exhibit resistance against insect pest *Spodoptera litura* and fungal pathogen *Phaeoisariopsis personata. Trans. Plant J.*, 2, 157–164, 2008.
66. Tiwari, S., Mishra, D.K., Chandrasekhar, K., Singh, P.K., Tuli, R., Expression of δ-endotoxin Cry1EC from an inducible promoter confers insect protection in peanut (*Arachis hypogaea* L.) plants. *Pest Manag. Sci.*, 67, 137–145, 2011.
67. Benchekroun, A., Michaud, D., Nguyen-Quoc, B., Overney, S., Desjardins, Y., Yelle, S., Synthesis of active oryzacystatin I in transgenic potato plants. *Plant Cell Rep.*, 14, 585–588, 1995.
68. Gatehouse, A.M., Norton, E., Davison, G.M., Babbé, S.M., Newell, C.A., Gatehouse, J.A., Digestive proteolytic activity in larvae of tomato moth, *Lacanobia oleracea*; Effects of plant protease inhibitors *in vitro* and *in vivo. J. Insect Physiol.*, 45, 545–558, 1999.
69. Mi, X., Liu, X., Yan, H., Liang, L., Zhou, X., Yang, J., Si, H., Zhang, N., Expression of the *Galanthus nivalis* agglutinin (GNA) gene in transgenic potato plants confers resistance to aphids. *C. R. Biol.*, 340, 7–12, 2017.
70. Hagh, Z.G., Rahnama, H., Panahandeh, J., Rouz, B.B., Jafari, K.M., Mahna, N., Green-tissue-specific, C4-PEPC-promoter-driven expression of Cry1Ab makes transgenic potato plants resistant to tuber moth (*Phthorimaea operculella*, Zeller). *Plant Cell Rep.*, 28, 1869–1879, 2009.
71. Outchkourov, N.S., Rogelj, B., Strukelj, B., Jongsma, M.A., Expression of sea anemone equistatin in potato. Effects of plant proteases on heterologous protein production. *Plant Physiol.*, 133, 379–390, 2003.
72. Li, X.B., Mao, H.Z., Bai, Y.Y., Transgenic plants of rutabaga (*Brassica napobrassica*) tolerant to pest insects. *Plant Cell Rep.*, 15, 97–101, 1995.
73. Kanrar, S., Venkateswari, J., Kirti, P., Chopra, V., Transgenic Indian mustard (*Brassica juncea*) with resistance to the mustard aphid (*Lipaphis erysimi* Kalt.). *Plant Cell Rep.*, 20, 976–981, 2002.
74. Sayyed, A.H., Schuler, T.H., Wright, D.J., Inheritance of resistance to Bt canola in a field-derived population of *Plutella xylostella*. *Pest Manag. Sci.*, 59, 1197–1202, 2003.
75. Wang, J., Chen, Z., Du, J., Sun, Y., Liang, A., Novel insect resistance in *Brassica napus* developed by transformation of chitinase and scorpion toxin genes. *Plant Cell Rep.*, 24, 549–555, 2005.
76. Meng, F., Li, Y., Zang, Z., Li, N., Xue, R., Cao, Y., Li, T., Zhou, Q., Li, W., Expression of the double-stranded RNA of the soybean pod borer

Leguminivora glycinivorella (Lepidoptera: Tortricidae) ribosomal protein P0 gene enhances the resistance of transgenic soybean plants. *Pest Manag. Sci.*, 73, 2447–2455, 2017.

77. Yu, H., Li, Y., Li, X., Wu, K., Arthropod abundance and diversity in transgenic Bt soybean. *Environ. Entomol.*, 43, 1124–1134, 2014.
78. Shade, R.E., Schroeder, H.E., Pueyo, J.J., Tabe, L.M., Murdock, L.L., Higgins, T.J.V., Chrispeels, M.J., Transgenic pea seeds expressing the α-amylase inhibitor of the common bean are resistant to bruchid beetles. *Nat. Biotechnol.*, 12, 793, 1994.
79. Schroeder, H.E., Gollasch, S., Moore, A., Tabe, L.M., Craig, S., Hardie, D.C., Chrispeels, M.J., Spencer, D., Higgins, T.J., Bean [alpha]-amylase inhibitor confers resistance to the pea weevil (*Bruchus pisorum*) in transgenic peas (*Pisum sativum* L.). *Plant Physiol.*, 107, 1233–1239, 1995.
80. Chakraborti, D., Sarkar, A., Mondal, H.A., Das, S., Tissue specific expression of potent insecticidal, *Allium sativum* leaf agglutinin (ASAL) in important pulse crop, chickpea (*Cicer arietinum* L.) to resist the phloem feeding *Aphis craccivora*. *Transgenic Res.*, 18, 529–544, 2009.
81. Lüthi, C., Álvarez-Alfageme, F., Ehlers, J.D., Higgins, T.J., Romeis, J., Resistance of αAI-1 transgenic chickpea (*Cicer arietinum*) and cowpea (*Vigna unguiculata*) dry grains to bruchid beetles (Coleoptera: Chrysomelidae). *Bull. Entomological Res.*, 103, 373–381, 2013.
82. Ganguly, M., Molla, K.A., Karmakar, S., Datta, K., Datta, S.K., Development of pod borer-resistant transgenic chickpea using a pod-specific and a constitutive promoter-driven fused cry1Ab/Ac gene. *Theor. Appl. Genet.*, 127, 2555–2565, 2014.
83. Dowd, P.F., Lagrimini, L.M., Nelsen, T.C., Relative resistance of transgenic tomato tissues expressing high levels of tobacco anionic peroxidase to different insect species. *Nat. Toxins*, 6, 241–249, 1998.
84. Reddy, K.R.K. and Rajam, M.V., Targeting chitinase gene of *Helicoverpa armigera* by host-induced RNA interference confers insect resistance in tobacco and tomato. *Plant Molec. Biol.*, 90, 281–292, 2016.
85. Yi, D., Yang, W., Tang, J., Wang, L., Fang, Z., Liu, Y., Zhuang, M., Zhang, Y., Yang, L., High resistance of transgenic cabbage plants with a synthetic cry1Ia8 gene from *Bacillus thuringiensis* against two lepidopteran species under field conditions. *Pest Manag. Sci.*, 72, 315–321, 2016.
86. Narendran, M., Deole, S.G., Harkude, S., Shirale, D., Nanote, A., Bihani, P., Parimi, S., Char, B.R., Zehr, U.B., Efficient genetic transformation of okra (*Abelmoschus esculentus* (L.) Moench) and generation of insect-resistant transgenic plants expressing the cry1Ac gene. *Plant Cell Rep.*, 32, 1191–1198, 2013.
87. Xiao, Y., Wang, K., Ding, R., Zhang, H., Di, P., Chen, J., Zhang, L., Chen, W., Transgenic tetraploid *Isatis indigotica* expressing Bt Cry1Ac and *Pinellia ternata* agglutinin showed enhanced resistance to moths and aphids. *Molec. Biol. Rep.*, 39, 485–491, 2012.

88. Liu, X., Chen, M., Onstad, D., Roush, R., Shelton, A.M., Effect of Bt broccoli and resistant genotype of *Plutella xylostella* (Lepidoptera: Plutellidae) on development and host acceptance of the parasitoid *Diadegma insulare* (Hymenoptera: Ichneumonidae). *Transgenic Res.*, 20, 887–897, 2011.
89. Zhao, J.Z., Collins, H.L., Tang, J.D., Cao, J., Earle, E.D., Roush, R.T., Herrero, S., Escriche, B., Ferré, J., Shelton, A.M., Development and characterization of diamondback moth resistance to transgenic broccoli expressing high levels of Cry1C. *Appl. Environ. Microbiol.*, 66, 3784–3789, 2000.
90. Markwick, N.P., Docherty, L.C., Phung, M.M., Lester, M.T., Murray, C., Yao, J.L., Mitra, D.S., Cohen, D., Beuning, L.L., Kutty-Amma, S., Christeller, J.T., Transgenic tobacco and apple plants expressing biotin-binding proteins are resistant to two cosmopolitan insect pests, potato tuber moth and lightbrown apple moth, respectively. *Transgenic Res.*, 12, 671–681, 2003.
91. Grace, L.J., Charity, J.A., Gresham, B., Kay, N., Walter, C., Insect-resistant transgenic *Pinus radiata*. *Plant Cell Rep.*, 24, 103–111, 2005.
92. Tabashnik, B.E., Dennehy, T.J., Sims, M.A., Larkin, K., Head, G.P., Moar, W.J., Carrière, Y., Control of resistant pink bollworm (*Pectinophora gossypiella*) by transgenic cotton that produces *Bacillus thuringiensis* toxin Cry2Ab. *Appl. Environ. Microbiol.*, 68, 3790–3794, 2002.
93. Mao, J., Burt, A.J., Ramputh, A.I., Simmonds, J., Cass, L., Hubbard, K., Miller, S., Altosaar, I., Arnason, J.T., Diverted secondary metabolism and improved resistance to European corn borer (*Ostrinia nubilalis*) in maize (*Zea mays* L.) transformed with wheat oxalate oxidase. *J. Agricult. Food Chem.*, 55, 2582–2589, 2007.
94. Kumar, M., Shukla, A.K., Singh, H., Tuli, R., Development of insect resistant transgenic cotton lines expressing cry1EC gene from an insect bite and wound inducible promoter. *J. Biotechnol.*, 140, 143–148, 2009.
95. Dunse, K.M., Stevens, J.A., Lay, F.T., Gaspar, Y.M., Heath, R.L., Anderson, M.A., Coexpression of potato type I and II proteinase inhibitors gives cotton plants protection against insect damage in the field. *Proc. Nat. Acad. Sci.*, 107, 15011–15015, 2010.
96. Vajhala, C.S., Sadumpati, V.K., Nunna, H.R., Puligundla, S.K., Vudem, D.R., Khareedu, V.R., Development of transgenic cotton lines expressing *Allium sativum* agglutinin (ASAL) for enhanced resistance against major sap-sucking pests. *PloS One*, 8, e72542, 2013.
97. Shukla, A.K., Upadhyay, S.K., Mishra, M., Saurabh, S., Singh, R., Singh, H., Thakur, N., Rai, P., Pandey, P., Hans, A.L., Srivastava, S., Expression of an insecticidal fern protein in cotton protects against whitefly. *Nat. Biotechnol.*, 34, 1046, 2016.
98. Han, Q., Wang, Z., He, Y., Xiong, Y., Lv, S., Li, S., Zhang, Z., Qiu, D., Zeng, H., Transgenic cotton plants expressing the HaHR3 gene conferred enhanced resistance to *Helicoverpa armigera* and improved cotton yield. *Int. J. Molec. Sci.*, 18, 1874, 2017.

99. Bi, R.M., Jia, H.Y., Feng, D.S., Wang, H.G., Production and analysis of transgenic wheat (*Triticum aestivum* L.) with improved insect resistance by the introduction of cowpea trypsin inhibitor gene. *Euphytica*, 151, 351–360, 2006.
100. Christy, L.A., Arvinth, S., Saravanakumar, M., Kanchana, M., Mukunthan, N., Srikanth, J., Thomas, G., Subramonian, N., Engineering sugarcane cultivars with bovine pancreatic trypsin inhibitor (aprotinin) gene for protection against top borer (*Scirpophaga excerptalis* Walker). *Plant Cell Rep.*, 28, 175–184, 2009.
101. Weng, L.X., Deng, H.H., Xu, J.L., Li, Q., Zhang, Y.Q., Jiang, Z.D., Li, Q.W., Chen, J.W., Zhang, L.H., Transgenic sugarcane plants expressing high levels of modified cry1Ac provide effective control against stem borers in field trials. *Transgenic Res.*, 20, 759–772, 2011.
102. Khan, M.S., Ali, S., Iqbal, J., Developmental and photosynthetic regulation of δ-endotoxin reveals that engineered sugarcane conferring resistance to 'dead heart' contains no toxins in cane juice. *Molec. Biol. Rep.*, 38, 2359–2369, 2011.
103. Kramer, K.J., Morgan, T.D., Throne, J.E., Dowell, F.E., Bailey, M., Howard, J.A., Transgenic avidin maize is resistant to storage insect pests. *Nat. Biotechnol.*, 18, 670, 2000.
104. Wang, Z., Zhang, K., Sun, X., Tang, K., Zhang, J., Enhancement of resistance to aphids by introducing the snowdrop lectin gene gna into maize plants. *J. Biosci.*, 30, 627–638, 2005.
105. Mao, Y.B., Cai, W.J., Wang, J.W., Hong, G.J., Tao, X.Y., Wang, L.J., Huang, Y.P., Chen, X.Y., Silencing a cotton bollworm P450 monooxygenase gene by plant-mediated RNAi impairs larval tolerance of gossypol. *Nat. Biotechnol.*, 25, 1307–1313, 2007.
106. Baum, J.A., Bogaert, T., Clinton, W., Heck, G.R., Feldmann, P., Ilagan, O., Johnson, S., Plaetinck, G., Munyikwa, T., Pleau, M., Vaughn, T., Control of coleopteran insect pests through RNA interference. *Nat. Biotechnol.*, 25, 1322, 2007.
107. Dowd, P.F., Johnson, E.T., Price, N.P., Enhanced pest resistance of maize leaves expressing monocot crop plant-derived ribosome-inactivating protein and agglutinin. *J. Agricult. Food Chem.*, 60, 10768–10775, 2012.
108. Osman, G.H., Assem, S.K., Alreedy, R.M., El-Ghareeb, D.K., Basry, M.A., Rastogi, A., Kalaji, H.M., Development of insect resistant maize plants expressing a chitinase gene from the cotton leaf worm, *Spodoptera littoralis*. *Scientific Rep.*, 5, 18067, 2015.
109. Schellenberger, U., Oral, J., Rosen, B.A., Wei, J.Z., Zhu, G., Xie, W., McDonald, M.J., Cerf, D.C., Diehn, S.H., Crane, V.C., Sandahl, G.A., A selective insecticidal protein from Pseudomonas for controlling corn rootworms. *Science*, 354, 634–637, 2016.
110. Wei, J.Z., O'Rear, J., Schellenberger, U., Rosen, B.A., Park, Y.J., McDonald, M.J., Zhu, G., Xie, W., Kassa, A., Procyk, L., Nelson, M., A selective insecticidal protein from *Pseudomonas mosselii* for corn rootworm control. *Plant Biotechnol. J.*, 16, 649–659, 2017.

111. Tu, J., Zhang, G., Datta, K., Xu, C., He, Y., Zhang, Q., Khush, G.S., Datta, S.K., Field performance of transgenic elite commercial hybrid rice expressing *Bacillus thuringiensis* δ-endotoxin. *Nat. Biotechnol.*, 18, 1101, 2000.
112. Ramesh, S., Nagadhara, D., Pasalu, I.C., Kumari, A.P., Sarma, N.P., Reddy, V.D., Rao, K.V., Development of stem borer resistant transgenic parental lines involved in the production of hybrid rice. *J. Biotechnol.*, 111, 131–141, 2004.
113. Zaidi, M.A., Ye, G., Yao, H., You, T.H., Loit, E., Dean, D.H., Riazuddin, S., Altosaar, I., Transgenic Rice Plants Expressing a Modified cry1Ca1 Gene are Resistant to *Spodoptera litura* and *Chilo suppressalis*. *Molec. Biotechnol.*, 43, 232–242, 2009.
114. Moghaieb, R.E., Transgenic rice plants expressing cry1Ia5 gene are resistant to stem borer (*Chilo agamemnon*). *GM Crops*, 1, 288–293, 2010.
115. Ling, F., Zhou, F., Chen, H., Lin, Y., Development of marker-free insect-resistant indica rice by *Agrobacterium tumefaciens*-mediated co-transformation. *Front. Plant Sci.*, 7, 1608, 2016.
116. Yang, Y.Y., Mei, F., Zhang, W., Shen, Z., Fang, J., Creation of Bt rice expressing a fusion protein of Cry1Ac and Cry1I-like using a green tissue-specific promoter. *J. Economic Entomol.*, 107, 1674–1679, 2014.
117. Rao, K.V., Rathore, K.S., Hodges, T.K., Fu, X., Stoger, E., Sudhakar, D., Williams, S., Christou, P., Bharathi, M., Bown, D.P., Powell, K.S., Expression of snowdrop lectin (GNA) in transgenic rice plants confers resistance to rice brown planthopper. *Plant J.*, 15, 469–477, 1998.
118. Foissac, X., Loc, N.T., Christou, P., Gatehouse, A.M., Gatehouse, J.A., Resistance to green leafhopper (*Nephotettix virescens*) and brown planthopper (*Nilaparvata lugens*) in transgenic rice expressing snowdrop lectin (*Galanthus nivalis* agglutinin; GNA). *J. Insect Physiol.*, 46, 573–583, 2000.
119. Saha, P., Chakraborti, D., Sarkar, A., Dutta, I., Basu, D., Das, S., Characterization of vascular-specific RSs1 and rolC promoters for their utilization in engineering plants to develop resistance against hemipteran insect pests. *Planta*, 226, 429–442, 2007.
120. Yarasi, B., Sadumpati, V., Immanni, C.P., Vudem, D.R., Khareedu, V.R., Transgenic rice expressing *Allium sativum* leaf agglutinin (ASAL) exhibits high-level resistance against major sap-sucking pests. *BMC Plant Biol.*, 8, 102, 2008.
121. Chandrasekhar, K., Vijayalakshmi, M., Vani, K., Kaul, T., Reddy, M.K., Phloem-specific expression of the lectin gene from *Allium sativum* confers resistance to the sap-sucker *Nilaparvata lugens*. *Biotechnol. Lett.*, 36, 1059–1067, 2014.
122. Duan, X.L., Li, X.G., Xue, Q.Z., Abo-El-Saad, M., Xu, D., Wu, R., Transgenic rice plants harboring an introduced potato proteinase inhibitor II gene are insect resistant. *Nat. Biotechnol.*, 14, 494–498, 1996.
123. Yoza, K.I., Imamura, T., Kramer, K.J., Morgan, T.D., Nakamura, S., Akiyama, K., Kawasaki, S., Takaiwa, F., Ohtsubo, K.I., Avidin expressed in transgenic

rice confers resistance to the stored-product insect pests *Tribolium confusum* and *Sitotroga cerealella. Biosci. Biotechnol. Biochem.*, 69, 966–971, 2005.
124. Choi, M.S., Kim, Y.H., Park, H.M., Seo, B.Y., Jung, J.K., Kim, S.T., Kim, M.C., Shin, D.B., Yun, H.T., Choi, I.S., Kim, C.K., Expression of BrD1, a plant defensin from *Brassica rapa*, confers resistance against brown planthopper (*Nilaparvata lugens*) in transgenic rices. *Molec. Cells*, 28, 131–137, 2009.
125. Zha, W., Peng, X., Chen, R., Du, B., Zhu, L., He, G., Knockdown of midgut genes by dsRNA-transgenic plant-mediated RNA interference in the hemipteran insect *Nilaparvata lugens. PloS One*, 6, e20504, 2011.
126. Pradhan, S., Chakraborty, A., Sikdar, N., Chakraborty, S., Bhattacharyya, J., Mitra, J., Manna, A., Gupta, S.D., Sen, S.K., Marker-free transgenic rice expressing the vegetative insecticidal protein (Vip) of *Bacillus thuringiensis* shows broad insecticidal properties. *Planta*, 244, 789–804, 2016.
127. Ji, H., Kim, S.R., Kim, Y.H., Suh, J.P., Park, H.M., Sreenivasulu, N., Misra, G., Kim, S.M., Hechanova, S.L., Kim, H., Lee, G.S., Map-based cloning and characterization of the BPH18 gene from wild rice conferring resistance to brown planthopper (BPH) insect pest. *Scientific Rep.*, 6, 34376, 2016.
128. Saha, P., Majumder, P., Dutta, I., Ray, T., Roy, S.C., Das, S., Transgenic rice expressing *Allium sativum* leaf lectin with enhanced resistance against sap-sucking insect pests. *Planta*, 223, 1329–1343, 2006.
129. Dubey, N.K., Mishra, D.K., Idris, A., Nigam, D., Singh, P.K., Sawant, S.V., Whitefly and aphid inducible promoters of *Arabidopsis thaliana* L. *J. Genet.*, 97, 109–119, 2018.
130. Ye, R., Huang, H., Yang, Z., Chen, T., Liu, L., Li, X., Chen, H., Lin, Y., Development of insect-resistant transgenic rice with Cry1C*-free endosperm. *Pest Manag. Sci.*, 65, 1015–1020, 2009.
131. Sharma, M., Charak, K.S., Ramanaiah, T.V., Agricultural biotechnology research in India: Status and policies. *Curr. Sci.*, 84, 297–302, 2003.
132. Grover, A. and Pental, D., Breeding objectives and requirements for producing transgenics for major field crops of India. *Curr. Sci.*, 84, 310–20, 2003.
133. Darbani, B., Farajnia, S., Toorchi, M., Zakerbostanabad, S., Noeparvar, S., Stewart, C.N., DNA-delivery methods to produce transgenic plants. *Biotechnology*, 7, 385–402, 2008.
134. Acharjee, S. and Sarmah, B.K., Biotechnologically generating 'super chickpea' for food and nutritional security. *Plant Sci.*, 207, 108–116, 2013.
135. Zhang, W., Dong, Y., Yang, L., Ma, B., Ma, R., Huang, F., Wang, C., Hu, H., Li, C., Yan, C., Chen, J., Small brown planthopper resistance loci in wild rice (*Oryza officinalis*). *Mol. Genet. Genomics*, 289, 373–382, 2014.
136. Chand, R. and Pal, S., Policy and technological options to deal with India's food surpluses and shortages. *Curr. Sci.*, 84, 388–398, 2003.
137. Puchta, H., Marker-free transgenic plants. *Plant Cell Tissue Org. Cult.*, 74, 123–134, 2003.
138. Yau, Y. and Stewart, C.N., Jr., Less is more: Strategies to remove marker genes from transgenic plants. *BMC Biotechnol.*, 13, 36, 2013.

139. Wu, J. and Tian, Y., Development of insect-resistant transgenic cotton with chimeric TVip3A accumulating in chloroplasts, in: *Transgenic Cotton*, pp. 247–58, Humana Press, Totowa, NJ, 2013.
140. Veluthambi, K., Gupta, A.K., Sharma, A., The current status of plant transformation technologies. *Curr. Sci.*, 84, 368–380, 2003.
141. Niu, L., Ma, Y., Mannakkara, A., Zhao, Y., Ma, W., Lei, C., Chen, L., Impact of single and stacked insect-resistant Bt-cotton on the honey bee and silkworm. *PLoS One*, 8, e72988, 2013.
142. Liu, P., He, X., Chen, D., Luo, Y., Cao, S., Song, H., Liu, T., Huang, K., Xu, W., A 90-day subchronic feeding study of genetically modified maize expressing Cry1Ac-M protein in Sprague–Dawley rats. *Food Chem. Toxicol.*, 50, 3215–3221, 2012.
143. Hendriksma, H.P., Härtel, S., Steffan-Dewenter, I., Testing pollen of single and stacked insect-resistant Bt-maize on *in vitro* reared honey bee larvae. *PLoS One.*, 6, e28174, 2011.
144. Freeman, S. and Mwang'ombe, A.W., Crop Protection through Pest-Resistant Genes. *Biotechnology-Volume VIII: Fundamentals in Biotechnology*, 8, 81, 2009.

3

Advances in Crop Improvement: Use of miRNA Technologies for Crop Improvement

Clarissa Challam*, N. Nandhakumar and Hemant Balasaheb Kardile

ICAR–Central Potato Research Station, Upper Shillong, Meghalaya, India

Abstract

Over the past 25 years, there have been several reports describing applications of microRNAs (miRNAs) as key mediators of post-transcriptional gene regulators in eukaryotic. miRNAs are class of noncoding small endogenous regulatory RNAs. In plants, miRNAs mediated gene silencing (fully or nearly complementary targets) through cleavage of target mRNA or translational repression. Although the biological functions of most miRNAs are still unknown, studies have helped to elucidate their participation in different developmental stages or signal transduction, diseases resistance, nutritional value, and metabolomics technologies in genetic engineering. Developments in the field of genetics and molecular biology such as microarray, gene transfer, RNAi, and CRISPR/Cas9 have led to significant boost for crop improvement. Since the first identification of miRNA genes, let-7 and lin-4 from *Caenorhabditis elegans*, thousands of miRNAs have been identified computationally and experimentally in both plants and animals. The current MiRBase entries for plants (viridiplantae), are 10,504 mature sequences. miRNAs have emerged as a promising approach for crop improvement, such as reverse genetics and genetic modification of several desirable plant traits.

Keywords: miRNA, siRNA, RNAi, transcription factor, post-translational, crop improvement

**Corresponding author*: clarissachallam@gmail.com

Rintu Banerjee, Garlapati Vijay Kumar, and S.P. Jeevan Kumar (eds.) OMICS-Based Approaches in Plant Biotechnology, (55–74) © 2019 Scrivener Publishing LLC

3.1 Introduction

Humans depend on plants both directly and indirectly. Plants have served not only for the primary human needs, viz., food, shelter, and clothing, but also resource other derivatives including resin, gum, oil, dyes, drugs, secondary metabolites, and fossil fuels. With the imminent doubling of the world population to 9 billion by 2050, our world communities are facing considerable pressure on food supplies (FAO). Seventy percent increase in food production will be required to feed the ever increasing population on already scarce agricultural resources, biotic and abiotic stresses, and climate changes. To overcome these problems, timely and appropriate techniques including tissue culture, mutagenesis and transformation have been used. Advance functional genomics studies give better understanding of plant genome and help in modifying it. RNA interference (RNAi), next-generation sequencing (NGS), nanotechnology, and CRISPR/Cas9 technology have become new promising techniques for improving crop according to future need.

RNAi is one of the major landmark discoveries in the history of molecular biology that has boosted our knowledge in functional genomics and opened up novel avenues of research that has an immense potential for crop improvement application. RNAi is a blanket term for the action of small interfering RNAs (siRNA) and microRNAs (miRNA). Many classes of small non-coding RNAs are known, of which miRNAs and siRNAs have been well defined and studied. Although both are functionally similar, they differ in their modes of biogenesis. miRNAs are class of small (21–24 nt), non-protein-coding RNA sequences that have important regulatory roles in eukaryotic gene expression. This chapter gives a brief account of plant miRNAs and their advance applications in crop improvement.

3.2 Discovery of miRNAs

The year 2018 marked the 25th anniversary of the first discovery of miRNA, *lin-4*, in the worm *Caenorhabditis elegans* back in 1993 through a forward genetic approach by two independent laboratories, the Ambros Lab and Ruvkun Lab [1, 2]. For nearly a decade, the genomics of this tiny regulatory molecule appeared simple until the discovery of another miRNA, *let-7*, also in *C. elegans* [3]. Later, during 2000–2003, plant miRNAs were also discovered through a combination of genetic screening and bioinformatics analyses [4, 5]. Over the years, significant miRNA discoveries have been made in the areas of plant science. To date, thousands of plant, animal, and virus miRNAs have been identified and registered in the miRBase database (www.mirbase.org). Until March 2018, the recent Release 22 of the miRBase database contains 38,589 entries representing hairpin precursor miRNAs, expressing 48,885 mature miRNA products, in 271 species. The maximum number of plant miRNAs has been discovered in *Medicago truncatula* (710), *Oryza sativa* (671), *Gylcine max* (685), *Picea abies* (594), *Populus trichocarpa* (364), and *Arabidopsis thaliana* (329).

3.3 Evolution and Organization of Plant miRNAs

The near-perfect complementary recognition for target as well as close evolutionary dynamics of miRNAs and their targets [6] suggests that the origin of plant miRNAs is linked to their cognate targets. Most plant miRNA genes are intergenic and are predominantly found at single loci in the genome. They are rarely arranged in tandem, although clustering has been observed in soybean [7]. The identification of many plant miRNAs has provided opportunity to understand the conservation of miRNAs over both large and small evolutionary distances. At one extreme in plant miRNA gene evolution, miRNA families (miR156, miR160, miR319, and miR390) are conserved in moss, indicating their very ancient origin. Also, most miRNA gene families have shown to be conserved among closely related species, and homologs were found among distantly related species. However, not all miRNAs are equally conserved and that ancient miRNA families have a higher level of conservation than new miRNA families [8]. In the single-cell algae, *Chlamydomonas*, none of the identified miRNAs seems conserved in multicellular plants [9], suggesting a convergent origin of the miRNA pathway in these two main groups of Viridiplantae. Some miRNA families evolved after the split between mosses and flowering plants but before the monocot and dicot divergence. Another extreme in plant miRNA gene evolution are miRNA families that were detected early on but typically represented by one copy number and not conserved in phylogenies [10]. Most of the newly identified non-conserved miRNAs in *Arabidopsis* and their predicted targets could possibly have much wider range of proteins covering most aspects of plant biology than currently known for conserved miRNAs [11]. Therefore, a comprehensive picture of the miRNAs evolution in plants can only be obtained when miRNAs from key representative species starting from unicellular plants to flowering plants are fully uncovered.

3.4 Identification of Plant miRNAs

Identification of plant miRNAs is based on their major characteristics: (1) small non-coding RNAs of ~20–22 nt in length; (2) precursors with a well-predicted secondary hairpin structure having lowest free energy (−32–57 kcal/mol), as predicted by mfold or other computational programs; and (3) evolutionarily conserved; in plants the mature miRNAs are usually conserved instead of miRNA precursors in animals. However, some of the above characteristics are not unique to miRNAs. For instance, tRNA also has stem-loop structures. Therefore, to avoid designating other small RNAs or fragments of other RNAs as miRNAs, Ambros *et al.*, [12] developed combined criteria (5) to identify new miRNAs, which include both biogenesis and expression criteria, neither of which on its own is sufficient for identifying a candidate gene as a new miRNA [13].

Currently, there are four approaches for identifying miRNAs in plants: genetic screening, direct cloning after isolation of small RNAs, computational strategy, and expressed sequence tag (ESTs) analysis. Initially, miRNAs were identified by genetic screening. For example, the discovery of miR319/JAW was in a developmental screen of activation-tagged *A. thaliana* lines [14]. This experimental approach is highly dominated by chance and the application is expensive and time consuming. To overcome some of the shortcomings, another experimental approach was described that involves direct cloning after isolation of small RNAs. Firstly, small RNA molecules are isolated by size fractionation and ligated at their 5' and 3' ends to RNA adapters. Finally, small RNAs are converted into cDNA, then amplified and sequenced. Because the method isolates only small RNAs, it is more efficient to obtain miRNAs than general genetic screening. Further, this method was refined by combining it with massively parallel signature sequencing (MPSS) to study *Arabidopsis* miRNAs. This method can also quantify miRNAs' abundance at the same time. However, the quantification of mature miRNAs is rather difficult due to their often low abundance, short length, and sequence homology between miRNA species. The third approach is the computational strategy. Bioinformatics methods have been valuable in the identification of many plant miRNAs based on genome sequence. Early predictor algorithms predict putative miRNAs in genome sequences by targeting secondary RNA structures that are evolutionary conserved between different species. Such approach allows filtering out many of the false-positive candidates, but does limit the discovery of novel miRNAs. Besides, some recently evolved miRNAs appear to be species-specific and are non-conserved. Hence, the concept of machine learning algorithms has been subsequently devised to identify non-conserved miRNAs. Several tools are currently available for finding miRNA genes and prediction of their putative targets in plants [15]. Till date, most sequence information (e.g., ESTs or GSS) used for computational prediction was sequenced by the traditional Sanger method. The development of NGS technologies now provides a rapid way to discover several nonconserved or less expressed miRNAs through cloning and deep sequencing of small RNA.

The fourth approach is the expressed sequence tag (EST) analysis. The approach combined EST analyses with various computational approaches for identification and characterization of plant miRNAs. This approach has proven to be an economically feasible alternative for species lacking draft genomes. This approach can also identify only conserved miRNA by searching for sequences containing mature miRNAs and check if these miRNAs bearing ESTs can form stable secondary stem-loop structures. miRNAs that are more likely nonconserved cannot be identified based on the EST approach. Also, some ESTs are unequally obtained from different tissues.

Thus, all the four existing approaches have different advantages and shortcomings. However, the advancement of technologies such as bioinformatics and NGS has allowed the identification of a large number of putative miRNAs in different plants. Computational algorithms have been adapted to harmonize experimental approaches directed at identifying real miRNAs in cells.

3.5 miRNA vs. siRNA

RNAs are classified into coding and non-coding RNAs. Coding RNAs comprise of messenger RNA (mRNAs), while ncRNAs are subdivided into ribosomal RNAs (rRNAs), transfer RNAs (tRNAs), and small RNAs (sRNAs). Among the sRNAs, small interfering RNAs (siRNAs) and microRNAs (miRNAs) have attracted much attention for their role in gene regulation, which makes them likely targets for crop improvement programs. miRNAs and siRNAs have similar physicochemical properties; both are processed by Dicer-like ribonucleases from long RNA precursors, and both regulate the target gene to produce a gene silencing effect. But their precursor structures, pathway of biogenesis, and modes of action are very distinct (Table 3.1).

Table 3.1 Comparison of miRNAs and siRNAs.

Property	miRNAs	siRNAs
Origin	Natural molecule: encoded by their cognate targets	Natural or a synthetic: encoded by transposons, viruses, heterochromatin
Prior to Dicer processing	Pre-miRNA contains 70–100 nt with interspersed mismatches and hairpin-shaped ssRNA	Double-stranded RNA that contains 30 to over 100 nt
Structure	20–22 bp RNA duplex with 2 nucleotide 3′ overhang	21–24 bp RNA duplex with 2 nucleotide 3′ overhang
Evolutionary conservation	Nearly always conserved in related organisms	Rarely conserved in related Organisms
Cellular role	Degradation of mRNA, translational repression	Degradation of mRNA, DNA methylation, histone modification
Gene regulation mechanism	Regulate genes with partial or full complementary to mRNA	Mediate the silencing of the target mRNA with full complementarity
Functions	Cell differentiation, cell development, regulation of developmental process, biotic and abiotic stress response	Defense against viruses and transposons and stress adaptation

3.6 Biogenesis of miRNAs and Their Regulatory Action in Plants

The biogenesis and regulatory action of miRNAs includes a series of steps. First, miRNA gene is transcribed into primary miRNAs (pri-miRNAs), which are imperfect long folded back structures, by RNA polymerase II (Pol II) enzymes [4, 13, 15]. Second, the pri-miRNA is processed by the RNAse III enzyme DICER-LIKE1 (DCL1) into a perfect or near-perfect stem-loop structure, called precursor miRNA (pre-miRNAs). Two precursor-processing pathways are proposed in this step for plant miRNA genes, viz., stem-to-loop and loop-to-stem processing. The former requires sequence and structure beyond the miRNA:miRNA* site, while the latter requires only the structure between the miRNA and miRNA* in order to excise the mature miRNA [16, 17]. The resulting miRNAs:miRNAs*duplex is then translocated from the nucleus into the cytoplasm by nucleocytoplasmic transporter 1 (HASTY) [15]. Unlike animal miRNAs, which end with free 2′, 3′ hydroxyl groups, plant miRNAs are methylated at the 2′-OH of the 3′-ends by HUA enhancer1 (HEN1), a small RNA methyl transferase [18, 19]. This is required to protect miRNAs from further modification or degradation [20]. The methylated miRNAs:miRNAs* duplex completes its assembly into the RNA Induced Silencing Complexes (RISCs), which contain Argonaut proteins (AGO1) as the core components [21]. The cytoplasm miRNAs are converted into single-strand mature miRNAs by helicases. When the small RNA guide in the RISCs pairs extensively to a target mRNA, the RISCs direct endonucleolytic cleavage of a single phosphodiester bond in the target mRNA at the nucleotides 10 and 11 of the miRNA or siRNA, resulting to target mRNA cleavage or translational repression (Figure 3.1).

Although the regulation of miRNAs is still unclear, reports have shown that miRNA biogenesis is under feedback mechanism. DCL1 gene in plants may be regulated by the status of miRNA biogenesis by two different mechanisms. First, DCL1 mRNA has a binding site for miR162, which leads to the cleavage of DCL1 mRNA [22]. Second, the 14th intron of the DCL1 mRNA also harbors the precursor to miR838, whose processing regulates the functional DCL1 mRNA. The main miRNA effector protein, AGO1 (Argonaut) gene, is also under the regulation by miR168 programmed, AGO1-catalyzed mRNA cleavage in *Arabidopsis* [23].

3.7 Application of miRNA for Crop Improvement

Global warming and climate change are major irreversible problems faced by our world communities, and result in more serious issues, including salinity, drought, and high and low temperature. To meet the growing challenges while feeding a rapidly growing world population, crop production should be significantly increased in a sustainable manner. miRNA is a promising approach for improving agronomical trait and tolerance to abiotic and biotic stresses. Increasing evidence has shown the miRNAs as an important player in response to these stresses (Table 3.2). Hence, the deployment of miRNAs in crop improvement via genetic engineering or marker-assisted selection can contribute toward achieving this goal.

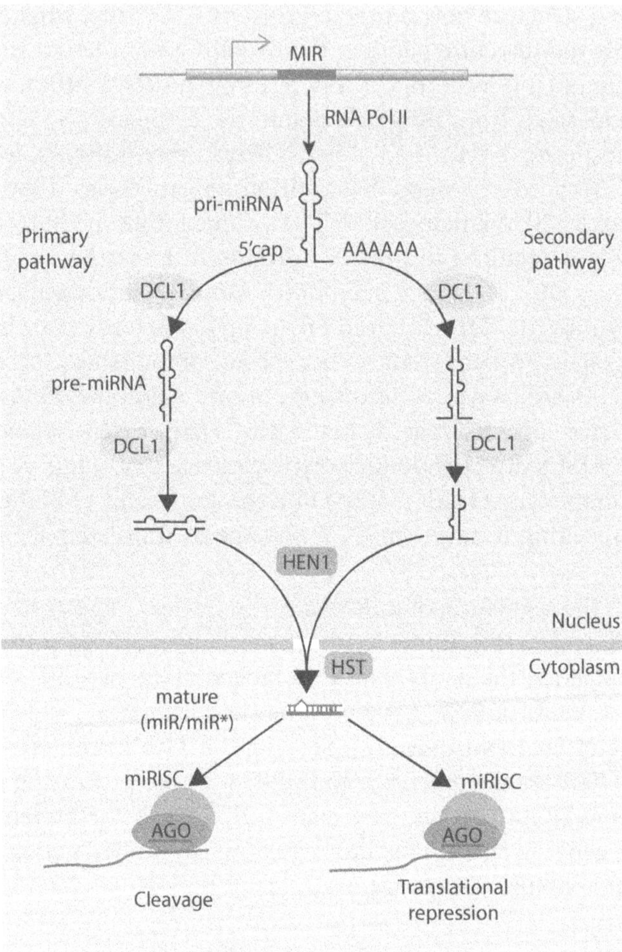

Figure 3.1 The plant miRNA biogenesis [17].

3.8 Concluding Remarks

Two decades of miRNA research in genome biology has given a better insight in understanding the transcriptional regulation mediated by miRNAs in plants; it is also clear that much is still to be discovered. Thorough understandings of the post-transcriptional and post-translational changes are equally important and are future interest. The miRNAs are considered as "master regulators" of gene expression and have been implicated in many important processes like growth, developmental, and biotic and abiotic stress responses. A combination of computational and experimental

Table 3.2 Plant miRNAs and their biological functions.

Species	miRNAs	Target gene	Biological functions	Reference
Agronomy crops				
Rice	miRNA160a, miRNA398b and miRNA769	ARF transcription factors	Resistance to blast disease	[24, 25]
	MiR769	*Nramp*	Resistance to blast disease	[24]
	miR393	*TIR1* and *AFB2*	Drought response, High tillering and early flowering	[26]
	miR820	*DRM2*	Resistance to salt and high temperature	[27]
	miR167	ARF transcription factors	Resistance to cold stress	[28]
	miR397	L-ascorbate oxidase	Heat stress response and adaptation	[28]
	miR1848	*CYP51G3*	Biosynthesis of phytosterol	[29]
Wheat	miR397, miR437	L-ascorbate oxidase	Growth and development	[30]
	miR395	ATP sulfurylase	Abiotic stress	

(*Continued*)

Table 3.2 Plant miRNAs and their biological functions. (*Continued*)

Species	miRNAs	Target gene	Biological functions	Reference
Maize	miR160	ARF transcription factors	Growth and development	[31]
	miR164	NAC1 transcription factors	Development of endosperm	[32]
	miR169	NF-YA transcription factors	Resistance to drought stress response	[33]
Barley	miR399b	Phosphatase transporter	Resistance to drought stress response	[34]
	miR396d	Growth factor	Development of seed	
Sugarcane	miR156	SBP/SPL transcription factors	Growth and development, stress response	[35]
	miR169	HAP12-CCAAT-box transcription factors	Resistance to salt	[36]
	miR399	Inorganic pyrophosphatase 2	Resistance to drought	[37]
Cotton	miR156	SBP/SPL transcription factors	Growth and development, stress response	[38]
	miR396	Callose synthase	Development of fiber	[39]
	miR167a	ARF transcription factors	Resistance to salt	[40]

(*Continued*)

Table 3.2 Plant miRNAs and their biological functions. (*Continued*)

Species	miRNAs	Target gene	Biological functions	Reference
Sorghum	miR156	SBP/SPL transcription factors	Growth and development, increased biomass metabolism	[41]
	miR169	NFY	Growth and development, drought response	[42]
	miR398	Selenium binding protein	Transportation	[43]
Horticultural crops 1. Vegetables				
Tomato	miR156/157	Squamosa promoter binding-like protein	Fruit development	[44]
	miR159	MYB transcription factor		
	miR160	*ARF16*		
	miR164	NAC domain protein		
	miR165/166	Class III homeodomain-leucine zipper		

(*Continued*)

Table 3.2 Plant miRNAs and their biological functions. (*Continued*)

Species	miRNAs	Target gene	Biological functions	Reference
	miR169	Nuclear transcription factor Y subunit A-3		
	miR170/171	GRAS family		
	miR173	Protein strawberry notch homolog 1		
	miR156	SBP-box gene	Fruit ripening	[45]
Sweet potato	miR828	MYB and TLD	Biosynthesis of phenylpropanoid and lignin	[46]
	miR156, miR162	Squamosa binding protein transcription factors	Storage root initiation and growth and development	[47]
Potato	miR172	*RAP1*	Development of flower	[48]
	miR473	Serine/threonine kinase-like	Metabolism	[49]
	miR475	Thioredoxin	Metabolism	
	miR160	ARF transcription factors	Growth and development	

(*Continued*)

Table 3.2 Plant miRNAs and their biological functions. (*Continued*)

Species	miRNAs	Target gene	Biological functions	Reference
Hot pepper	miR156	BP transcription factors	Flowering time	[50]
	miR159/319	MYB and TCP transcription factor	Formation of meristem and seed development	
	miR164	*ORESARA1*	Cell death	
	miR395	Sulfate transporter	Sulfur assimilation pathway	
Cassava	miR164	NAC transcription factors	Tolerant to drought	[51]
2. Fruit crops				
Grapevine	miR160, miR167	ARF transcription factors	Growth and development, stress response	[52]
	miR393, miR394	F-box	Growth and development, stress response	[53]
Apple	miR159	MYB transcription factors	Production of anthocyanin and color formation	[54]

(*Continued*)

Table 3.2 Plant miRNAs and their biological functions. (*Continued*)

Species	miRNAs	Target gene	Biological functions	Reference
	miR828	MYB transcription factors	Signaling pathways and metabolic and biological process	
	miR858	*TAS4*	Signaling pathways and metabolic and biological process	
	miR169a, miR160e miR167b,g	ARF transcription factors	Resistance to fire blight	[55]
Pear	miR160a	ARF16, ARF17, and ARF18	Development of fruit	[56]
Citrus	miRNA156/157	Squamosa promoter binding protein (SBP)-like genes	Development of early flower	[57]
	miRNA160, miRNA167	ARF 8 and ARF 10	Hormone signal	
	miR393	TIR1, ARF, and AFB	Adaptive responses of leaf to B-deficiency	[58]
	miR165	Homeo-domain leucine zipper and HD-Zip protein	Growth and development, stress response, root absorption	[59]

(*Continued*)

Table 3.2 Plant miRNAs and their biological functions. (*Continued*)

Species	miRNAs	Target gene	Biological functions	Reference
Banana	miR156/157	SPL transcription factors	Fruit development and ripening	[60]
	miR319m	GAMYB	Regulating flower formation	
	miR156	*SPL16*	Regulating root and fruit	
3. Plantation crops				
Coffee	miR167	ARF transcription factors	Growth and development, stress response	[61]
	miR393	Transport inhibitor-like protein, DNA-binding proteins, GRR1-like protein	Chitin, cold, salt stress, and water deprivation	[62]
	miR390	*TAS3*	Growth and development, cellular signaling pathways	[61]
Tea	miR156	SBP/SPL transcription factors	Plant growth, development Cold stress response	[63]
	miR171	GRAS family transcription factors	Growth and development, stress response	
	miR408	Plastocyanin-like	Cold stress	[64]

approaches will therefore allow more plant miRNA discoveries in the future, which were then limited to few model species such as *Arabidopsis*, rice, etc. This will gradually reveal the panorama of the miRNAs and further help to understand the evolutionary aspects, functions, structure, and regulation of miRNAs as well as for fine-tuning the regulatory processes.

References

1. Lee, R.C., Feinbaum, R.L., Ambros, V., The *C. elegans* heterochronic gene lin-4 encodes small RNAs with antisense complementarity to lin-14. *Cell*, 75, 843–854, 1993.
2. Wightman, B., Ha, I., Ruvkun, G., Posttranscriptional regulation of the heterochronic gene lin-14 by lin-4 mediates temporal pattern formation in *C. elegans*. *Cell*, 75, 855–862, 1993.
3. Reinhart, B.J., Slack, F.J., Basson, M., Pasquinelli, A.E., Bettinger, J.C., Rougvie, A.E. *et al.*, The 21-nucleotide let-7 RNA regulates developmental timing in *Caenorhabditis elegans*. *Nature*, 403, 6772, 901–6, 2000.
4. Reinhart, B.J., Weinstein, E.G., Rhoades, M.W., Bartel, B., Bartel, D.P., MicroRNAs in plants. *Genes Dev.*, 6, 13, 1616–1626, 2002.
5. Rhoades, M.W., Reinhart, B.J., Lim, L.P., Burge, C.B., Bartel, B., Bartel, D.P., Prediction of plant microRNA targets. *Cell*, 110, 4, 513–520, 2002.
6. Zhao, M., Meyers, B.C., Cai, C. *et al.*, Evolutionary patterns and coevolutionary consequences of MIRNA genes and MicroRNA targets triggered by multiple mechanisms of genomic duplications in soybean. *Plant Cell*, 2015.
7. Zhang, J.F., Yuan, L.J., Shao, Y., Du, W., Yan, D.W., Lu, Y.T., The disturbance of small RNA pathways enhanced abscisic acid response and multiple stress responses in *Arabidopsis* plant. *Cell Env.*, 31, 562–574, 2008.
8. Axtell, M.J. and Bowman, J.L., Evolution of plant microRNAs and their targets. *Trends Plant Sci.*, 13, 343–349, 2008.
9. Molnár, A., Schwach, F., Studholme, D.J., Thuenemann, E.C., Baulcombe, D.C., miRNAs control gene expression in the single-cell alga *Chlamydomonas reinhardtii*. *Nature*, 447, 1126–1129, 2007.
10. Zhang, B., Pan, X., Cannon, C.H., Cobb, G.P., Anderson, T.A., Conservation and divergence of plant microRNA genes. *Plant J.*, 46, 243–259, 2006.
11. Voinnet, O., Origin, biogenesis, and activity of plant microRNAs. *Cell*, 136, 4, 669–687, 2009.
12. Ambros, V., Bartel, B., Bartel, D.P., Burge, C.B., Carrington, J.C., Chen, X., Dreyfuss, G., Eddy, S.R., Griffiths-Jones, S., Marshall, M., Matzke, M., Ruvkun, G., Tuschl, T., A uniform system for microRNA annotation. *RNA*, 9, 277–279, 2003.

13. Xie, Z., Johansen, L.K., Gustafson, A.M., Kasschau, K.D., Lellis, A.D., Zilberman, D., Jacobsen, S.E., Carrington, J.C., Genetic and functional diversification of small RNA pathways in plants. *PLoS Biol.*, 2, 5, E104, 2004.
14. Palatnik, J.F., Allen, E., Wu, X., Schommer, C., Schwab, R., Carrington, J.C., Weigel, D., Control of leaf morphogenesis by microRNAs. *Nature*, 425, 6955, 257–263, 2003.
15. Bologna, N.G., Mateos, J.L., Bresso, E.G., Palatnik, J.F., A loop-to-base processing mechanism underlies the biogenesis of plant microRNAs miR319 and miR159. *EMBO J.*, 28, 23, 3646–3656, 2009.
16. Song, L., Axtell, M.J., Fedoroff, N.V., RNA secondary structural determinants of miRNA precursor processing in *Arabidopsis. Curr. Biol.*, 20, 1, 37–41, 2010.
17. Jiandong, D., Shuigeng, Z., Jihong, G., Finding MicroRNA targets in plants: Current status and perspectives. *Gen. Prot. Bioin.*, 10, 5, 264–275, 2012.
18. Yang, L., Liu, Z., Lu, F., Dong, A., Huang, H., SERRATE is a novel nuclear regulator in primary microRNA processing in *Arabidopsis. Plant J.*, 47, 841–850, 2006.
19. Yu, B., Yang, Z., Li, J., Minakhina, S., Yang, M., Padgett, R.W., Steward, R., Chen, X., Methylation as a crucial step in plant microRNA biogenesis. *Science*, 307, 932–935, 2005.
20. Li, J., Yang, Z., Yu, B., Liu, J., Chen, X., Methylation protects miRNAs and siRNAs from a 30-end uridylation activity in *Arabidopsis. Curr. Biol.*, 15, 1501–1507, 2005.
21. Park, M.Y., Wu, G., Gonzalez-Sulser, A., Vaucheret, H., Poethig, R.S., Nuclear processing and export of microRNAs in Arabidopsis. *Proc. Natl. Acad. Sci., USA*, 102, 3691–3696, 2005.
22. Xie, Z., Kasschau, K.D., Carrington, J.C., Negative feedback regulation of Dicer-Like1 in Arabidopsis by microRNA-guided mRNA degradation. *Curr. Biol.*, 13, 784–789, 2003.
23. Vaucheret, H., Vazquez, F., Crete, P., Bartel, D.P., The action of Argonaute1 in the miRNAs pathway and its regulation by the miRNAs pathway are crucial for plant development. *Genes Dev.*, 18, 1187–119, 2004.
24. Campo, S., Peris-Peris, C., Siré, C., Moreno, A.B., Donaire, L., Zytnicki, M., Notredame, C., Llave, C., San Segundo, B., Identification of a novel microRNA (miRNA) from rice that targets an alternatively spliced transcript of the Nramp6 (Natural resistance-associated macrophage protein 6) gene involved in pathogen resistance. *New Phytol.*, 199, 212–227, 2013.
25. Liu, W., Liu, J., Ning, Y., Ding, B., Wang, X., Wang, Z., Wang, G.L., Recent progress in understanding PAMP- and effector-triggered immunity against the rice blast fungus *Magnaporthe oryzae. Mol. Plant*, 6, 605–620, 2013.
26. Zhou, L., Liu, Y., Liu, Z., Kong, D., Duan, M., Luo, L., Genome-wide identification and analysis of drought-responsive microRNAs in *Oryza sativa. J. Exp. Bot.*, 61, 4157–4168, 2010.

27. Sharma, N., Panchal, S., Sanan-mishra, N., Protocol for artificial microRNA mediated over-expression of miR820 in indica rice. *Am. J. Plant Sci.*, 6, 1951–1961, 2015.
28. Jeong, D.H., Park, S., Zhai, J., Gurazada, S.G., De Paoli, E., Meyers, B.C. *et al.*, Massive analysis of rice small RNAs: Mechanistic implications of regulated microRNAs and variants for differential target RNA cleavage. *Plant Cell*, 23, 4185–4207, 2011.
29. Xia, K., Ou, X., Tang, H., Wang, R., Wu, P., Jia, Y., Wei, X., Xu, X., Seung-Hye, K., Kim, S., Zhang, M., Rice microRNA osa-miR1848 targets the obtusifoliol 14α-demethylase gene OsCYP51G3 and mediates the biosynthesis of phytosterols and brassinosteroids during development and in response to stress. *New Phytol.*, 208, 2015.
30. Han, J., Kong, M.L., Xie, H., Sun, Q.P., Nan, Z.J., Zhang, Q.Z. *et al.*, Identification of miRNAs and their targets in wheat (*Triticum aestivum* L.) by EST analysis. *Genet. Mol. Res.*, 12, 3793–3805, 2013.
31. Gu, Y., Liu, Y., Zhang, J., Liu, H., Hu, Y., Du, H. *et al.*, Identification and characterization of microRNAs in the developing maize endosperm. *Genomics*, 102, 472–478, 2013.
32. Ding, H., Gao, J., Luo, M., Peng, H., Lin, H., Yuan, G. *et al.*, Identification and functional analysis of miRNAs in developing kernels of a viviparous mutant in maize kernel. *Crop J.*, 1, 115–126, 2013.
33. Sheng, L., Chai, W., Gong, X., Zhou, L., Cai, R., Li, X. *et al.*, Identification and characterization of novel maize miRNAs involved in different genetic background. *Int. J. Biol. Sci.*, 11, 781–793, 2015.
34. Shuzuo, L.V., Nie, X., Le, W., Du, X., Biradar, S.S., Jia, X. *et al.*, Identification and characterization of microRNAs from barley (*Hordeum vulgare* L.) by high-throughput sequencing. *Int. J. Biol. Sci.*, 13, 2973–2984, 2012.
35. Zanca, A.S., Vicentini, R., Ortiz-Morea, F.A., Bem, L.E.D., da Silva, M.J., Vincentz, M. *et al.*, Identification and expression analysis of microRNAs and targets in the biofuel crop sugarcane. *BMC Plant Biol.*, 10, 260, 2010.
36. Carnavale-Bottino, M., Rosario, S., Grativol, C., Thiebaut, F., Rojas, C.A., Farrineli, L. *et al.*, High-throughput sequencing of small RNA transcriptome reveals salt stress regulated microRNAs in sugarcane. *PLoS ONE*, 8, e59423, 2013.
37. Ferreira, T.H., Gentile, A., Vilela, R.D., Costa, G.G., Dias, L.I., Endres, L., microRNAs associated with drought response in the bioenergy crop sugarcane (*Saccharum spp.*). *PLoS ONE*, 7, e46703, 2012.
38. Wang, Q. and Zhang, B., MicroRNAs in cotton: An open world needs more exploration. *Planta*, 241, 1303–1312, 2015.
39. Zhang, Q., Strategies for developing green super rice. *Proc. Natl. Acad. Sci. USA*, 104, 16402–16409, 2007.
40. Yin, Z.J., Li, Y., Yu, J.W., Liu, Y.D., Li, C.H., Han, X.L. *et al.*, Difference in miRNA expression profiles between two cotton cultivars with distinct salt sensitivity. *Mol. Biol. Rep.*, 39, 4961–4970, 2012.

41. Katiyar, A., Smita, S., Chinnusamy, V., Pandey, D.M., Bansal, K.C., Identification of miRNAs in sorghum by using bioinformatics approach. *Plant Signal. Behav.*, 7, 246–259, 2012.
42. Paterson, A.H., Bowers, J.E., Bruggmann, R., Dubchak, I., Grimwood, J., Gundlach, H. *et al.*, The Sorghum bicolor genome and the diversification of grasses. *Nature*, 457, 551–556, 2009.
43. Du, J.F., Wu, Y.J., Fang, X.F., Xia, C.J., Liang, Z., Heng, T.S., Prediction of sorghum miRNAs and their targets with computational methods. *Chin. Sci. Bull.*, 55, 1263–1270, 2010.
44. Moxon, S., Jing, R., Szittya, G., Schwach, F., Rusholme, P.R.L., Moulton, V., Dalmay, T., Deep sequencing of tomato short RNAs identifies microRNAs targeting genes involved in fruit ripening. *Genome Res.*, 18, 1602–1609, 2008.
45. Mohorianu, I., Schwach, F., Jing, R., Lopez-Gomollon, S., Moxon, S., Szittya, G., Sorefan, K., Moulton, V., Dalmay, T., Profiling of short RNAs during fleshy fruit development reveals stage-specific sRNAome expression patterns. *Plant J.*, 67, 232–246, 2011.
46. Lin, J.S., Lin, C.C., Lin, H.H., Chen, Y.C., Jeng, S., MicroR828 regulates lignin and H2O2 accumulation in sweet potato on wounding. *New Phytol.*, 196, 427–440, 2012.
47. Sun, R., Guo, T., Cobb, J., Wang, Q., Zhang, B., Role of microRNAs during flower and storage root development in sweet potato. *Plant Mol. Biol. Rep.*, 33, 1731–1739, 2015.
48. Zhang, Y.C., Yu, Y., Wang, C.Y., Li, Z.Y., Liu, Q., Xu, J. *et al.*, Overexpression of microRNA OsmiR397 improves rice yield by increasing grain size and promoting panicle branching. *Nat. Biotechnol.*, 31, 848–852, 2013.
49. Din, M., Barozai, M.Y.K., Baloch, I.A., Identification and functional analysis of new conserved microRNAs and their targets in potato (*Solanum tuberosum* L.). *Turk. J. Bot.*, 38, 1199–1213, 2014.
50. Hwang, D.-G., Park, J.H., Lim, J.Y., The Hot Pepper (*Capsicum annuum*) MicroRNA transcriptome reveals novel and conserved targets: A foundation for understanding MicroRNA functional roles in hot pepper. Unver, T., (Ed.) *PLoS ONE*, 8, 5, e64238, 2013.
51. Patanun, O., Lertpanyasampatha, M., Sojikul, P., Viboonjun, U., Narangajavana, J., Computational identification of microRNAs and their targets in cassava (*Manihot esculenta* Crantz.). *Mol. Biotechnol.*, 53, 257–269, 2013.
52. Wang, C., Shangguan, L., Kibet, K.N., Wang, X., Han, J., Song, C. *et al.*, Characterization of microRNAs identified in a table grapevine cultivar with validation of computationally predicted grapevine miRNAs by miR-RACE. *PLoS ONE*, 6, e21259, 2011.
53. Han, J., Fang, J., Wang, C., Yin, Y., Sun, X., Leng, X. *et al.*, Grapevine microRNAs responsive to exogenous Gibberellin. *BMC Genomics*, 15, 111, 2014.
54. Xia, R., Zhu, H., Hong, An, Y., Beers, E.P., Liu, Z., Apple miRNAs and tasiRNAs with novel regulatory networks. *Genome Biol.*, 13, R47, 2012.

55. Kaja, E., Szcześniak, M.W., Jensen, P.J., Axtell, M.J., McNellis, T., Makałowska, I., Identification of apple miRNAs and their potential role in fire blight resistance. *Tree Genet. Genomes*, 11, 812, 2015.
56. Wu, J., Wang, D.F., Liu, Y.F., Wang, L., Qiao, X., Zhang, S.L., Identification of miRNAs involved in pear fruit development and quality. *BMC Genomics*, 15, 953, 2014.
57. Song, C., Fang, J., Li, X., Liu, H., Thomas Chao, C., Identification and characterization of 27 conserved microRNAs in citrus. *Planta*, 230, 671–685, 2009.
58. Lu, Y.B., Yang, L.T., Qi, Y.P., Li, Y., Li, Z., Chen, Y.B. *et al.*, Identification of boron-deficiency-responsive microRNAs in *Citrus sinensis* roots by Illumina sequencing. *BMC Plant Biol.*, 14, 123, 2014.
59. Song, Y., Zhang, Y.M., Liu, J., Wang, C.Z., Liu, M.Y., Feng, S.Q. *et al.*, Comparison of gibberellin acid content and the genes relatived to GA biosynthesis between 'Changfu 2' apple (*Malus domestica* Borkh.) and its spur sport. *Sci. Agric. Sin.*, 45, 2668–2675, 2012.
60. Chai, J., Feng, R., Shi, H., Ren, M., Zhang, Y., Wang, J., Bioinformatic identification and expression analysis of banana MicroRNAs and their targets. Alvarez, M.L., (Ed.) *PLoS ONE*, 10, 4, e0123083, 2015.
61. Chaves, S.S., Fernandes-Brum, C.N., Silva, G.F.F., Ferrara-Barbosa, B.C., Paiva, L.V., Nogueira, F.T.S. *et al.*, New insights on coffea miRNAs: Features and evolutionary conservation. *Appl. Biochem. Biotechnol.*, 177, 879–908, 2015.
62. Akter, A., Islam, M.M., Mondal, S.I., Mahmud, Z., Jewel, N.A., Ferdous, S. *et al.*, Computational identification of miRNA and targets from expressed sequence tags of coffee (*Coffea arabica*). *Saudi J. Biol. Sci.*, 21, 3–12, 2014.
63. Zhu, Q.W. and Luo, Y.P., Identification of miRNAs and their targets in tea (*Camellia sinensis*). *Biomed. Biotechnol.*, 14, 916–923, 2013.
64. Zhang, Y., Zhu, X., Chen, X. *et al.*, Identification and characterization of cold-responsive microRNAs in tea plant (*Camellia sinensis*) and their targets using high-throughput sequencing and degradome analysis. *BMC Plant Biol.*, 14, 271, 2014.

4

Gene Discovery by Forward Genetic Approach in the Era of High-Throughput Sequencing

Vivek Thakur[1]* and Samart Wanchana[2]

[1]SBDM Unit, 300 Building, International Crops Research Institute for the Semi-Arid Tropics (ICRISAT), Patancheru, Hyderabad, India
[2]National Center for Genetic Engineering and Biotechnology (BIOTEC), National Science and Technology Development Agency (NSTDA), Khlong Luang, Pathum Thani, Thailand

Abstract
Forward genetic experiments are one of the most common approaches for discovery of a gene function, which involves mutagenesis to screen for a feature of interest, and investigate the causal locus/gene. With advancement in sequencing technologies, the mapping of locus can now be simultaneously done by whole-genome sequencing of mapping populations. The chapter describes the strategy for mapping-by-sequencing and various factors such as effect of sequencing coverage, pool size, etc., that must be considered in performing that. Toward the end, the currently available tools have also been mentioned.

Keywords: Mapping-by-sequencing, gene discovery, whole-genome sequencing, functional genomics, mutagenesis

4.1 Introduction

Advancements in DNA sequencing technology have revolutionized the field of genomics such that we now have draft genome sequence of a large number of prokaryotic and eukaryotic species including plants. With the

**Corresponding author*: vivek22@gmail.com

Rintu Banerjee, Garlapati Vijay Kumar, and S.P. Jeevan Kumar (eds.) OMICS-Based Approaches in Plant Biotechnology, (75–90) © 2019 Scrivener Publishing LLC

sequencing technology, the computational methods for their assembly and annotation have also seen tremendous improvement [1]. However, functional information of a large fraction of genes, particularly of newly sequenced genomes, remains either unknown or is based largely on computational evidences, such as sequence homology, conserved domains, etc., which may not have high accuracy [2]. Gene functions are generally discovered through forward and reverse genetic experiments; recent high-throughput approaches such as differential gene expression study using transcriptomes and genome-wide association study (GWAS) have also emerged as alternates [3].

Under the forward genetics approach, molecular lesions are induced in the genome of individual(s) to generate new phenotypes [4], followed by mapping the causal gene or locus. Whereas a reverse genetics approach involves characterization of a gene of interest for its functional role, and tools like TILLING [5] and T-DNA insertional mutagenesis [6] provide its implementation at high-throughput scale. The conventional mapping of causal locus in a forward genetic approach is a two-step process involving generation of mapping population followed by genotyping; and both have been time consuming and labor intensive [7]. However, with the availability of high-throughput sequencing technologies and/or genomic information, mapping is now preferably done by sequencing, which is much faster and relatively easier [8]. In this chapter, we will review the concept, methods, and tools of mapping by sequencing. We will also discuss the challenges often faced during such analysis, and their most appropriate solutions.

4.2 Mutagens Differ for Type and Density of Induced Mutations

Mutagenesis involves generation of mutants by treatment of zygote or embryo or seeds with a mutagen of choice. The advantage of using germ cells or embryo is that the descendant of mutated cells will be homogeneous for the induced mutations. The mutagen treatment (dose and/or duration) is generally based on the LD50 value, which is the amount of mutagen resulting in 50% lethality.

There are varieties of mutagens that have been used for the purpose of mutagenesis, and they differ for the nature of lesion induced in the DNA. Among the chemical mutagens, ethyl methanesulfonate (EMS) and N-Nitroso-N-methylurea (NMU) are the two most commonly used. The chemical mutagens typically induce point mutations that are exploited for mapping analysis as genetic markers. EMS, for example, causes point mutations in the DNA such that cytocines are substituted by thymines (C->T), which in turn causes substitution of the paired bases on the reverse strand (guanines to adenines, G->A) [9]. Physical mutagens include the high-energy radiations [10], and gamma-ray is one of the most commonly used physical mutagens. Unlike chemical mutagens, the physical mutagens primarily cause double-strand breaks in DNA leading to a variety of lesions like deletions of variable size, inversion, translocation, and SNPs [11].

4.3 High-Throughput Sequencing is Getting Better and Cheaper

The last 10–15 years has witnessed tremendous development in the sequencing technology, hence, the name next-generation sequencing (NGS). The advancements have been in terms of increase in read length, drop in cost of per base, volume of data sequenced per unit time, etc. Among various NGS platforms, Hiseq/Miseq/Nextseq from Illumina [12], 454 from Roche [13], and Ion PGM/Proton from Thermo-Fisher [14] are commonly used, and they generate single- or paired-end reads of size between 100 and 500 bp. These platforms find use in diverse applications such as, whole-genome sequencing/resequencing, transcriptome sequencing, metagenomics, and bisulfite sequencing. Some recent sequencing technologies from Pacific Biosciences (PacBio) [15] and Oxford nanopore [16] have substantially improved the read length measuring up to 5–6 kb and 1 Mb, respectively. With this advancement, the computational tools have also improved in terms of accuracy, speed, and hardware requirements.

When it comes to using NGS for mapping genomic locus underlying mutant phenotype, the coverage of sequence data is of higher importance than the read length. This is because such analysis typically involves resequencing of pooled samples, so in order to sequence DNA of individuals that were pooled, it would require a modest coverage to estimate the true allele frequency (details in Section 4.7 on effect of coverage on mapping).

4.4 Mapping-by-Sequencing

In a typical design, seeds are first mutagenized, and the mutant plants (M_1) grown from these mutated seeds, which will be heterozygous for the induced mutations, are self-pollinated to raise M_2 plants. In M_2 stage, the induced mutations (alleles) will segregate and form new allelic combinations, thus rendering them suitable for screening. The mutants may be further selfed to ensure its genetic stability and/or increase homozygosity (Figure 4.1).

In order to map the genetic locus underlying the mutant phenotype, a mapping population is required, theoretically containing shuffled combinations of induced mutations.

To generate a mapping population, the mutant line is either back-crossed with its progenitor or out-crossed with a polymorphic parent to obtain F1 plants. The F1 plants are then selfed to raise a segregating population with wild-type and mutant-type phenotypes approximately in the ratio of 3:1 (Figure 4.1).

The principle behind mapping the causal locus (i.e., underlying the mutant phenotype) is that all mutant F2 lines would share the causal mutation, present in homozygous state, whereas the wild-type F2 lines will either be absent for the causal mutation or be in the heterozygous state. For the mutant F2, a region of genome present in all lines implies that the causal mutation will show linkage with the neighboring markers, given the random nature of recombination events.

Mapping-by-sequencing is done by bulking the DNA of a significant sample size of mutant and wild-type F2 lines followed by their whole-genome sequencing at modest genome coverage (Figure 4.1). The alleles present in each bulked sample are identified on the basis of variations between the wild-type/reference sequence and homologous reads aligning uniquely against the wild-type/reference genome sequence. Depending upon the nature of mutagen used, these variants could be base substitutions, indels, and sometimes structural changes. Once the variants are identified, their allele frequency is obtained, which will be indicative of frequency of allele in the bulked population. Theoretically, the homologous reads from a given region are expected to have originated from all individuals that were bulked; however, it may be partly true in practice, and the allele frequency may deviate a little depending on the number of homologous reads.

A plot of allele frequency of the variants from both bulked populations along the genome is a comprehensive way to examine whether a given allele is present in all individuals (resulting in allele frequency of 1) and also shows linkage with neighboring markers (resulting in a trend line where the allele frequency declines from the peak value on both sides) (Figure 4.2). The pattern of allele frequency in the region harboring causal mutation will be very contrasting between mutant and wild-type samples.

In practice, such analysis may end up with more than one allele qualifying for the criteria mentioned above, thus making all such variants as

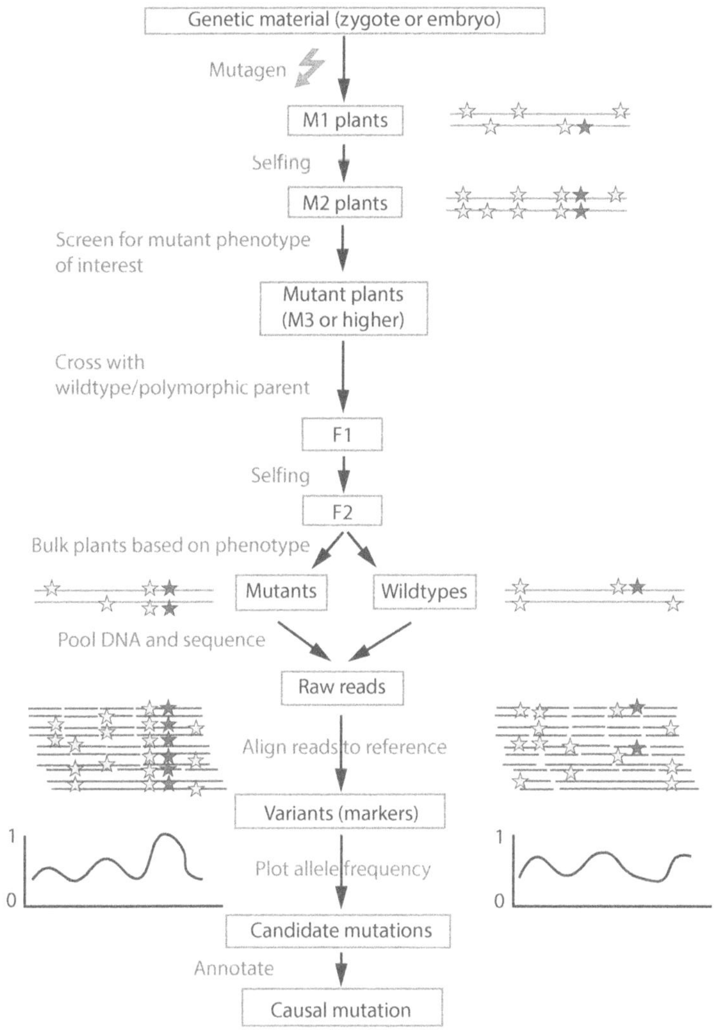

Figure 4.1 Schematic flowchart of mapping by sequencing involving isogenic mapping population (from back-cross). The homologous chromosome pairs are indicated by red lines; the mutations induced by mutagen are indicated by stars (the causal ones are in blue shade while the rest in yellow).

candidates for causal mutation. However, there exists an easy solution for this problem: annotate the variants to examine which candidates have the potential to cause functional abnormalities such as, truncated protein, amino acid change in functionally important motifs, etc. Often this helps in filtering the candidates; however, if there still remains more than one candidate causing potential functional changes, then additional information

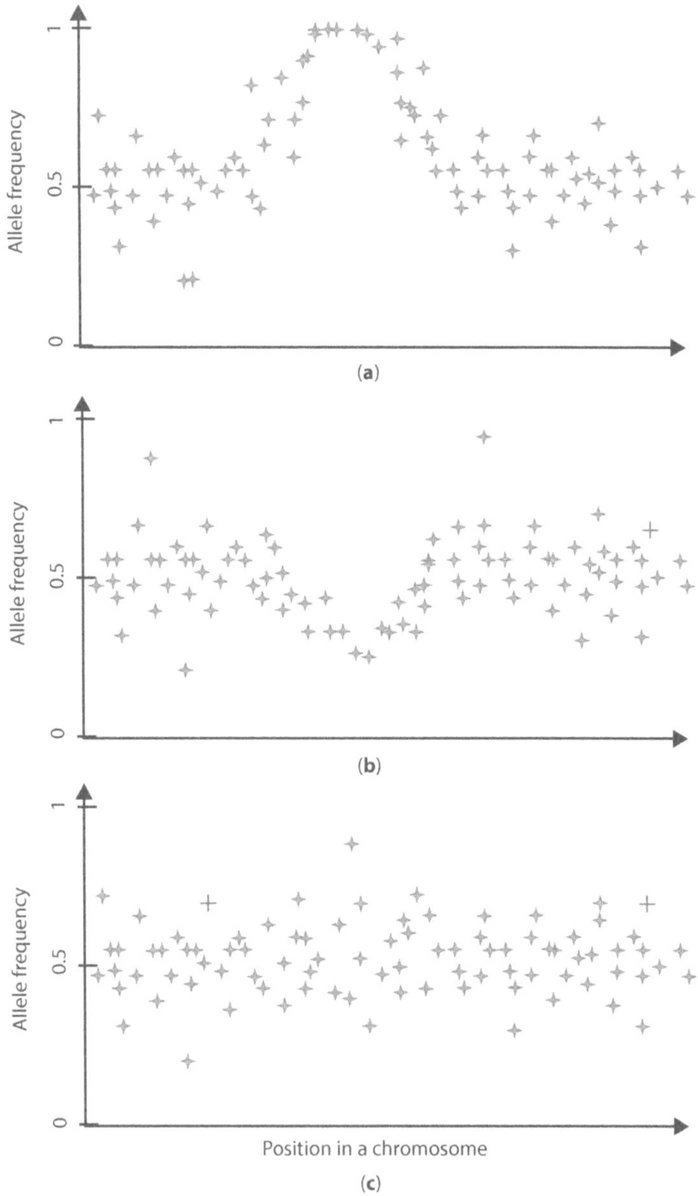

Figure 4.2 Schematic plots of allele frequency of induced mutations along a chromosome, and various likely patterns (in the case of isogenic mutant population). (a) Plot for the mutant-pool sample, having allele(s) with frequency of 1 along with linked alleles, indicating genetic association of the locus with the phenotype, (b) plot of allele frequency in same chromosome but for the nonmutant (wild-type) pool sample, and (c) plot showing behavior of induced mutations in loci not associated with the phenotype.

may be needed to discover causal mutation. The additional information could be ontology data (molecular function, biological process, cellular location), expression of gene of interest in relevant tissue/phenotype, etc.

The last aspect with such analysis is validation, which may be important if the evidences supporting the causal mutation are poor. Sometimes, bioinformatics pipelines for variant discovery may provide spurious candidates, so if their quality is poor, then it would be good practice to confirm the variant through amplicon sequencing in either mutant parent or F2 segregants. If the function of gene carrying the causal mutation is unknown, then their functional validation through knock-down or over-expression adds value to the overall analysis.

4.5 Different Mapping Populations for Specific Need

In the first report of mapping-by-sequencing, the mapping population was obtained by out-crossing the mutant to a polymorphic wild-type line (genetically diverged from the mutant progenitor) followed by selfing, which was reported as SHOREmap method [8]. Out-crossing resulted in introduction of a large number of natural polymorphisms in the recombinant genomes, and thereby provided a strong basis for allele frequency estimations. In this method, sequencing reads are aligned to the genome sequence of the wild-type parental line (progenitor) to search for regions with a high frequency of wild-type specific SNPs (Table 4.1). Crossing the mutant to a diverged strain, however, poses challenge(s) by altering and interfering with the mutant phenotype, which may result into incorrect phenotyping/pooling, eventually leading to a considerably larger than expected mapping interval [17–19].

As an alternate, mapping population can also be generated by back-crossing the mutant with the wild-type parent, which was reported as MutMap approach [18] (Figure 4.1). A similar method involving EMS mutagenesis for generation of isogenic mapping population was reported in the same year by Hartwig *et al.*, [19] and by Zhu *et al.*, [20]. The MutMap approach uses SNPs incorporated by mutagenesis as markers to look for the region having causal mutation. Unlike the phenotyping issues in SHOREmap, the F2 progeny here shows unequivocal segregation between the mutant and wild-type phenotypes. Besides, consideration of relatively fewer SNPs in MutMap leads to better alignment between genome sequences and lower noise in SNP calling (Table 4.1). The two approaches also differ in size of F2 progeny required for bulking: MutMap can be performed with as low as 20 [18], whereas application of SHOREmap typically requires 500 progenies to be bulked [8].

Table 4.1 Differences among three mapping populations typically used for mapping-by-sequencing.

Mapping population types→	Out-cross to distant line	Back-cross using progenitor	Segregating population at M3 generation
Implementing methods	SHOREmap	MutMap	MutMap+
Phenotyping	Could be challenging	Only mutant and WT phenotypes are likely	Only mutant and WT phenotypes are likely
Generation suitable for bulking	F2	BC1F2	M3
Typical bulk size	20	500	20–40
Markers used for mapping	SNPs distinct between two parents	Mutagen-induced SNPs showing MAF = 1 in mutant bulk	Mutagen-induced SNPs differing in MAF between the two bulks

There are some specific cases where crossing, either with distant line or wild-type parent, for generation of mapping population is practically difficult (for example, sterile F1). In such scenario, mapping cannot be performed in the absence of mapping population. To address this issue, a method namely MutMap+ was reported [21]. This method is based on selfing of heterozygous plants showing wild-type phenotype, whose M2 progeny segregates for wild-type and mutant phenotype in the ratio of 3:1. The M3 population, segregating for the target mutation, can be bulked based on phenotype and subjected for whole-genome sequencing. Thus, the population needed to map the locus is available over a shorter time span than other two approaches discussed above (Table 4.1).

Under the MutMap+ approach, the reads are aligned to the reference sequence of the parental genotype and mutant allele frequency (MAF) along the genome is plotted. To rule out the noncausal SNPs (those that become fixed to homozygous state in the M2 generation and thus are shared by all M3 plants), the allele frequency plots of both the wild-type and mutant bulks are compared (Table 4.1).

Regions showing MAF = 1 by random fixation of SNPs in the M2 generation should be shared between the two bulks, whereas the region having the causal mutation should be specific to the mutant bulk only.

4.6 Effect of Mutagen Type on Mapping

As discussed earlier, the chemical and physical mutagens differ in the nature of lesions they induce in the genomic DNA. If there is *a priori* information about the nature of mutations induced, as in the case of ethyl methanesulfonate (EMS), then it makes it much easier to distinguish induced mutations from the spurious ones during the data analysis step (i.e., variant filtering). In contrast, the gamma rays can induce a range of mutation type (i.e., SNPs, small indels, and structural variations), and each type needs to be examined for causality. Besides nature of mutations induced, the mutagens also differ in terms of density of induced mutations; the density for EMS was much higher than that from gamma-rays [22, 23]. Mapping causal locus using mutant populations with shallow density of induced mutations is much more challenging [23], as with fewer mutations, it becomes difficult to accurately pick the linkage of alleles with the one with MAF = 1. Secondly, low density of induced mutations implies requirement of large M_1 population for screening, if the purpose is to cover almost an entire set of genes present.

4.7 Effect of Bulk Size and Sequencing Coverage on Mapping

While designing a mapping experiment, two factors, namely, size of bulk and sequencing coverage, play very crucial roles. As the size of bulk declines, the length of mapped locus increases, resulting in too many candidate mutations to characterize. Moreover, the *chance occurrence* of allele frequency reaching its maximum value of 1 in mutant pool also increases. The underlying reason is that a genome may have several thousands of induced SNPs, and thus such patterns can be observed mere by chance, unless there is enough statistical power to rule out the noise. So it is critical that the bulk size should be well above the threshold.

James *et al.*, [24] examined trade-off between bulk size and sequencing coverage for both out-cross and back-cross populations. For the out-cross population, the bulk size has *large* effect on the mapped interval size than the sequencing coverage [24]. For example, at an average genome-wide coverage of 15×, pools with 50 and 200 recombinants yielded an average interval size of 783 (± 567) kb and 381 (± 222) kb, respectively.

The study recommended a pool size of 150 and sequencing coverage of >25 as optimal. Whereas, for the back-cross population, both factors (i.e., pool size and coverage) had high effects on the mapping interval size and the number of candidate mutations (Figure 4.3). As evident from the figure, at 25× coverage, some improvement is observed when the pool size increases from 10 to 20, but remains the same thereafter. At 50× coverage, further improvement is observed till the pool size is increased to 40. Further increasing the coverage to 100× shows even further improvements in terms of reduction in the number of candidate mutations (Figure 4.3). While relatively smaller pool size (50) was found optimum, however, a higher sequencing coverage (optimally 50×) was recommended to ensure a reasonable interval size [24].

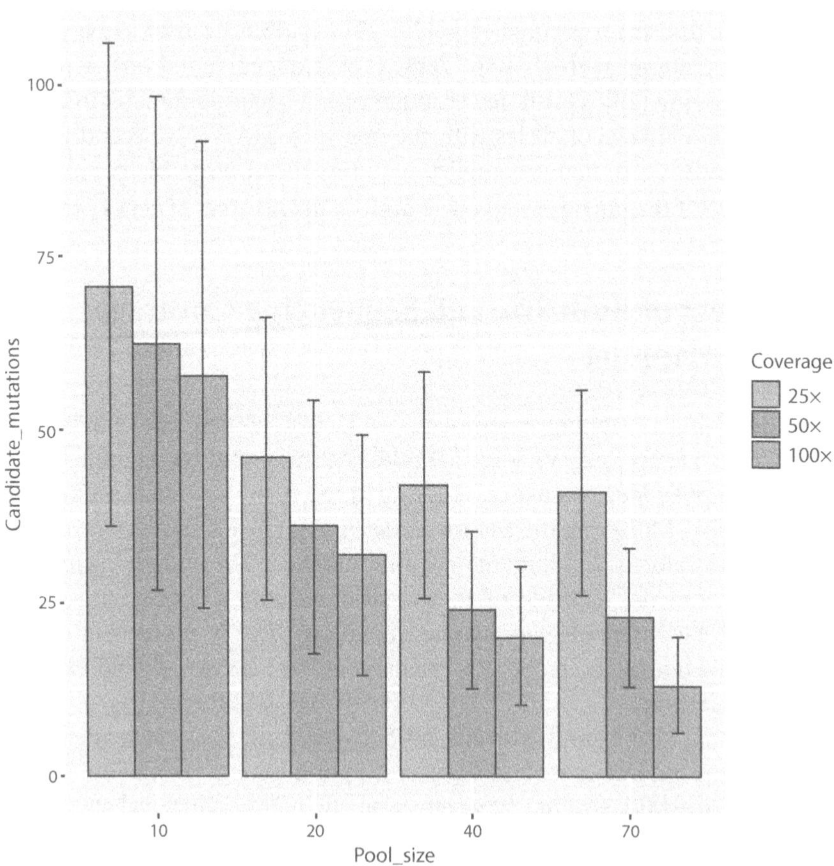

Figure 4.3 Effect of pool size (or bulk size) and sequencing coverage on the interval size of mapped locus (redrawn from Ref. [24]). Increase in value of both of these factors led to reduction in the number of candidate mutations. The coverage of 50× and pool size of 50 was reported as optimal by James *et al.*, [24].

4.8 Challenges in Variant Calling

Irrespective of the population used for mapping, SNPs are required as markers to map the causal mutation/locus. In the era of sequencing, the SNPs are identified while looking for variations between aligned reads and the reference sequence. While the strategies for alignment of shorter reads have improved a lot in terms of speed as well as accuracy, still a large number of SNPs are wrongly called [25]. Sometimes reads actually do not belong to a particular region where they get aligned as they somehow qualify the alignment score, thereby leading to false SNP calls. On the other hand, reads may contain some wrong base calls, which may be present despite quality control measures, thus contributing to the pool of false SNP calls. Such false positives often appear as variants with very low allele frequency (<0.25) [23, 26], which is unlikely to be observed either from homozygous or heterozygous genotypes.

There are some tools such as Coval [25] that try to address such issues. There are some easy strategies to reduce the number of falsely called SNPs such as the following: (a) by setting off the alignment of clipped reads unless the purpose is to discover indels and structural variants, (b) by ensuring that alignment quality of aligned reads is above a threshold, and (c) there should be a minimum number of reads to support any genotype call, etc. Discovery of true cases of structural variants is even more challenging than that of SNPs, and overlapping set of variants from multiple programs is often reliable.

4.9 Cases Where Genome Sequence is either Unavailable or Highly Diverged

The mapping-by-sequencing methods discussed above require resequencing followed by alignment of reads on to reference sequence to reconstruct the whole genome. Thereby, the analysis becomes limited to those species for which a reference genome sequence is already available. Despite the revolution in genomics, reference sequences of a number of crops, in particular, the orphan crops, are either under progress or are unavailable. Moreover, for species with existing reference sequence, even mapping-by-sequencing can become technically challenging if the individual of interest has diverged a lot. For instance, if causal mutation lies within large structural changes in the genome of such individuals, then their identification using resequencing data would be very challenging. To address this issue, an approach, namely, *Needle in K-mer Stack* (NIKS), was developed by Nordström *et al.*, [27], which can discover the causal mutation without aligning reads onto the reference genome sequence.

NIKS implementation also requires deep sequencing of mutant and wild-type bulks obtained from an isogenic mapping population after backcross, but the mapping strategy differs from existing methods [27]. First of all, k-mer frequency is obtained from each sequence data, followed by their distribution to identify k-mers that are highly infrequent, which could be an indication of sequencing errors. The filtered k-mers are then merged to form seeds, and the latter are compared between the two samples to identify variations (i.e., mutant alleles and their frequencies) between the samples. This approach can discover only point mutations or very small changes. Relatively larger changes can be discovered by examining if a seed from one sample is mapped to paired seeds from another sample, indicating a deletion. Finally, NIKS generates local *de novo* assemblies to extend the sequences associated with the mutated site resulting into contigs much longer in size. These contigs allow the users for functional annotation to examine the putative effect of mutations either by *de novo* gene predictions or alignments against known gene annotations.

4.10 Bioinformatics Tools for Mapping-by-Sequencing Analysis

In the last decade, along with the development of methods for mapping-by-sequencing, a set of software tools or programs were also developed to help the research community to do the analysis, in particular, to those having little computational experience of processing whole-genome sequencing data. The SHOREmap method is also available as an open source tool, and its current release is SHOREmap v3.x [28, 29]. Similarly, for the MutMap and MutMap+ methods, Abe *et al.*, [18] and Fekih *et al.*, [21] also developed separate pipelines available publicly [30]. The NIKS tool meant for reference-free mapping-by-sequencing is also available as a pipeline [31].

Alternatively, those having basic computational skill can set up their own pipeline by a set of individual bioinformatics programs/applications. Here we briefly describe the one suitable for Illumina paired-end sequencing of an isogenic mapping population. Once the sequencing data of relevant samples (both pools and/or wild-type line) are available, the same should be filtered/trimmed for low base quality and adaptors using tools like Trimmomatic [32], FASTX [33], etc. The clean reads can then be aligned onto reference genome using short read aligners such as BWA [34] or

Bowtie [35]. In case wild-type line has also been sequenced, then the reads of same should first be aligned to the reference sequence, and the consensus sequence can be obtained using Samtools suite [36] or Pilon [37] or GATK [38]. This personal genome should be used as reference to align reads from segregant pools. The alignment should then be processed for calling variants using GATK [38] or Samtools suite [36]. The called variants are generally stored in VCF format and should be filtered to remove those that have too low variant quality or read depth or genotype quality. The allele frequency of filtered variants can be plotted using any plotting software. The locus harboring causal mutation can be identified based on the plots, and the candidate variants should then be annotated using tools like SNPeff [39] or Annovar [40].

Acknowledgments

VT is supported by the Department of Biotechnology (DBT)–Ramalingaswami Fellowship. VT would like to thank SBDM unit, ICRISAT, for hosting his research activities.

References

1. Yandell, M. and Ence, D., A beginner's guide to eukaryotic genome annotation. *Nat. Rev. Genet.*, 13, 5, 329, 2012.
2. Rhee, S.Y. and Mutwil, M., Towards revealing the functions of all genes in plants. *Trends Plant Sci.*, 19, 4, 198, 2014.
3. Cavanagh, C., Morell, M., Mackay, I., Powell, W., From mutations to MAGIC: Resources for gene discovery, validation and delivery in crop plants. *Curr. Opin. Plant Biol.*, 11, 215–221, 2008.
4. Page, D.R. and Grossniklaus, U., The art and design of genetic screens: *Arabidopsis thaliana*. *Nat. Rev. Genet.*, 3, 124–136, 2002.
5. McCallum, C.M., Comai, L., Greene, E.A., Henikoff, S., Targeted screening for induced mutations. *Nat. Biotechnol.*, 18, 455–457, 2000.
6. Krysan, P.J., Young, J.C., Sussman, M.R., T-DNA as an insertional mutagen in *Arabidopsis*. *Plant Cell*, 11, 12, 2283–2290, 1999.
7. Schneeberger, K., Using next-generation sequencing to isolate mutant genes from forward genetic screens. *Nat. Rev. Genet.*, 15, 10, 662, 2014.
8. Schneeberger, K., Ossowski, S., Lanz, C., Juul, T., Petersen, A.H., Nielsen, K.L., Jorgensen, J.E., Weigel, D., Andersen, S.U., SHOREmap: Simultaneous mapping and mutation identification by deep sequencing. *Nat. Methods*, 6, 550, 2009.

9. Greene, E.A., Codomo, C.A., Taylor, N.E., Henikoff, J.G., Till, B.J., Reynolds, S.H., Enns, L.C., Burtner, C., Johnson, J.E., Odden, A.R., Comai, L., Henikoff, S., Spectrum of chemically induced mutations from a large-scale reverse-genetic screen in *Arabidopsis*. *Genetics*, 164, 731–740, 2003.
10. Muller, H.J., Artificial transmutation of the gene. *Science*, 66, 84–87, 1927.
11. Morita, R., Kusaba, M., Iida, S., Yamaguchi, H., Nishio, T., Nishimura, M., Molecular characterization of mutations induced by gamma irradiation in rice. *Genes Genet. Syst.*, 84, 5, 361, 2009.
12. Illumina sequencing platforms, https://www.illumina.com/systems/sequencing-platforms.html.
13. Roche sequencing solutions, https://sequencing.roche.com/en.html.
14. Thermo Fisher Scientific sequencing platforms, https://www.thermofisher.com/in/en/home/brands/ion-torrent.html.
15. Pacific Biosciences sequencing platform, https://www.pacb.com/.
16. Nanopore sequencing platforms, https://nanoporetech.com/.
17. Greenberg, M.V., Ausin, I., Chan, S.W., Cokus, S.J., Cuperus, J.T., Feng, S., Law, J.A., Chu, C., Pellegrini, M., Carrington, J.C., Jacobsen, S.E., Identification of genes required for *de novo* DNA methylation in *Arabidopsis*. *Epigenetics*, 6, 344–354, 2011.
18. Abe, A., Kosugi, S., Yoshida, K., Natsume, S., Takagi, H., Kanzaki, H., Matsumura, H., Yoshida, K., Mitsuoka, C., Tamiru, M., Innan, H., Cano, L., Kamoun, S., Terauchi, R., Genome sequencing reveals agronomically important loci in rice using MutMap. *Nat. Biotech.*, 30, 174, 2012.
19. Hartwig, B., Velikkakam James, G., Konrad, K., Schneeberger, K., Turck, F., Fast isogenic mapping-by-sequencing of EMS-induced mutant bulks. *Plant Physiol.*, 160, 591, 2012.
20. Zhu, Y., Mang, H.-G., Sun, Q., Qian, J., Hipps, A., Hua, J., Gene discovery using mutagen-induced polymorphisms and deep sequencing: Application to plant disease resistance. *Genetics*, 192, 1, 139, 2012.
21. Fekih, R., Takagi, H., Tamiru, M., Abe, A., Natsume, S., Yaegashi, H., Sharma, S., Sharma, S., Kanzaki, H., Matsumura, H., Saitoh, H., Mitsuoka, C., Utsushi, H., Uemura, A., Kanzaki, E., Kosugi, S., Yoshida, K., Cano, L., Kamoun, S., Terauchi, R., MutMap+: Genetic mapping and mutant identification without crossing in rice. *PLoS ONE*, 8, 7, e68529, 2013.
22. Shirasawa, K., Hirakawa, H., Nunome, T., Tabata, S., Isobe, S., Genome-wide survey of artificial mutations induced by ethyl methanesulfonate and gamma rays in tomato. *Plant Biotechnol. J.*, 14, 1, 51–60, 2016.
23. Rizal, G., Thakur, V., Dionora, J., Karki, S., Wanchana, S., Acebron, K., Larazo, N., Garcia, R., Mabilangan, A., Montecillo, F., Danila, F., Mogul, R., Pablico, P., Leung, H., Langdale, J.A., Sheehy, J., Kelly, S., Quick, W.P., Two forward genetic screens for vein density mutants in sorghum converge on a cytochrome P450 gene in the brassinosteroid pathway. *Plant J.*, 84, 2, 257, 2015.

24. James, G.V., Patel, V., Nordström, K.J., Klasen, J.R., Salomé, P.A., Weigel, D., Schneeberger, K., User guide for mapping-by-sequencing in *Arabidopsis*. *Genome Biol.*, 14, 6, R61, 2013.
25. Kosugi, S., Natsume, S., Yoshida, K., MacLean, D., Cano, L., Kamoun, S., Terauchi, R., Coval: Improving alignment quality and variant calling accuracy for next-generation sequencing data. *PLoS One*, 8, 10, e75402, 2013.
26. Rizal, G., Karki, S., Thakur, V., Wanchana, S., Alonso-Cantabrana, H., Dionora, J., Sheehy, J.E., Furbank, R., Von Caemmerer, S., Quick, W.P., A Sorghum (*Sorghum bicolor*) mutant with altered carbon isotope ratio. *PLoS One*, 12, 6, e0179567, 2017.
27. Nordström, K.J.V., Albani, M.C., Velikkakam James, G., Gutjahr, C., Hartwig, B., Turck, F., Paszkowski, U., Coupland, G., Schneeberger, K., Mutation identification by direct comparison of whole-genome sequencing data from mutant and wild-type individuals using k-mers. *Nat. Biotech.*, 31, 325, 2013.
28. SHOREmap pipeline, http://bioinfo.mpipz.mpg.de/shoremap/.
29. Sun, H. and Schneeberger, H., SHOREmap v3.0: Fast and accurate identification of causal mutations from forward genetic screens, in: *Plant Functional Genomics: Methods and Protocols*, 2nd Edition, J.M. Alonso and A.N. Stepanova (Eds.), pp. 381–395, Springer, New York, 2015.
30. MutMap & MutMap+ pipeline, http://genome-e.ibrc.or.jp/home/bioinformatics-team/mutmap.
31. NIKS pipeline, https://sourceforge.net/projects/niks/.
32. Trimmomatic tool for preprocessing of sequence data, http://www.usadellab.org/cms/?page=trimmomatic.
33. FASTX toolkit, http://hannonlab.cshl.edu/fastx_toolkit/.
34. Burrows-Wheeler Aligner (BWA), http://bio-bwa.sourceforge.net/.
35. Bowtie2 aligner, http://bowtie-bio.sourceforge.net/bowtie2/index.shtml.
36. Samtools/Bcftools, http://www.htslib.org/.
37. Pilon, http://software.broadinstitute.org/software/pilon/.
38. The Genome Analysis Tool-Kit, https://software.broadinstitute.org/gatk/.
39. SNPeff tool for SNP annotation, http://snpeff.sourceforge.net/index.html.
40. Annovar tool for variant annotation, http://annovar.openbioinformatics.org.

5

Functional Genomics of Thermotolerant Plants

Nagendra Nath Das

Department of Biotechnology, Berhampur University, Berhampur, Odisha, India

Abstract

In this chapter, critical outline of various aspects involved in functional genomic studies about thermotolerant plants has been described. While thermotolerant shrubs and trees have been noticeably well documented in regard to functional genomics, thermotolerant vegetative plants and herbs are put in the back foot direction. Desert plants pave the way for continuous exploitation of functional genomic research activities. More impetus is given for thermotolerant plants, which possess economical outputs, e.g., thyme, date palm, sunflower, juniper, and eucalyptus. This chapter also addresses unavailability of functional genomics data on some quintessential desert plants having importance on evolutionary point of view analysis.

Keywords: Functional genomics, gene expression and regulation, thermotolerant plants, next-generation sequencing

5.1 Introduction

Functional genomics is a molecular biology approach to decipher the dynamic aspects of data produced by genomic and transcriptomic projects, which are used to explain the workings of complex biological processes such as regulation of gene expression, gene–protein interactions, gene transcription, and translation mechanisms. With advent of a compilation of forward and reverse genetic approaches and rapid

Email: my.dear.sir@gmail.com

Rintu Banerjee, Garlapati Vijay Kumar, and S.P. Jeevan Kumar (eds.) OMICS-Based Approaches in Plant Biotechnology, (91–104) © 2019 Scrivener Publishing LLC

evolution of genome-editing methods, it provides platform for expediting gene function characterizations at a faster time scale [1, 2]. Most tools and techniques used in functional genomics are as follows: (a) Microarray, (b) SAGE, (c) RNAi, (d) VIGS, (e) Insertional mutagenesis, (f) NGS Techniques, (g) qPCR, (h) Transgenic knockouts, (i) ChIP-seq, (j) MSIA, (k) FISH, (l) ELISA, (m) CRISPR system, (n) GSEA, (o) Ingenuity Pathway analysis software, (p) GC-MS, and (q) TILLING. Applications of functional genomics in various categories are as follows: (a) drug target identification, (b) drug resistance, (c) host–pathogen interactions, (d) biological pathway analysis, (e) discovery of alleles/QTLs in association with biotic or abiotic stress tolerance mechanisms, (f) development of novel high-throughput markers, (g) epigenetics, and (h) development of high-density linkage map [3]. Gene driven and phenotype driven are two key significant approaches available for studies in functional genomics. Production of high-temperature transgenic plants has been achieved technologically through the source of various mesophilic plants (Figure 5.1), whereas in case of thermotolerant plants, there are scanty reports of literature data available. Rice 2020 project set up by the International Rice Functional Genomics Steering Committee (IRFGSC) becomes one of the prominent developments regarding the combat of food-security issues [4]. SAGE technique is applied to study host–pathogen interaction and abiotic stress in plants. VIGS and TILLING are among the reverse genetic approaches used for study of

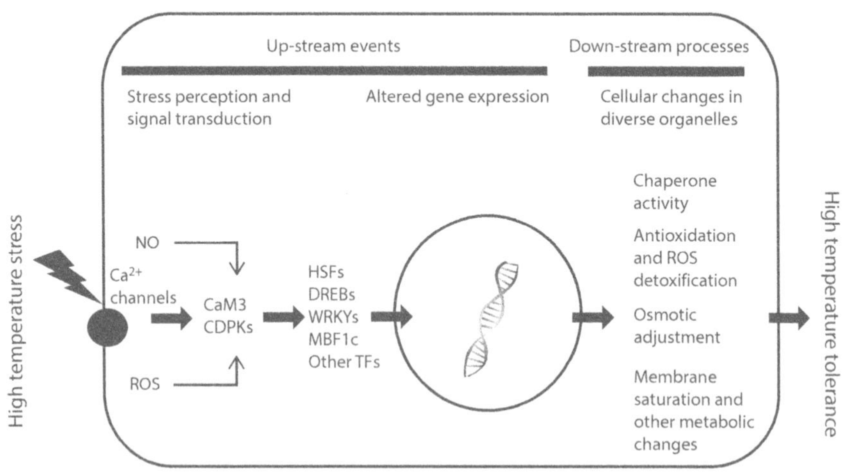

Figure 5.1 Model summarizing the up-stream and down-stream changes governing high-temperature stress tolerance in plants. (Adapted from Grover *et al.*, *Plant Science*, 2013.)

stress genes and phenotypic screening. NGS technology helps to study about genome-wide association and high-throughput markers. QTLs are helpful in detection of polymorphic markers by genotyping and high-density linkage mapping. DNA base-specific fluorescence banding types, e.g., FISH technique, are used for analysis of mapping of chromosome specific markers in plants, which is of high importance in plant molecular systematics, detection of hybrid nature, analysis of somaclonal variations, and diversity analysis.

5.2 Functional Genomics in Plants

Most mesophilic plants already used in the study of functional genomics are as follows: (a) *Arabidopsis*, (b) *Oryza sativa*, (c) *Brassica* sp., (d) *Solanum lycopersicon*, (e) soya bean, and (f) *Triticum* sp. Microarray and gene silencing approaches have been used widely to study functional genomics of these plants. Further developments in genome annotation of aforementioned plants came through NGS technologies [5–11]. Various strategies are employed at the whole plant and cellular and subcellular levels for genetic improvement of plants with temperature stress tolerance. Glycine betaine and proline are two compatible osmolytes that play an important role in maintenance of thermotolerance and were found to be absent in these plants, which are reengineered for production through application of functional genomics tools [12–14]. Cross-talk of phytohormones and secondary metabolites confers thermotolerance in plants, e.g., Brassinosteroids in tomato and oilseed rape [15], ethylene in rosemary [16], abscisic acid in *Agrostis palustris* plants [17], phenylalanine ammonia lyase in watermelon [18], and isoprenoids in different plants [19]. Plants themselves can alter intensive and extensive factors when they act on the abiotic environment surroundings. Some crop plant groups (e.g., fruit crops, forage, and polyploidy crops) exhibit unique breeding problems that require functional genomics strategies to employ, i.e., marker-assisted breeding, transgenic breeding, and genomics-aided breeding. Functional genomic studies in plants were done usually by *Agrobacterium* mediated or particle bombardment methods in association with other techniques, e.g., single cell expression profiling for transcript and protein analyses, AFLP-based RNA fingerprinting for novel variants, SuperSAGE being the most advanced transcriptome technology for functional genomics, CAGE to DeepCAGE for high-throughput transcription start site and promoter identification for gene network analysis, microarrays as a tool to decipher transcriptomes in symbiotic interactions, genome-wide analysis of mRNA expression by fluorescent differential display, real-time quantitation of microRNAs by TaqMan microRNA assays, VIGS for model and other crop plants, RNA interference, TILLING as a reverse genetics tool, transposon tagging in cereal crops, fast neutron mutagenesis, bioinformatics tools to discover co-expressed genes in plants, nanotechnologies and fluorescent proteins application, and next-generation sequencing technologies [20]. The availability of whole-genome sequences paves the way for gene discovery and functional genomics. Specific traits became molecular markers in functional genomics of plants, such as the following: (a) biochemical candidate, (b) positional candidate, and (c) expression candidate [21].

The main goals of functional genomics in plants are as follows: (a) reverse genetics methods for study of loss of function and (b) over-expression of normal gene in transgenic plants for study of gain of function [22]. A major paradigm shift occurred by the implication of gene silencing approaches and recently CRISPR system in crop plants. RNAi, which employs gene knock-out or knock-down phenomenon, can be performed in different ways, i.e., (a) promoter-region directed silencing and (b) coding region-directed silencing. This tool has also been used for analysis of plant metabolic pathways. However, RNAi technique has drawbacks, which suggest that it cannot be applied for mutants producing sterile phenotypes, and the exact sequence of the target gene is required. Multiple transgenic events should be generated and analyzed for conformity that each transgenic line is stably silencing its target, because some of the genes may be resistant to silencing due to undetermined reasons. Tropical crop plants are one of the world's most valuable assets and plant breeders have little opportunity to discover the value of genomics for their improvement efforts [23].

Applications of RNAi technique in some agriculturally significant plants have been illustrated in Table 5.1 for their stable or transient transformations.

5.3 Thermotolerant Plants

Plants that survive in extreme temperature environments (range from 40°C to 60°C) are categorized as thermotolerant in nature. Thermotolerant plants are classified as follows: (a) thermotolerant shrubs, (b) thermotolerant trees, (c) thermotolerant perennials, (d) thermotolerant annuals,

Table 5.1 RNAi technique application in some agriculturally significant plants. (Adapted from *RNAi for functional genomics in plants*, Briefings in functional genomics, 9, 2, 2010.)

Plant	Target gene	Stable "or" transient transformation	Reference
Tomato	ACC oxidase	Stable	[24]
	DET1	Stable	[25]
Wheat	Zinc finger, CCT domain	Stable	[26]
	protein	Stable	[27]
	MADS, K-box domain	Stable	[28]
	protein	Stable	[28]
	Starch branching enzyme IIa	Stable	[29]
	Starch branching enzyme IIb	Stable	[29]
	Transmembrane protein	Stable	[30]
	Phytoene desaturase	Stable	[31]
	NAM/NAC transcription		
	factor		
	Seed storage protein		
Potato	Rar1 (required for Mla12	Transient	[32]
	resistance)	Stable	[33]
	Granule bound starch		
	synthase I		
Coffee bean	CaMXMT1	Stable	[34]
Cotton	Δ12-desaturase	Stable	[35]
	Δ9-desaturase	Stable	[35]
	Myb transcription factor	Stable	[36]
Soybean	24 kDa Oleosin A	Stable	[37]
Rice	Hemeoxygenase	Stable	[38]
	RAC GTPase 1	Stable	[38]
	RAC GTPase 3	Stable	[38]
	RAC GTPase 4	Stable	[38]
	Putative PolyA binding	Stable	[38]
	protein	Stable	[38]
	Putative CBS domain	Stable	[38]
	protein		
	Putative ribosomal protein		
	L5		

(*Continued*)

Table 5.1 RNAi technique application in some agriculturally significant plants. (Adapted from *RNAi for functional genomics in plants*, Briefings in functional genomics, 9, 2, 2010.) (*Continued*)

Plant	Target gene	Stable "or" transient transformation	Reference
Maize	Male sterility factor 45 dihydroflavenol reductase Cytochrome P450	Stable Stable Stable	[39] [39] [39]
Tobacco	Polyphenol oxidase Chalcone isomerase	Stable Stable	[35]
Barley	GAMyb transcription factor Slender protein ABA-inducible kinase	Transient Transient Transient	[40] [40] [40]

(e) thermotolerant herbs, (f) thermotolerant vegetable plants, and (g) thermotolerant desert plants [41–43]. In xeric environments, the common feature observed in plants is their outer appearance, especially modifications in stalk, leaves, and flowers. The list of thermotolerant plants is given in Table 5.2.

Table 5.2 List of thermotolerant plants region-wise.

List of thermotolerant shrubs	Area/Region
(a) Butterfly bush (b) Sumac (c) Juniper (d) Rosemary (e) *Callistemon* sp. (f) Spirea	All mostly occurred in tropical and subtropical desert regions.
List of thermotolerant trees	
(a) Acacia (b) Cedar (c) Pine (d) Eucalyptus	
List of thermotolerant perennials	
(a) Echinacea (b) Blanket flower	

(*Continued*)

Table 5.2 List of thermotolerant plants region-wise. (*Continued*)

List of thermotolerant annuals	Area/Region
(a) Cosmos (b) Sunflower (c) Asparagus	
List of thermotolerant herbs	
(a) Lavender (b) Thyme (c) Rosemary	
List of thermotolerant vegetable plants	
(a) Cherokee purple tomato (b) Eggplants (c) Chili peppers (d) Melons (e) Swiss chard (f) Lantana	
List of thermotolerant Sahara region desert plants	
(a) African peyote cactus (b) Doum palm (c) Red acacia tree (d) Poisonous houseplant (e) *Ephedra alata* (f) *Daisylirion wheeleri* (g) Gourd (h) Sedge (j) Desert sage (k) Bunchgrass (unavailability of genomic data)	
List of thermotolerant other desert region plants	**Area/Region**
(a) Cactus plants: Acuna Cactus, Barrel Cactus, Cholla Cactus, Desert Christmas Cactus, Datil Yucca, Engelmann's Hedgehog Cactus, Organ Pipe Cactus, Prickly Pear Cactus, Saguaro Cactus, Soaptree Yucca, Thornber's Pincushion Cactus, Old Man Cactus (unavailability of genomic data)	All mostly occurred in tropical desert regions. San Bernardino County, California, United States.

(*Continued*)

Table 5.2 List of thermotolerant plants region-wise. (*Continued*)

List of thermotolerant other desert region plants	Area/Region
(b) Bastard toadflax (c) Date palm (d) Cottonwood (e) Desert saltgrass (f) Joshua tree (g) Saxaul (h) Mesquites	Native to the southern Great Plains of the United States (Oklahoma, New Mexico, Texas) and also to northeastern Mexico.
(i) *Ephedra antisyphilitica* Berland. ex C.A.Mey. (Mormon Tea, an IUCN Red List of Threatened Species plant, unavailability of genomic data)	Regions of North America.
(j) Poison ivy (k) Smoke tree (l) Ocotillo	Native to Namibian desert and can live up to 1500 years
(m) Four-wing saltbush (unavailability of genomic data) (n) Desert willow elephant tree	Found within the Karoo desert of South Africa
(o) *Welwitschia mirabilis* (unavailability of genomic data) (p) Baseball plant (*Euphorbia obesa)* (q) *Hydrona africana*	Native to Africa.

5.4 Studies on Functional Genomics of Thermotolerant Plants

Plants possess basal and acquired thermotolerance, which indicates that different genes contribute thermotolerance at different stages of their life cycles [44, 45]. Acquired thermotolerance in plants is done through engineering of chaperone activity, levels of compatible osmolytes, antioxidation and detoxification pathways, membrane and other metabolic processes, transcription factor genes, and up-stream signaling components [46]. Recently, in *Eucalyptus* plants, it was observed that long cold exposure induces transcriptional and biochemical remodeling of xylem secondary cell wall, and CBF transcription factors are master players in the trade-off between growth and stress resistance. A large amount of genomic data are being explored about eucalyptus plant and its thermotolerance capabilities at the phenological and anatomical level [47–50]. Sunflower (*Helianthus annuus* L.) whole-genome analysis provides high-throughput insight of oil metabolism, flowering, and Asterid evolution mechanisms [51]. Regarding the thermotolerant date palm plant, vast amount of genetic manipulation data are available due to its commercial importance [52]. Symbiotic genes and characterization of functional traits of *Mesorhizobium* sp. strains are associated with *Acacia* plant being characterized by phylogenetic tools [53]. Soybean Genomics Executive Committee (SoyGEC) maintained one database (www.soyabase.org) for soya bean thermotolerant plant improvement [54]. Joshua Tree Genome Project reveals genomic organization and function. It is a rare tree of Tikaboo Valley, Nevada (USA) and has evolutionary importance, being thermotolerant in nature [55]. Rosemary (*Rosmarinus officinalis* L.), an economically important thermotolerant species, has therapeutic potential to treat Alzheimer's disease [56]. Sequencing and de novo assembly of the *Toxicodendron radicans* (poison ivy) transcriptome revealed conserved and unique transcripts among Anacardiaceae species [57]. Saguaro Cactus plant (*Carnegiea gigantea*) acts as model species for study of novel adaptations in diverse groups of cacti in the future [58]. Complete chloroplast genome of *Juniperus cedrus*, a thermotolerant plant, reveals phylogenetic analysis among species wise [59]. Absorption of arsenic, sulfur, and phosphorus in *Chilopsis linearis* (desert willow tree) revealed about accumulation, speciation, and coordination in an inbred line [60]. *Lophophora williamsii* plant (African peyote cactus) de novo sequencing and transcriptome analysis infers about putative genes involved in mescaline biosynthesis [61].

5.5 Concluding Remarks

As global climate temperature is rising year wise, the amount of temperature stress is also increasing in plants. This is because plants must adapt themselves toward global climate change at the abiotic and biotic level for their sustainability. Therefore, this phenomenon puts high impetus on implementation of biotechnological tools to improve their sustainability and production rate, so that the challenge on food security can be solved at a faster scale.

Thermotolerant plants can play a vital role in tracing out of genetic data on temperature tress mechanisms, which will ultimately address new insights on the discovery of novel genetic processes and products in other plants or crops. I, myself, as an author, hope that this chapter will facilitate details about the functional genomic role in case of thermotolerant plants.

Abbreviations

RNAi, RNA interference; VIGS, virus-induced gene silencing; SAGE, Serial Analysis Gene Expression; NGS, next-generation sequencing; qPCR, quantitative polymerase chain reaction; ChIP-seq, chromatin immunoprecipitation sequencing; MSIA, mass spectrometric immunoassay; FISH, fluorescent *in situ* hybridization; ELISA, enzyme-linked immunosorbent assay; CRISPR system, Clustered Regularly Interspaced Short Palindromic Repeats system; GSEA, Gene Set Enrichment Analysis; GC-MS, gas chromatography–mass spectrometry; TILLING, targeting-induced local lesions in genomes; QTL, quantitative trait loci; GWAS, genome-wide association study; IUCN, International Union for Conservation of Nature.

References

1. Colebatch, G., Trevaskis, B., Udvardi, M., Functional genomics: Tools of the trade. *New Phytol.*, 153, 27–36, 2002.
2. Sachse *et al.*, High-throughput RNA interference strategies for target discovery and validation by using synthetic short interfering RNAs: Functional genomics investigations of biological pathways. *Methods Enzymol.*, 392, 242–277, 2005.
3. Xin, C., Eric, J., Siu Tim, C., New tools for functional genomic analysis. *Drug Discov. Today*, 14, 15–16, 754–760, 2009.
4. Zhang, Q., Li, J., Xue, Y., Han, B., Deng, X.W., Rice 2020: A call for an international coordinated effort in rice functional genomics. *Mol. Plant*, 1, 5, 715–9, 2008.
5. Debmalya, B., Muhammad Sarwar, K., Eric, D. (Eds.), *Plant Omics: The Omics of Plant Science*, pp. 1–839, Springer India, 2015.
6. Guan, M., Li, X., Guan, C., Microarray analysis of differentially expressed genes between *Brassica napus* strains with high- and low-oleic acid contents. *Plant Cell Rep.*, 31, 929–943, 2012.
7. Igarashi, K., Kazama, T., Motomura, K., Toriyama, K., Whole genomic sequencing of RT98 mitochondria derived from *Oryza rufipogon* and Northern blot analysis to uncover a cytoplasmic male sterility-associated gene. *Plant Cell Physiol.*, 54, 237–243, 2013.
8. La, H., Ding, B., Mishra, G.P., Zhou, B., Yang, H., BellizziMdel, R., Chen, S., Meyers, B.C., Peng, Z., Zhu, J.K., Wang, G.L., A 5-methylcytosine DNA glycosylase/lyase demethylates the retrotransposon Tos17 and promotes its transposition in rice. *Proc. Natl. Acad. Sci. USA*, 108, 15498–15503, 2011.
9. Matsumura, H., Nasir, K.H.B., Yoshida, K., Ito, A., Kahl, G., Kruger, D.H., Terauchi, R., SuperSAGE array: The direct use of 26 base-pair transcript tags in oligonucleotide arrays. *Nat. Methods*, 3, 469–474, 2006.

10. Zhang, C., Grosic, S., Whitham, S.A., Hill, J.H., The requirement of multiple defense genes in soybean Rsv1-mediated extreme resistance to soybean mosaic virus. *Mol. Plant Microbe Interact.*, 25, 1307–1313, 2012.
11. Zha, W., Peng, X., Chen, R., Du, B., Zhu, L., He, G., Knockdown of midgut genes by dsRNA-transgenic plant-mediated RNA interference in the hemipteran insect *Nilaparvata lugens*. *PLOS One*, 6, 5, e20504, 2011.
12. Sakamoto, A. and Murata, N., The role of glycine betaine in the protection of plants from stress: Clues from transgenic plants. *Plant Cell Environ.*, 25, 163–171, 2002.
13. Quan, R., Shang, M., Zhang, H., Zhao, Y., Zhang, J., Engineering of enhanced glycine betaine synthesis improves drought tolerance in maize. *Plant Biotech. J.*, 2, 477–486, 2004.
14. Kavi Kishore, P.B., Sangam, S., Amrutha, R.N., Laxmi, P.S., Naidu, K.R., Rao, K.R.S.S., Rao, S., Reddy, K.J., Theriappan, P., Sreenivasulu, N., Regulation of proline biosynthesis, degradation, uptake and transport in higher plants: Its implications in plant growth and abiotic stress tolerance. *Curr. Sci.*, 88, 424–438, 2005.
15. Dhaubhadel, S., Chaudhary, S., Dobinson, K.F., Krishna, P., Treatment with 24-epibrassinolide, a brassinosteroid, increases the basic thermotolerance of *Brassica napus* and tomato seedlings. *Plant Mol. Biol.*, 40, 333–342, 1999.
16. Munne-Bosch, S., Lopez-Carbonell, M., Alegre, L.A., Van Onckelen, H.A., Effect of drought and high solar radiation on 1-aminocyclopropane-1-carboxylic acid and abscisic acid in *Rosmarinus officinalis* plants. *Physiol. Plant*, 114, 380–386, 2002.
17. Larkindale, J. and Huang, B., Effects of abscisic acid, salicylic acid, ethylene and hydrogen peroxide in thermotolerance and recovery for creeping bent grass. *Plant Growth Regul.*, 47, 17–28, 2005.
18. Rivero, R.M., Ruiz, J.M., Garcia, P.C., Lopez-Lefebre, L.R., Sanchez, E., Romero, L., Resistance to cold and heat stress: Accumulation of phenolic compounds in tomato and watermelon plants. *Plant Sci.*, 160, 315–321, 2001.
19. Loreto, F., Forster, A., Durr, M., Csiky, O., Seufert, G., On the monoterpene emission under heat stress and on the increased thermotolerance of leaves of *Quercus ilex* L. fumigated with selected monoterpenes. *Plant Cell Environ.*, 21, 101–107, 1998.
20. Kahl, G. and Meksem, K., *The Handbook of Plant Functional Genomics: Concepts and Protocols*, pp. 1–577, Wiley-VCH Verlag GmbH & Co. KGaA, Weinheim, Germany, 2008.
21. Kole, C. and Abbott, A.G., *Principles and Practices of Plant Genomics - Volume 3: Advanced Genomics*, pp. 1–614, Science Publishers, P.O. Box 699, Enfield, NH 03748, USA, 2010.
22. Abdeeva *et al.*, Transgenic Plants as a Tool for Plant Functional Genomics, in: *Transgenic Plants - Advances and Limitations*, Y. Ozden Çiftçi (Ed.), pp. 1–478, InTech, 2012.

23. McGinnis, K.M., RNAi for functional genomics in plants. *Brief Funct. Genom.*, 9, 2, 2010.
24. Xiong *et al.*, Different effects on ACC oxidase gene silencing triggered by RNA interference in transgenic tomato. *Plant Cell Rep.*, 23, 9, 639–646, 2005.
25. Davuluri *et al.*, Fruit specific RNAi-mediated suppression of DET1 enhances carotenoid and flavonoid content in tomatoes. *Nat. Biotechnol.*, 23, 7, 890–895, 2005.
26. Yan *et al.*, The wheat VRN2 gene is a flowering repressor down-regulated by vernalization. *Science*, 303, 5664, 1640–1644, 2004.
27. Loukoianov *et al.*, Regulation of VRN-1 vernalization genes in normal and transgenic polyploid wheat. *Plant Physiol.*, 138, 4, 2364–2373, 2005.
28. Regina *et al.*, High-amylose wheat generated by RNA interference improves indices of large bowel health in rats. *Proc. Natl. Acad. Sci. USA*, 103, 10, 3546–3551, 2006.
29. Travella, S., Klimm, T.E., Keller, B., RNA interference based gene silencing as an efficient tool for functional genomics in hexaploid bread wheat. *Plant Physiol.*, 142, 1, 6–20, 2006.
30. Uauy *et al.*, A NAC Gene regulating senescence improves grain protein, zinc, and iron content in wheat. *Science*, 314, 5803, 1298–1301, 2006.
31. Yue *et al.*, Generation of transgenic wheatlines with altered expression levels of 1Dx5 high-molecular weight gluten in subunit by RNA interference. *J. Cereal Sci.*, 47, 2, 153–161, 2007.
32. Bhaskar *et al.*, Agrobacterium-mediated transient gene expression and silencing: A rapid tool for functional gene assay in potato. *PLoS One*, 4, 6, e5812, 2009.
33. Heilersig *et al.*, Post-transcriptional gene silencing of GBSSI in potato: Effects of size and sequence of the inverted repeats. *Plant Mol. Biol.*, 60, 5, 647–662, 2006a.
34. Ogita *et al.*, Application of RNAi to confirm theobromine as the major intermediate for caffeine biosynthesis in coffee plants with potential for construction of decaffeinated varieties. *Plant Mol. Biol.*, 54, 6, 931–941, 2004.
35. Wesley, S.V., Helliwell, C.A., Smith, N.A., Construct design for efficient, effective and high throughput gene silencing in plants. *Plant J.*, 27, 6, 581–590, 2001.
36. Machado *et al.*, The MYB transcription factor GhMYB25 regulates early fibre and trichome development. *Plant J.*, 59, 1, 52–62, 2009.
37. Schmidt, M.A. and Herman, E.M., Suppression of soybean oleosin produces micro-oil bodies that aggregate into oil body/ER complexes. *Mol. Plant*, 1, 6, 910–924, 2008.
38. Okano, Y., Miki, D., Shimamoto, K., Small interfering RNA(siRNA) targeting of endogenous promoters induces DNA methylation, but not necessarily gene silencing in rice. *Plant J.*, 53, 1, 65–77, 2008.
39. Cigan, A.M., Unger-Wallace, E., Haug-Collet, K., Transcriptional gene silencing as a tool for uncovering gene function in maize. *Plant J.*, 43, 6, 929–940, 2005.

40. Zentella, R., Yamauchi, D., Ho, T.H., Molecular dissection of the gibberellin/abscisic acid signaling pathways by transiently expressed RNA interference in barley aleurone cells. *Plant Cell*, 14, 9, 2289–2301, 2002.
41. Gardenerdy, A., Comprehensive List of Desert Plants No One Ever Gave You Before, https://gardenerdy.com/desert-plants-list, 2018.
42. The Spruce, 11 Best Heat Tolerant Shrubs, The 13 Best Heat-Tolerant Trees, https://www.thespruce.com/, 2018.
43. Gill, D., The Times–Picayune garden columnist, Crops That Take the Heat, http://www.nola.com/homegarden/index.ssf/2009/07/post_27.html, 2018.
44. Bell, A. and and Bachman, S., *Ephedra antisyphilitica* Berland. ex C.A.Mey. The IUCN Red List of Threatened Species, http://www.iucnredlist.org/details/201712/0, 2011.
45. Larkindale, J. and Vierling, E., Core genome responses involved in acclimation to high temperature. *Plant Physiol.*, 146, 748–761, 2008.
46. von Koskull-Doring, P., Scharf, K.D., Nover, L., The diversity of plant heat stress transcription factors. *Trends Plant Sci.*, 12, 452–457, 2007.
47. Grover, A., Dheeraj, M., Manisha, N., Lavania, D., Generating high temperature tolerant transgenic plants: Achievements and challenges. *Plant Sci.*, 205–206, 38–47, 2013.
48. Ployet, R., Soler, M., Carocha, V., Ladouce, N., Alves, A., Rodrigues, J.C., Harvengt, L., Marque, C., Teulières, C., Grima-Pettenati, J., Mounet, F., Long cold exposure induces transcriptional and biochemical remodelling of xylem secondary cell wall in *Eucalyptus*. *Tree Physiol.*, 1–14, 2017.
49. Nguyen, H.C., Cao, P.B., San Clemente, H., Ployet, R., Mounet, F., Ladouce, N., Harvengt, L., Marque, C., Teulieres, C., Special trends in CBF and DREB2 groups in *Eucalyptus gunnii* vs *Eucalyptus grandis* suggest that CBF are master players in the trade-off between growth and stress resistance. *Physiol. Plant*, 159, 445–467, 2017.
50. Plasencia, A., Soler, M., Dupas, A., Ladouce, N., Silva-Martins, G., Martinez, Y., Lapierre, C., Franche, C., Truchet, I., Grima-Pettenati, J., Eucalyptus hairy roots, a fast, efficient and versatile tool to explore function and expression of genes involved in wood formation. *Plant Biotechnol. J.*, 14, 1381–1393, 2016.
51. Helene *et al.*, The sunflower genome provides insights into oil metabolism, flowering and Asterid evolution. *Nature*, 000, 1–14, 2017.
52. Jain, S.M., Al-Khayri Jameel, M., Johnson, D.V. (Eds.), *Date Palm Biotechnology*, pp. 1–743, Springer Science+Business Media B.V., 2011.
53. Diouf *et al.*, Phylogenetic analyses of symbiotic genes and characterization of functional traits of *Mesorhizobium* spp. strains associated with the promiscuous species *Acacia seyal* Del. *J. Appl. Microbiol.*, 108, 818–830, 2010.
54. Soybean Genomics Executive Committee, https://soybase.org/resources/soygec.php, 2018.
55. The Joshua Tree Genome Project, http://joshuatreegenome.org/, 2018.

56. Habtemariam, S., The therapeutic potential of rosemary (*Rosmarinus officinalis*) diterpenes for Alzheimer's disease. *Evid. Based Complement Alternat. Med.*, 2680409, 2016.
57. Weisberg *et al.*, Sequencing and de novo assembly of the *Toxicodendron radicans* (poison ivy) transcriptome. *Genes*, 8, 317, 2017.
58. Saguaro Genome Shows Why Many Cactus Species Are Hard to Classify, https://news.azpm.org/p/news-articles/2017/11/7/119597-saguaro-genome-shows-why-many-cactus-species-are-hard-to-classify/, 2018.
59. Guo, Q., Bianba, D., Zheng, W., Characterization of the complete chloroplast genome of *Juniperus cedrus* (Cupressaceae). *Mitochondrial DNA*, 1–2, 2015.
60. Gardea-Torresdey *et al.*, Accumulation, speciation, and coordination of arsenic in an inbred line and a wild type cultivar of the desert plant species *Chilopsis linearis* (desert willow). *Phytochemistry*, 70, 540–545, 2009.
61. Ibarra-Laclette *et al.*, De novo sequencing and analysis of *Lophophora williamsii* transcriptome, and searching for putative genes involved in mescaline biosynthesis. *BMC Genomics*, 161–14, 657, 2015.

Part 2
METABOLOMICS

6

A Workflow in Single Cell-Type Metabolomics: From Data Pre-Processing and Statistical Analysis to Biological Insights

Biswapriya B. Misra

Center for Precision Medicine, Department of Internal Medicine, Section on Molecular Medicine, Wake Forest School of Medicine, Medical Center Boulevard, Winston-Salem, NC, USA

Abstract

Metabolomics is a promising approach for the quantification of metabolically relevant biomarkers in cells, tissues, organs, whole organisms, and biofluids for disease identification and prediction. Hyphenated technologies such as liquid chromatography–mass spectrometry (LC-MS) and gas chromatography–mass spectrometry (GC-MS) platforms have helped generate copious amount of metabolomics data available for further fine-grained analysis. With the growing importance of cell-type resolution for deciphering cellular mechanisms pertaining to biological events, single cell types such as the stomatal guard cells lend themselves for analyses of the stress-responsive metabolic changes when challenged with specific stressors. However, challenges remain with metabolomics data processing and more especially with the statistical treatment of the data. Using computational, bioinformatics, and statistical approaches, and a dataset that explored the guard cell metabolome's response to a abiotic stress treatment (bicarbonate), the results are presented reflecting steps from data treatment to biological interpretation. The insights provided in this study would excite future investigators conducting metabolomics studies to consider the statistical pre-treatment parameters such as normalization, transformation, and scaling more rationally while performing data analysis.

Keywords: Metabolomics, single cell-type, mass spectrometry, workflow, statistical, R, multivariate

Email: bbmisraccb@gmail.com

Rintu Banerjee, Garlapati Vijay Kumar, and S.P. Jeevan Kumar (eds.) OMICS-Based Approaches in Plant Biotechnology, (107–128) © 2019 Scrivener Publishing LLC

6.1 Introduction

Metabolomics refers to the systematic identification and quantification of metabolites (i.e., small molecules in the molecular weight range of 50–2000 Da) present in biological systems (cells, tissues, organs, biofluids, or whole organisms) or environmental samples. Gas chromatography–mass spectrometry (GC-MS) is suitable for the analysis of a broad range of polar metabolites with an order of magnitude increase in metabolite detectability over other methodologies used in metabolomics [1]. Using GC-MS as a tool for metabolomics studies, the underlying mechanisms of several biotic and abiotic stress responses in whole plants [2], tissues [3], and cell-types [4] have been revealed. Similarly, liquid chromatography–mass spectrometry (LC-MS) platforms have been identified as corner stones of discovery and validation metabolomics studies in plants [5]. Thus, the role of combined analysis by GC-MS and LC-MS datasets for comprehensive plant metabolomics is underscored [6].

Paired stomatal guard cells are a plant single cell-type on the leaf surfaces that control the opening of a pore known as the stoma or stomate, thus providing a channel for photosynthetic gas exchange (carbon dioxide, CO_2 and oxygen, O_2), transpirational release of water in land plants, and as a defense against pathogens. Thus, the guard cell physiology (biology) forms an integral part of understanding of biotic and abiotic stress responses in plants. Abscisic acid (ABA), CO_2, and red and blue light mediated stomatal movements have generated immense interest since time immemorial. Moreover, the guard cell metabolome is more or less catalogued, which indicates the role of primary metabolism (simple carbohydrates, amino acids, and citric acid cycle, glycolysis) in stomatal function and movement, reactive oxygen species in stomatal signaling, lipid signaling in movement and development of stomata, the role of phytohormones in stomatal functioning, and the unique sulfur- and nitrogen-rich metabolites in guard cells [7]. In order to reveal the role of CO_2 as a cue for stomatal movements such as opening and closing, it is important to understand the underlying biochemical or metabolic changes using *in vitro* model systems such as isolated and enriched guard cells as well as the use of gaseous CO_2 or its mimic in, i.e., bicarbonate (HCO_3^-). In one such metabolomics study, Misra *et al.*, [8] showed HCO_3-induced significant biochemical changes in a time-dependent manner in guard and mesophyll cells in canola (*Brassica napus*). Moreover, this study formed the basis of several studies concerned with abiotic stress (such as CO_2 response) and guard cell/stomata metabolism [9–11]. Using this publicly archived dataset, I set out to explore the effect of statistical treatment on interpretation of the biological interpretation, which is the metabolic response of plant cell-types to bicarbonate.

6.2 Methods and Data

6.2.1 Source of Data

The dataset used in this study was retrieved from MetaboLights database i.e., accession ID MTBLS309 available here: https://www.ebi.ac.uk/metabolights/MTBLS309, and based on an earlier publication [8].

6.2.2 Processing of Raw Mass Spectrometry Data

The GC-MS data (.RAW data files) were pre-processed, cleaned, aligned, and processed using AMDIS. Libraries consulted were as follows: MSRI spectral libraries of Golm Metabolome Database available from Max Planck Institute for Plant Physiology, Golm, Germany (http://csbdb.mpimp-golm.mpg.de/csbdb/gmd/gmd.html) by matching the mass spectra (Kopka *et al.*, [30]), NIST Mass Spectral Reference Library (NIST11/2011; National Institute of Standards and Technology, USA), and the Mass Bank of North America (MoNA) (http://mona.fiehnlab.ucdavis.edu/).

The LC-MS datasets (AB SCIEX .wiff and .wiffscan files) from the targeted data analysis were analyzed using the settings and parameters as described in Misra *et al.*, [8] using the vendor's raw data to ABF file format that were read using MS-DIAL suite as described [12, 13]. The ABF file converter is available: http://www.reifycs.com/english/AbfConverter/.

The filtered and combined raw GC-MS and LC-MS datasets comprised of data from four technical replicates and *287* manually curated metabolites. The base peak areas of the mass fragments (m/z) were normalized using median normalization and $\log_2$ transformation. Peak areas were normalized by dividing each peak area value by the area of the internal standard(s) for a given sample.

6.2.3 Statistical Analyses

Statistical analyses were performed using the statistical software R (version 3.5.1, R Development Core Team 2017, http://www.R-project.org) [14, 15]. Transformations of continuous variables were performed using DeviumWeb [16]. Normalized, transformed, imputed, outlier removed, and scaled peak areas representative of relative metabolite abundance are presented. Metabolite abundance differences were significant when $P < 0.05$.

Normalization was performed using the Normalyze R stand-alone tool Normalyzer (version 1.1.1) [17]. The normalization procedures performed on the datasets included $\log_2$, locally estimated scatterplot smoothing (LOESS), robust linear regression (Rlr), variance stabilization normalization (Vsn), total intensity (TI), median intensity (MI), average intensity (AI), quantile, and NormFinder (NF).

Principal components analysis (PCA) was performed using DeviumWeb [16], where the output consisted of score plots to visualize the contrast between different samples. The data were scaled to unit variance without any transformation. Partial least-squares discriminate analysis (PLS-DA) was used to highlight differences between the metabolic phenotypes of the groups (control, C and treatment, T) used in the study.

6.2.4 Pathway Enrichment and Clustering Analysis

Pathway enrichment was performed using MetaboAnalyst (www.Metaboanalyst.ca) [18].

6.3 Results

6.3.1 Design of the Study and Data Analysis

Based on the available data, with multifactorial experimental design, an attempt was made to find the optimal data treatment strategies. Thus, sample type (control C, treatment T), normalization (sum S, median M, quantile Q), transformation ($\log_2$ L, cube root C), and scaling (mean centering M, auto scaling A, Pareto scaling P, range scaling R) and all their possible combinations were chosen to treat the same datasets iteratively. This workflow design is depicted in Figure 6.1. The data file consisted of 48 (samples) by 287 (metabolite IDs) data matrix, and unpaired with 6 groups.

6.3.2 The Guard Cell Metabolomics Dataset

The unique study design of the original study of Misra *et al.*, [8] relied on comparison of both guard and mesophyll cells. However, for the simplicity of data treatment and comprehension of the readers for reduced complexity, only the guard cell metabolomics data were selected. Two treatments (control [C] vs. treatment/bicarbonate [HCO_3^-] [T]) and time-course

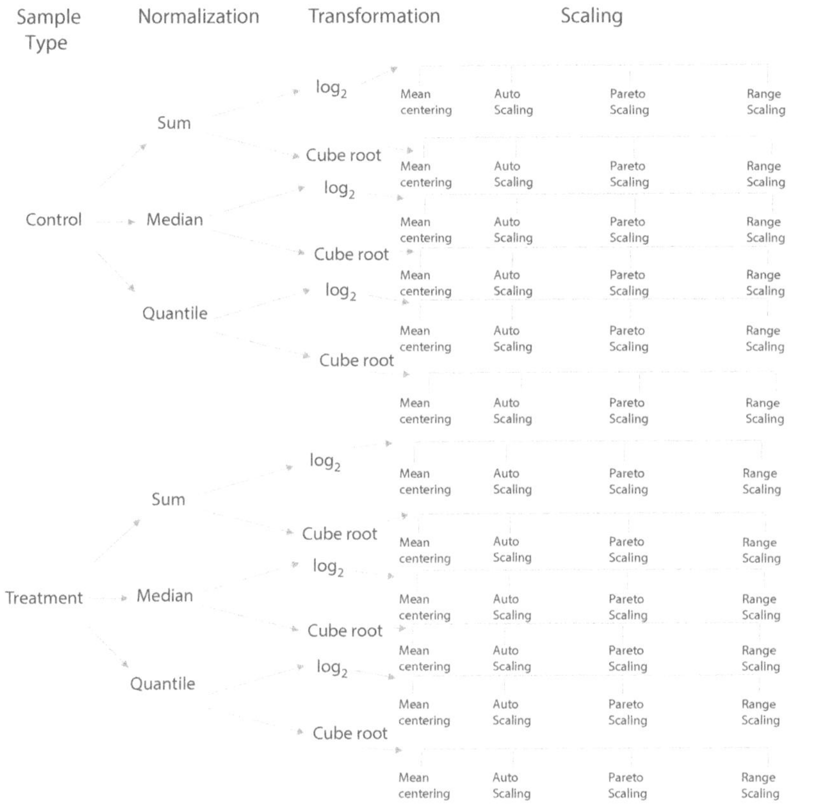

Figure 6.1 Summarized statistical workflow (normalization, transformation, and scaling strategies) addressed in this chapter. Both C and T datasets were subjected to normalization (sum, median, and quantile), followed by transformation (log_2 and cube root), followed by scaling (mean centering, auto scaling, Pareto scaling, and range scaling) methods. All such combinations were tested individually prior to using the data for PCA and PLS-DA for discrimination between time points.

treatments (0, 5, 15, 30, 60, 120 minutes; 6 time points) provided multifactorial experimental design for these metabolomics datasets. The extracted datasets displayed the relative quantitative abundance for 287 metabolites in this study.

The 287 metabolites were mapped onto over 59 different metabolic pathways identified using the publicly available Kyoto Encyclopedia of Genes and Genomes (KEGG) metabolic pathways, presented in a network view (Figure 6.2). These enriched pathways include urea cycle (nine metabolites), taurine and hypotaurine metabolism, glutathione metabolism, betaine metabolism, beta-alanine metabolism, malate-aspartate shuttle, histidine metabolism, methionine metabolism, glutamate metabolism, galactose

Figure 6.2 Displaying the metabolic pathway enrichment analysis output for the metabolites covered in these two datasets using a network view. Darker color nodes (red circles) represent a higher number of associated metabolites to a given pathway, as opposed to lighter color nodes (circles). Edges (lines), whenever present, only connect the related biological pathways in this network, and their lengths do not represent anything.

metabolism, aspartate metabolism, glycine, serine, and threonine metabolism, phenylalanine and tyrosine metabolism, cysteine metabolism, and arginine and proline metabolism among others.

6.3.3 Multivariate Analysis for Insights into Data Pre-Processing

Results from PCA (Tables 6.1 and 6.2) indicate that for the treatment (HCO_3^-) group (T), the combinations of SCM, QCM, MCM, SCP, and MCP performed as the top 5 based on the cumulative PC scores of the first two components (PC1, PC2), which were 69.5, 69.2, 64.3, 58.4, and 56.7%, respectively. This suggests that, for this single cell-type metabolomics datasets, the M and S normalizations followed by C transformation work well with M and P scaling. The worst performers (bottom 5) were the combinations of QLA, MCR, QCR, MCA, and QCA as indicated from the cumulative PC scores of 30.8, 30.2, 30.2, 29.3, and 28.7%, respectively. The results indicate that A and R scaling methods were unhelpful in discrimination of datasets, and more so in combination of C transformation and Q normalization.

For supervised PLS-DA, the results were comparable with QCM, SCM, MCM, SCP, and QCP performing better explaining the variations at 67.4, 66.8, 59, 56.1, and 53.8%, respectively, based on the first two PCs (PC1, PC2). These indicate the superiority of discrimination after pre-processing of the metabolomics datasets with either Q or S normalizations, followed by C transformation M or P scaling. For PLS-DA, the QLA, SLR, QCA, SCA, and SLA combinations performed the worst, explaining only 27.2, 26.5, 25.8, 25.2, and 24.9% variations, respectively, based on the first two PCs. This typically suggested that all transformations were negatively affected with A and R scaling, more specifically with S and Q normalizations.

Thus, it can be clearly concluded that M, S, and Q normalizations with C transformation worked better with M and P scaling methods for this multitreatment, small powered single cell-type metabolomics dataset. Both PCA and PLS-DA based best and worst case scenarios for the T group are shown in Figure 6.3.

Further, it was interesting to observe the effect of data treatment on the outcome of the statistical results. For instance, using ANOVA, the significantly changed metabolites were obtained for the best performing data pretreatment strategies such as MCP, SCP, MCM, QCM, and SCM, which revealed 51, 45, 51, 45, and 45, respectively, significantly ($P < 0.001$) changed metabolites across the five time points in the T datasets. Nonetheless, all five of them shared 40 common metabolites. Further exploration revealed that both MCP and MCM shared 11 metabolites that were absent for the SCP and SCM combination, which in turn shared only 5 unique metabolites.

Table 6.1 Data showing the effect of various normalization, transformation, and scaling strategies on the contribution of PC1 and PC2 in PCA and PLS-DA, and the summed PC values (%ages) in case of the C (control) dataset. The data treatment naming convention follows: normalization (sum S, median M, quantile Q), transformation (log_2 L, cube root C), and scaling (mean centering M, auto scaling A, Pareto scaling P, range scaling R).

Treatment	**PC1 in PCA**	**PC2 in PCA**	**PCA Sum [PC1, PC2]**	**PC1 in PLS-DA**	**PC2 in PLS-DA**	**PLS-DA Sum [PC1, PC2]**
CSCM	49.7	25.5	**75.2**	42.4	31.7	**74.1**
CMCM	45.8	25.1	**70.9**	36.1	34.5	**70.6**
CSCP	43.8	18	**61.8**	39.1	22.4	**61.5**
CMCP	42.6	18.2	**60.8**	37.9	21.5	**59.4**
CQCM	**37.6**	**22.3**	**59.9**	**33.1**	**23.1**	**56.2**
CQCP	31.9	17.4	**49.3**	28.1	17.4	**45.5**
CMLM	26.1	10.8	**36.9**	23	9.5	**32.5**
CSLM	24.5	11.2	**35.7**	21.6	8.9	**30.5**
CQLM	23.2	11	**34.2**	19.6	10.4	**30**
CMLP	22.4	11.4	**33.8**	20	7.1	**27.1**
CSLP	20.9	12.7	**33.6**	18.6	6.7	**25.3**
CSLA	19.2	14	**33.2**	16.9	6.1	**23**

(*Continued*)

Table 6.1 Data showing the effect of various normalization, transformation, and scaling strategies on the contribution of PC1 and PC2 in PCA and PLS-DA, and the summed PC values (%ages) in case of the C (control) dataset. The data treatment naming convention follows: normalization (sum S, median M, quantile Q), transformation (log_2 L, cube root C), and scaling (mean centering M, auto scaling A, Pareto scaling P, range scaling R). (*Continued*)

Treatment	PC1 in PCA	PC2 in PCA	PCA Sum [PC1, PC2]	PC1 in PLS-DA	PC2 in PLS-DA	PLS-DA Sum [PC1, PC2]
CSCA	20.5	12	**32.5**	18.7	9.7	**28.4**
CSLR	18.6	13.7	**32.3**	16.1	6.3	**22.4**
CSCR	19.9	12.2	**32.1**	18.4	10.1	**28.5**
CQLP	20.5	10.9	**31.4**	17.8	7.8	**25.6**
CMLA	19.8	11	**30.8**	17.4	6.2	**23.6**
CMLR	19.6	10.9	**30.5**	16.7	6.8	**23.5**
CMCR	19.2	11.3	**30.5**	16.7	6.1	**22.8**
CMCA	18.7	11.5	**30.2**	16.3	6.3	**22.6**
CQLA	18.3	10.4	**28.7**	15.9	6.6	**22.5**
CQLR	18.2	10.5	**28.7**	15.4	7.1	**22.5**
CQCR	17.1	11.2	**28.3**	14.5	6.7	**21.2**
CQCA	16.6	10.7	**27.3**	13.9	6.8	**20.7**

Table 6.2 Data showing the effect of various normalization, transformation, and scaling strategies on the contribution of PC1 and PC2 in PCA and PLS-DA, and the summed PC values (%ages) in case of the T (treatment) dataset. The data treatment naming convention follows: normalization (sum S, median M, quantile Q), transformation ($\log_2$ L, cube root C), and scaling (mean centering M, auto scaling A, Pareto scaling P, range scaling R).

Treatment	PC1 in PCA	PC2 in PCA	PCA Sum [PC1, PC2]	PC1 in PLS-DA	PC2 in PLS-DA	PLS-DA Sum [PC1, PC2]
TSCM	49.9	19.6	**69.5**	49.7	17.1	**66.8**
TQCM	46	23.2	**69.2**	45.9	21.5	**67.4**
TMCM	40.7	23.6	**64.3**	40.4	18.6	**59**
TSCP	42	16.4	**58.4**	41.8	14.3	**56.1**
TMCP	36.5	20.2	**56.7**	36	13	**49**
TQCP	36.3	19.4	**55.7**	35.8	18	**53.8**
TMLM	30.7	12.5	**43.2**	30.7	11.8	**42.5**
TSLM	29	13.2	**42.2**	28.9	11	**39.9**
TQLM	28.5	12.1	**40.6**	28.4	11	**39.4**
TSLP	22.2	17.6	**39.8**	22.1	8.6	**30.7**
TSLR	21.3	18.3	**39.6**	18.3	8.2	**26.5**

(Continued)

Table 6.2 Data showing the effect of various normalization, transformation, and scaling strategies on the contribution of PC1 and PC2 in PCA and PLS-DA, and the summed PC values (%ages) in case of the T (treatment) dataset. The data treatment naming convention follows: normalization (sum S, median M, quantile Q), transformation (log_2 L, cube root C), and scaling (mean centering M, auto scaling A, Pareto scaling P, range scaling R). (*Continued*)

Treatment	PC1 in PCA	PC2 in PCA	PCA Sum [PC1, PC2]	PC1 in PLS-DA	PC2 in PLS-DA	PLS-DA Sum [PC1, PC2]
TSCR	21.1	18.1	**39.2**	18.5	8.7	**27.2**
TSLA	22.1	17.1	**39.2**	17.1	7.8	**24.9**
TSCA	20.8	16.9	**37.7**	17.1	8.1	**25.2**
TMLP	24.4	11.8	**36.2**	24.4	10	**34.4**
TQLP	23.3	10.9	**34.2**	23.2	8.4	**31.6**
TMLR	21.4	11.6	**33**	21.4	9.6	**31**
TMLA	20.3	12.1	**32.4**	20.2	9.9	**30.1**
TQLR	20.6	11.2	**31.8**	20.6	8.2	**28.8**
TQLA	19.4	11.4	**30.8**	19.4	7.8	**27.2**
TMCR	19.4	10.8	**30.2**	19.4	9.5	**28.9**
TQCR	19.2	11	**30.2**	19.2	8.7	**27.9**
TMCA	18.3	11	**29.3**	18.2	9.1	**27.3**
TQCA	17.8	10.9	**28.7**	17.7	8.1	**25.8**

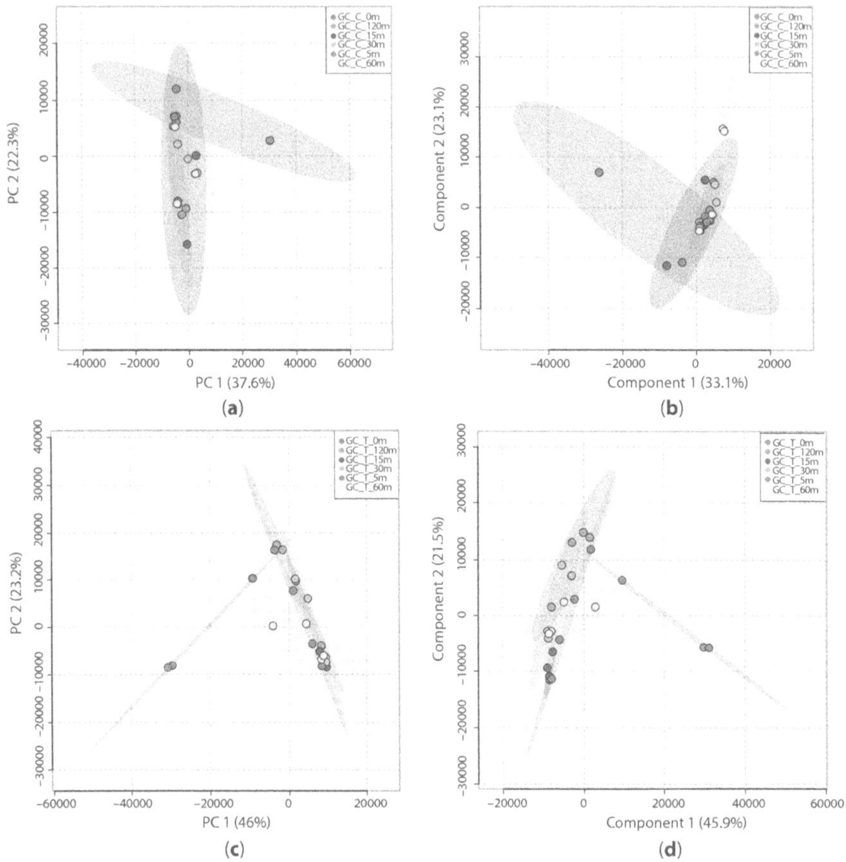

Figure 6.3 Multivariate analysis displaying the best combinations of normalization, transformation, and scaling for the dataset. (a) Q-C-M output for PCA for C datasets, (b) Q-C-M output for PLS-DA for C-dataset, (c) S-C-M output for PCA for T-dataset, and (D) T-Q-C-M output for PLS-DA for T-dataset.

The 11 metabolites include galacturonic acid, succinic acid, 2,4-dihydroxyacetophenone, rosmarinic acid, jasmonic acid, sphingosine, capsaicin, naringin, O-phosphoserine, myricetin, and ethylene glycol, whereas the 5 shared between the SCP and SCM combinations were threonine, cellobiose, diethanolamine, homocysteine, and mannitol. Moreover, the median *P*-values for these five treatment strategies were 1.13E-04, 1.41E-05, 1.13E-04, 1.41E-05, and 1.37E-04, respectively, indicating a better and comparable performance for MCP and MCM over the other combinations, where QCM performed the worst, indicating a greater variance stabilizing effect for M and C combinations (Figure 6.4).

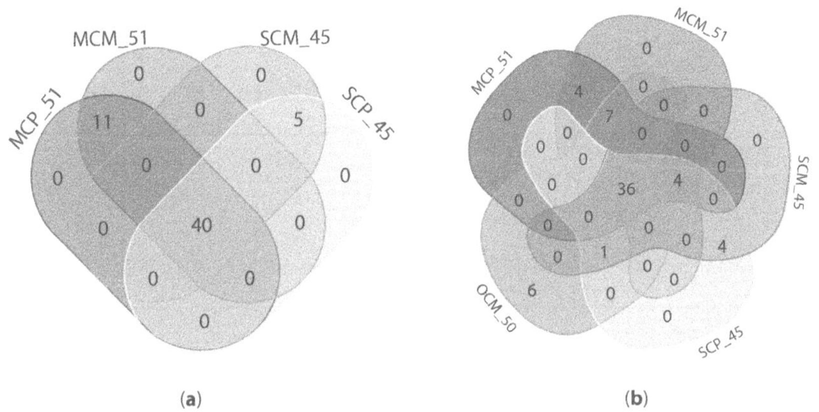

Figure 6.4 (a) A four-way Venn diagram showing the unique and shared metabolites between M-C-P, M-C-M, S-C-M, and S-C-P strategies with core (shared) 40 metabolites showing significant changes ($P < 0.001$) using ANOVA tests. (b) A five-way Venn diagram showing unique and shared metabolites between M-C-P, M-C-M, S-C-M, S-C-P, and a fifth, Q-C-M strategies with core 40 metabolites showing significant changes ($P < 0.001$) using ANOVA tests.

6.3.4 Effect of Data Normalization Methods

The control (C) and treatment (T) datasets were further subjected to various normalization methods independently. These methods were $\log_2$, LOESS, robust linear regression (Rlr), variance stabilization normalization (Vsn), total intensity (TI), median intensity (MI), average intensity (AI), quantile, and NormFinder (NF), among others, by using the Normalyze R tool.

Results from the boxplots that demonstrate the whisker alignment and lengths suggest that "Q" normalization works better for this dataset (Figure 6.5). Multidimensional scaling (MDS), an alternative to factor analysis, aims to place each object in N-dimensional space such that the between-object distances are well preserved.

An MDS analysis indicated that the Q-N method in C dataset revealed at least two outliers within the 0 min time point data (Figure 6.6A), whereas all time-points (samples) were clustered together in absence of any treatment or perturbation, indicating little or no biological variations. On the other hand, in T dataset, expectedly, the four data points (samples) from 0 min clearly stood out from the rest of treatment time points (5, 15, 30, 60, 120 min), which all received HCO_3^- treatment (Figure 6.6B). Thus, again, Q normalization was robust enough to not only specify the outliers but also discriminated control from treatment samples in this study.

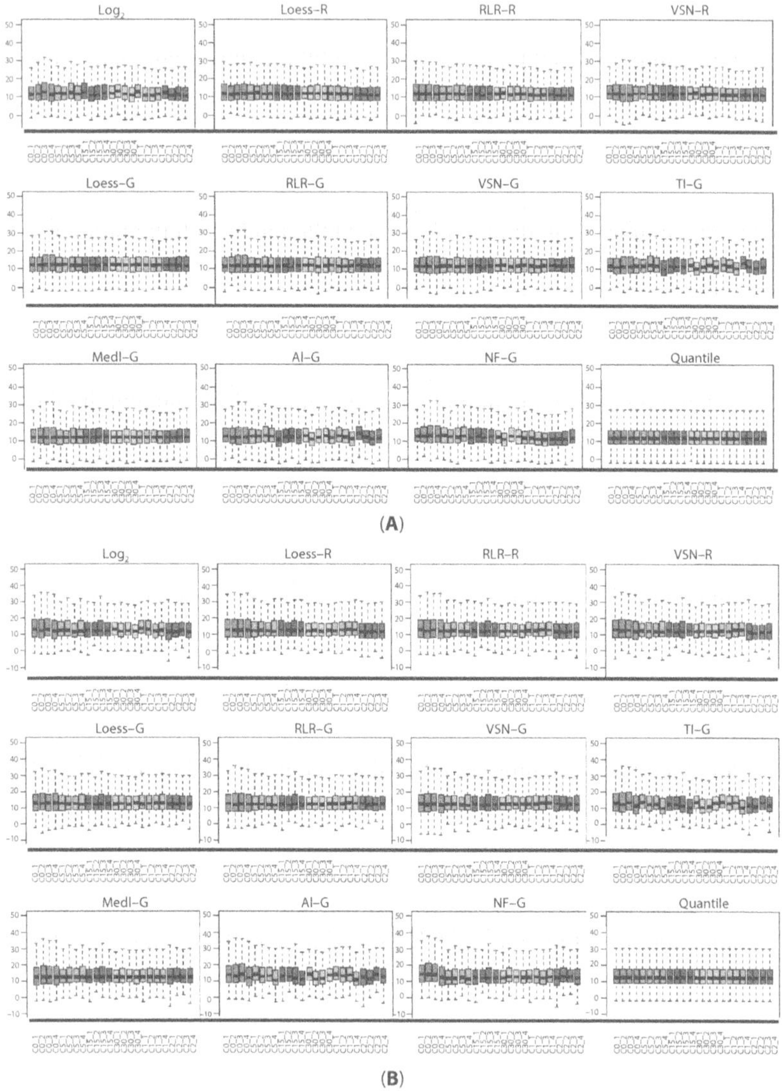

Figure 6.5 Box whisker plots showing the output of 12 normalization methods for (A) C dataset and (B) T dataset. The 12 normalization methods are $\log_2$, LOESS-R, RLR-R, VSN-R, LOESS-G, RLR-G, VSN-G, TI-G, MedI-G, AI-G, NF-G, and Quantile.

A closer look into the pooled intragroup coefficient of variation (PCV) and pooled intragroup median absolute deviation (PMAD) painted interesting pictures on these datasets. For instance, in the PCV analysis for C data, there is very little variation in samples, as evident from different normalization methods owing to lack of any treatment, and that time has no

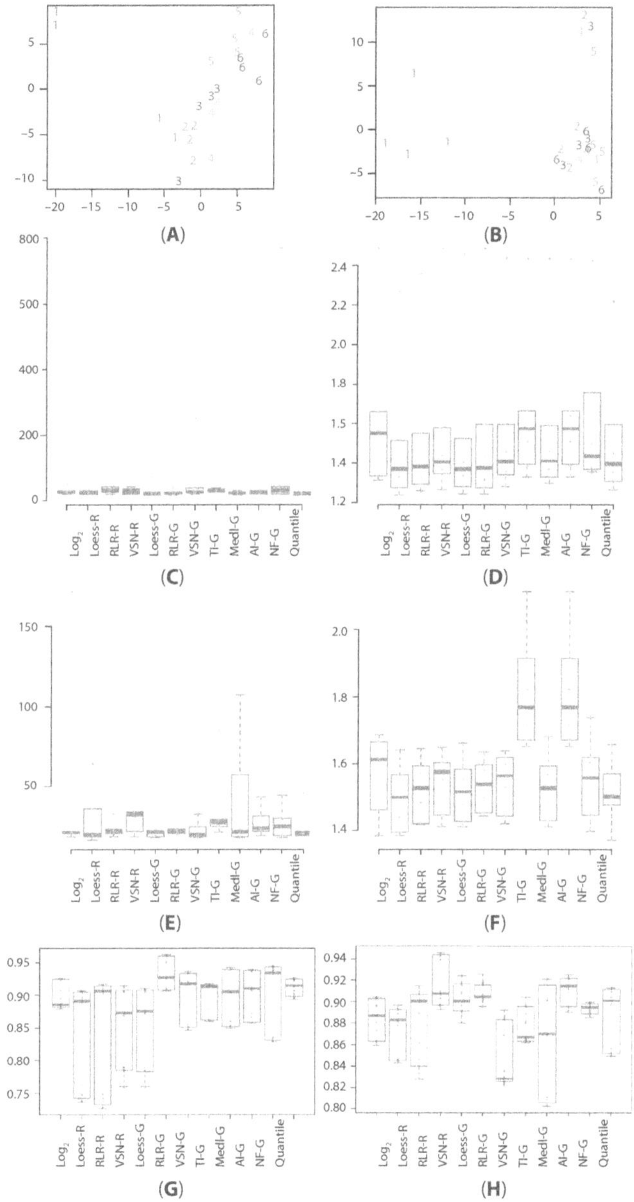

Figure 6.6 MDS plots: (A) control, (B) treatment; where numbers 1–6 represent the six time-points, (C) PCV for 12 normalization methods performed on C dataset, (D) PMAD for 12 normalization methods performed on C dataset, (E) PCV for 12 normalization methods for T dataset, and (F) PMAD for 12 normalization methods for T dataset. (G) Pearson correlation coefficients for 12 normalization methods for C dataset, and (H) Pearson correlation coefficients for 12 normalization methods for T dataset.

or very little effect on the metabolome changes (Figure 6.6C). Similarly, the absence of significant whiskers in the plots indicates comparable performance of the normalization methods for C datasets, although Quantile and LOESS methods performed better (Figure 6.6D). In case of T datasets, interestingly, Quantile, log_2, and Rlr methods worked best as far as PCV is concerned (Figure 6.6E) and Quantile and LOESS handled the intragroup PMAD better (Figure 6.6F). Further, the results of the correlation analysis using Pearson correlation coefficients that measure linear relationships (in this case, intragroup for the data points [samples]) indicated that for C data, the Pearson correlation was rather homogenous and was distributed in the range of 0.87–0.95 (Figure 6.6G). Moreover, the distribution of the coefficients was tighter in case of Quantile and log_2 followed by Rlr methods. On the other hand, for the T datasets, the Pearson correlation coefficients were more unevenly distributed and were spread in the range 0.82–0.92 (Figure 6.6H), indicating that the intragroup correlations were lower for these datasets. Moreover, Vsn normalization performed the worst of all methods, whereas NF, Rlr, and LOESS handled the data better where the T group consisted of 0 min untreated and five other treated sample sets.

6.4 Discussion

Normalization is widely recognized as an integral part of data processing [19] and ≥19 methods are utilized for MS-based metabolomics data [20–22]. Although a lot of online tools (such as MetaboAnalyst [18]) offer such normalization options, they do not provide a service with performance evaluation—a bottleneck in such approaches. Up to 24 methods in total are available as popular tools for MS-based metabolomics data normalization including Auto Scaling, Cross-contribution Compensating Multiple standard Normalization (CCMN), Contrast, Cubic Splines, Cyclic Loess, EigenMS, Level Scaling, Linear Baseline Scaling, Log-transform, Mean Normalization, Median Normalization, mass spectra total useful signal (MSTUS), normalization using optimal selection of multiple internal standards (NOMIS), Pareto Scaling, Power Scaling, Probabilistic Quotient Normalization (PQN), Quantile, Range Scaling, remove unwanted variation-2 (RUV-2), RUV-random, single internal standard (SIS), Total Sum, Vast Scaling, and variance stabilizing normalization (VSN) [23]. Moreover, the approaches depend on whether quality control (QC) samples are taken into consideration or in first place; they were made available to a data analyst. Moreover, with more scientific research groups sticking to the FAIR (Findable; Accessible; Interoperable; Reusable) guiding principles [24] for scientific data management and stewardship, a lot of -omics and large-scale datasets are being made available to the public. In this dataset that we used, MTBLS309 [8], the raw datasets were made available to the research and data analysis community, but at that time did not provide the QC datasets thereby not allowing this analysis to take advantage of the QC samples for consideration of normalization analysis. However, with the use of internal standards (ribitol for GC-MS) and two standards for the LC-MS analyses (lidocaine and camphor-10-sulfenic acid), the datasets provided a QC measure to check for inter-sample variabilities.

For omics-scale data such as metabolomics, the data do not conform to assumptions of normality and/or homoscedasticity or homogeneity of variance. Hence, the data warrant a transformation procedure to improve their normality. Commonly discussed traditional transformations include adding constants, cube root, square root, converting to logarithmic (e.g., base 10, natural $\log_2$) scales, inverting and reflecting, and applying trigonometric transformations such as sine wave transformations [25]. There exist a multitude of normalization methods, such as linear regression normalization (Rlr, RlrMA, RlrMACyc), local regression normalization (LoessF, LoessCyc), and VSN that are derived from microarray/transcriptomics, as well as EigenMS, other proprietary methods, in addition to the popularly used simpler methods such as quantile and median. Quantile normalization forces the distributions of the samples to be the same on the basis of the quantiles of the samples by replacing each point of a sample with the mean of the corresponding quantile [26]. Median normalization is based on the assumption that the samples of a dataset are separated by a constant as it scales the samples so that they have the same median [27].

The purpose of these transformations was to find simple data transformation that stabilizes the variance in metabolomics datasets obtained by combining from LC-MS and GC-MS platforms. This investigation used multivariate analysis to analyze data from well designed experiments with variance structures that allow for heterogeneity of variance between treatments. Recently, a treatise discussed the transformation, normalization, and batch effect in the analysis of mass spectrometry-based data for omics studies such as metabolomics and proteomics [28]. We clearly observed that normalization improved the discrimination when compared with non-normalized data. The nature of bias might be very different for different datasets, and the normalization methods may not make too stringent assumptions on the type of bias or else the methods become too specific for a dataset. Thus, a quantile method that does not make strict assumptions about the nature of the bias is a very generic method [27]. Also, for label-free proteomics datasets, it was shown that the median and quantile normalization methods were on par with most of the normalization methods [27].

In addition, observing the density, MDS, and MeanS plots, and the overall distribution for skewness, replicate clustering, and outliers, and checking out the variance as independent of the mean provide more QC measures on a given dataset. However, the exclusiveness and platform specificity of metabolites identified and quantified are challenges that typical metabolomics studies suffer from. For instance, Zhu and Assmann [11] observed that no metabolites were common to GC-MS and LC-MS run in positive mode, and only 5 metabolites were common to LC-MS in positive mode and negative mode out of a total of 390 metabolites quantified in the study. Recent studies [8, 29] reported 268 and 358 metabolites, respectively, in the metabolite profiling of guard cell enriched epidermal peels from *B. napus*, in studies of metabolome responses to HCO_3^- and elevated CO_2, whereas the ABA-responsive metabolome showed an overlap of 62 and 74 metabolites with those datasets [11]. In addition, another study reported 411 metabolites changing in response to low CO_2 [29].

6.5 Conclusion

In the current investigation, which is designed as a proof of concept for single cell-type metabolomics (i.e., guard cells), data analysis using a plethora of data treatment and the effects thereof was conducted. Although the benefits of specific normalization, transformation, and scaling strategies were discussed with this typical plant cell-type metabolomics datasets consisting of time-course data, treatments, and possible outliers from a combined LC-MS and GC-MS approach, the results need to be taken with a grain of salt given that other metabolomics platforms and datasets from other biological sources and experiments might quickly throw off these conclusions. One of the primary focus of this exercise was to demonstrate the feasibility of obtaining publicly archived metabolomics datasets for reanalysis (reuse and re-purpose) for understanding and development of the statistical tools and workflows in metabolomics, as well as generation of functional genomics insights with choice of optimal methods of normalization, transformation, and scaling prior to downstream multivariate analyses and providing biological answers.

Conflicts of Interest

The author declares no conflicts of interest.

Acknowledgment

The author would like to acknowledge colleagues for helpful discussions during preparation of the manuscript and MetaboLights for helping in archiving the author's datasets among so many other metabolomics datasets generated worldwide. A big thanks to the author's previous mentor, Professor Sixue Chen, for archiving the metabolomics datasets for reuse and thus promoting by the FAIR (findable, accessible, interoperable, and reusable) principles.

References

1. Fiehn, O., Metabolomics by gas chromatography–mass spectrometry: Combined targeted and untargeted profiling. *Curr. Prot. Molec. Biol.*, 30–4, 2016.
2. Nikiforova, V.J., Kopka, J., Tolstikov, V., Fiehn, O., Hopkins, L., Hawkesford, M.J., Hesse, H., Hoefgen, R., Systems rebalancing of metabolism in response to sulfur deprivation, as revealed by metabolome analysis of *Arabidopsis* plants. *Plant Physiol.*, 138, 304–318, 2005.
3. Wiklund, S., Johansson, E., Sjöström, L., Mellerowicz, E.J., Edlund, U., Shockcor, J.P., Gottfries, J., Moritz, T., Trygg, J., Visualization of GC/TOF-MS-based metabolomics data for identification of biochemically interesting compounds using OPLS class models. *Anal. Chem.*, 80, 115–122, 2008.
4. Jin, X., Wang, R.S., Zhu, M., Jeon, B.W., Albert, R., Chen, S., Assmann, S.M., Abscisic acid–responsive guard cell metabolomes of *Arabidopsis* wild-type and gpa1 G-protein mutants. *Plant Cell*, 25, 4789–4811, 2013.
5. Xiao, J.F., Zhou, B., Ressom, H.W., Metabolite identification and quantitation in LC-MS/MS-based metabolomics. *Trends Anal. Chem.*, 32, 1–14, 2012.
6. t'Kindt, R., Morreel, K., Deforce, D., Boerjan, W., Van Bocxlaer, J., Joint GC–MS and LC–MS platforms for comprehensive plant metabolomics: Repeatability and sample pre-treatment. *J. Chromatogr. B*, 877, 3572–3580, 2009.
7. Misra, B.B., Acharya, B.R., Granot, D., Assmann, S.M., Chen, S., The guard cell metabolome: Functions in stomatal movement and global food security. *Front. Plant Science*, 6, 2015.
8. Misra, B.B., de Armas, E., Tong, Z., Chen, S., Metabolomic responses of guard cells and mesophyll cells to bicarbonate. *PLoS One*, 10, e0144206, 2015.
9. Misra, B.B., Yin, Z., Geng, S., De Armas, E., Chen, S., Metabolomic responses of *Arabidopsis* suspension cells to bicarbonate under light and dark conditions. *Sci. Rep.*, 6, 35778, 2016.
10. Geng, S., Yu, B., Zhu, N., Dufresne, C., Chen, S., Metabolomics and proteomics of *Brassica napus* guard cells in response to low CO_2. *Front. Mol. Biosci.*, 4, 51, 2017.

11. Zhu, M. and Assmann, S.M., Metabolic signatures in response to abscisic acid (ABA) Treatment in *Brassica napus* guard cells revealed by metabolomics. *Sci. Rep.*, 7, 12875, 2017.
12. Tsugawa, H., Cajka, T., Kind, T., Ma, Y., Higgins, B., Ikeda, K., Kanazawa, M., VanderGheynst, J., Fiehn, O., Arita, M., MS-DIAL: Data-independent MS/MS deconvolution for comprehensive metabolome analysis. *Nat. Methods*, 12, 523–526, 2015.
13. Lai, Z., Tsugawa, H., Wohlgemuth, G., Mehta, S., Mueller, M., Zheng, Y., Ogiwara, A., Meissen, J., Showalter, M., Takeuchi, K., Kind, T., Identifying metabolites by integrating metabolome databases with mass spectrometry cheminformatics. *Nat. Methods*, 15, 53–56, 2017.
14. Team, R.C., *R: A language and environment for statistical computing [Internet]*, R Foundation for Statistical Computing, Vienna, Austria, 2017.
15. Sokal, R.R. and Rohlf, F.J., *Biometry: The principles and practice of statistics in biological research*, W.H. Freeman and Company 337, New York, 1995.
16. Grapov, D., DeviumWeb: version 0.3.2. ZENODO, 2014, https://github.com/dgrapov/DeviumWeb.
17. Chawade, A., Alexandersson, E., Levander, F., Normalyzer: A tool for rapid evaluation of normalization methods for omics data sets. *J. Proteome Res.*, 13, 6, 3114–3120, 2014.
18. Xia, J., Sinelnikov, I.V., Han, B., Wishart, D.S., MetaboAnalyst 3.0—Making metabolomics more meaningful. *Nucleic Acids Res.*, 43, W251–W257, 2015.
19. De Livera, A.M., Olshansky, M., Speed, T.P., Roessner, U., Dias, D.A., Statistical analysis of metabolomics data, in: *Metabolomics Tools for Natural Product Discovery: Methods and Protocols*, pp. 291–307, Springer Science+Business Media, LLC, 1055, 2013.
20. De Livera, A.M., Dias, D.A., De Souza, D., Rupasinghe, T., Pyke, J., Tull, D., Roessner, U., McConville, M., Speed, T.P., Normalizing and integrating metabolomics data. *Anal. Chem.*, 84, 10768–10776, 2012.
21. De Livera, A.M., Sysi-Aho, M., Jacob, L., Gagnon-Bartsch, J.A., Castillo, S., Simpson, J.A., Speed, T.P., Statistical methods for handling unwanted variation in metabolomics data. *Anal. Chem.*, 87, 3606–3615, 2015.
22. Warrack, B.M., Hnatyshyn, S., Ott, K.H., Reily, M.D., Sanders, M., Zhang, H., Drexler, D.M., Normalization strategies for metabonomic analysis of urine samples. *J. Chromatogr. B Analyt. Technol. Biomed. Life Sci.*, 877, 547–552, 2009.
23. Li, B., Tang, J., Yang, Q., Li, S., Cui, X., Li, Y., Chen, Y., Xue, W., Li, X., Zhu, F., NOREVA: Normalization and evaluation of MS-based metabolomics data. *Nucleic Acids Res.*, 45, W1, W162–W170, 2017.
24. Wilkinson, M.D., Dumontier, M., Aalbersberg, I.J., Appleton, G., Axton, M., Baak, A., Blomberg, N., Boiten, J.W., da Silva Santos, L.B., Bourne, P.E., Bouwman, J., The FAIR Guiding Principles for scientific data management and stewardship. *Sci. Data*, 3, 160018, 2016.

25. Osborne, J.W., Improving your data transformations: Applying the Box–Cox transformation. *Pract. Assess. Res. & Eval.*, 15, 1–9, 2010.
26. Bolstad, B.M., Irizarry, R.A., Åstrand, M., Speed, T.P., A comparison of normalization methods for high density oligonucleotide array data based on variance and bias. *Bioinformatics*, 19, 185–193, 2003.
27. Välikangas, T., Suomi, T., Elo, L.L., A systematic evaluation of normalization methods in quantitative label-free proteomics. *Brief. Bioinform.*, bbw095, 2016.
28. Mertens, B.J., Transformation, normalization, and batch effect in the analysis of mass spectrometry data for omics studies, in: *Statistical analysis of proteomics, metabolomics, and lipidomics data using mass spectrometry*, pp. 1–21, Springer International Publishing, 2017.
29. Geng, S., Misra, B.B., Armas, E., Huhman, D.V., Alborn, H.T., Sumner, L.W., Chen, S., Jasmonate-mediated stomatal closure under elevated CO_2 revealed by time-resolved metabolomics. *Plant J.*, 88, 947–962, 2016.
30. Kopka, J., Schauer, N., Krueger, S., Birkemeyer, C., Usadel, B., Bergmüller, E., Dörmann, P., Weckwerth, W., Gibon, Y., Stitt, M., Willmitzer, L., GMD@CSB. DB: The Golm metabolome database. *Bioinformatics*, 21, 8, pp. 1635–1638, 2005.

7

Metabolite Profiling and Metabolomics of Plant Systems Using ^{1}H NMR and GC-MS

Manu Shree, Maneesh Lingwan and Shyam K. Masakapalli*

School of Basic Sciences, Indian Institute of Technology Mandi, Himachal Pradesh, India

Abstract

Metabolites are the small molecules (typically <1000 Dalton) that play critical role as intermediates of metabolism. The metabolome comprises of all the metabolites present in a cellular system at a particular time and space. These variations in the metabolome due to genetic, environmental, and other physiological parameters are captured using metabolomics, an approach that often involves chemometric (multivariate) statistical analysis. The current chapter introduces the routine practices to undertake metabolite profiling and metabolomics analysis in plant systems using proton nuclear magnetic resonance (^{1}H NMR) spectroscopy and gas chromatography–mass spectroscopy (GC-MS). In particular, emphasis is laid on the general experimental workflow, metabolite data acquisition and their analysis for metabolite profiling, statistical analysis for metabolomics using MetaboAnalyst, a free web-based tool, and applications with relevance to phytochemical diversity, nutritional biology, and medicine.

Keywords: NMR, GC-MS, metabolomics, MetaboAnalyst, PCA, PLS-DA, metabolite profiling, phytochemicals

7.1 Introduction

Metabolic systems biology explains the precise metabolic phenotypes of cellular systems along with the bottlenecks of their physiology with applications in various fields such as agriculture, synthetic biology, metabolic

**Corresponding author*: shyam@iitmandi.ac.in

Rintu Banerjee, Garlapati Vijay Kumar, and S.P. Jeevan Kumar (eds.) OMICS-Based Approaches in Plant Biotechnology, (129–144) © 2019 Scrivener Publishing LLC

engineering, and health [1–3]. The techniques at the core of metabolic systems biology are metabolite profiling, metabolomics, and fluxomics [4, 5]. Metabolome or metabolite profiling deals with studying the complete set of metabolites in a cellular system. Metabolomics involves subjecting the metabolite profiles to multivariate statistical analysis to capture the extent of variations among the cellular systems or tissues that may differ either due to genetic, environmental, or physiological conditions. For example, the effect of biotic or abiotic stress on the metabolome of the biological system can be understood. Metabolomics provides valuable insights into the systems biological features and has wide applications such as biomarker discovery, etc. With the progress achieved in the last decade, the experimental design, the sample extraction methods and their processing, the sensitivity of the analytical platforms used, the analysis of the metabolite profiles, and the statistical approaches used became critical factors for reliable biological interpretation.

Defining the metabolome of plant systems is of interest for agronomic and industrial applications [6–9]. Major approaches currently being used are metabolite fingerprinting, metabolite profiling, and metabolomics via targeted or untargeted analysis [10–12]. Metabolite profiling and metabolomics of different plants species could shed light on their physiology, the dynamic behavior of the metabolic networks, in addition to identifying several primary and secondary metabolites of interest [13, 14]. It has been established that comprehensive metabolic analysis largely depends on the experimental design, sensitivity of the analytical platforms (such as NMR, HPLC, GC-MS, LC-MS/MS), and data handling (binning, baseline corrections, and statistical analysis) [8, 14–16]. The modern techniques that are routinely used for metabolite profiling are mainly ^{1}H NMR and mass spectroscopy (MS). Owing to the low sensitivity of NMR, it is highly suitable to analyze the highly abundant metabolites in a complex mixture [17]. Mass spectrometry is more powerful compared to NMR in terms of better sensitivity and high resolving power. Among the mass spectrometry approaches available for metabolomics, GC-MS is suitable mainly for volatile compounds analysis in complex samples, whereas LC-MS is widely used for nonvolatile and temperature-sensitive metabolites. In addition, the availability of EI-based mass spectral libraries is the major advantage for GC-MS-based metabolomics [18].

Here we introduce the general methodology and critical steps routinely recommended to analyze the high-throughput metabolic data from typical plant systems using ^{1}H NMR and GC-MS (Figure 7.1). The discussion and references cited will serve as first-hand information to guide researchers performing metabolite profiling and metabolomics of plants.

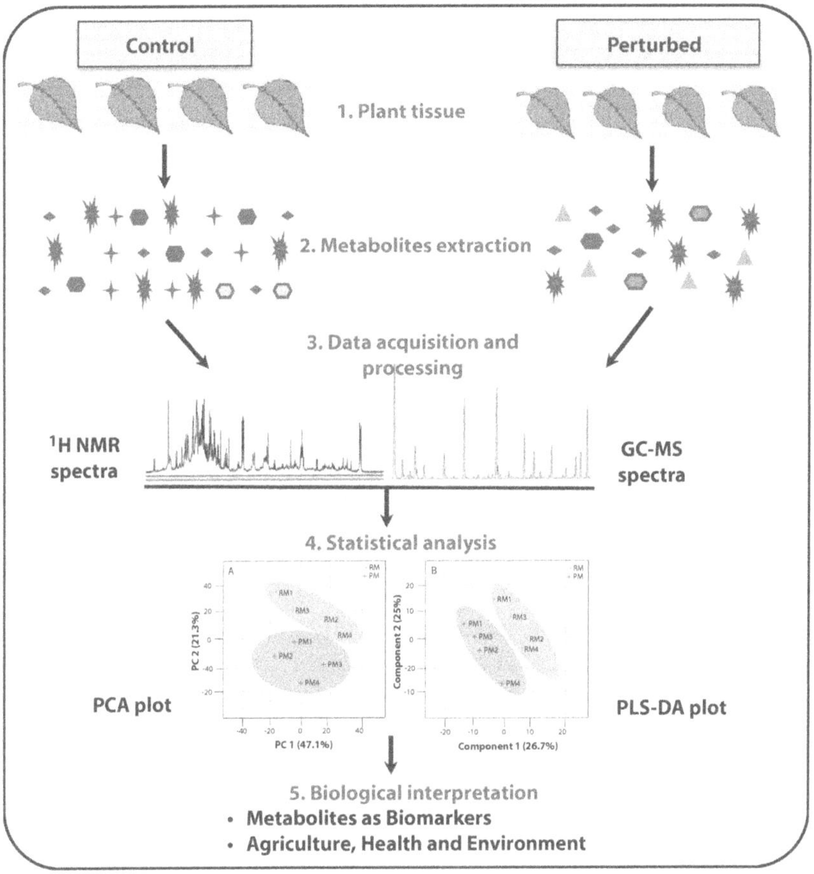

Figure 7.1 A typical experimental workflow showcasing the metabolomics of plants. The first and foremost step while conducting any metabolic study is to collect plant tissues of interest for metabolite extraction. Different solvents individually or in combination (methanol, chloroform, acetone, and water) are routinely used for the process of extraction. The extracted metabolites are then identified (metabolic profiling) using analytical platforms (GC-MS, ^{1}H NMR, etc.). In order to achieve comprehensive biological interpretation, multivariate statistical analysis using supervised (PCA: principal component analysis) and unsupervised methods (PLS-DA: partial least squares projections to latent structures—discriminant analysis) is routinely followed.

7.2 Materials and Methods

The workflows that can be adapted for analysis of plant extracts using ^{1}H NMR and GC-MS are presented below. The routine materials used for metabolite extraction are Milli-Q water, HPLC-grade methanol, and

HPLC-grade chloroform. During metabolite analysis using NMR, deuterated solvents like deuterated methanol (MeOD), deuterated chloroform ($CDCl_3$), and deuterated water (D_2O) along with internal standards like DSS (4,4-dimethyl-4-silapentane-1-sulfonic acid) and TMS (tetramethylsilane) are used. For GC-MS-based metabolite profiling different derivatization reagents are used including N-methyl-N-trimethylsilyl trifluoroacetamide (MSTFA), methoxamine hydrochloride (MeOX-HCl), pyridine, etc. The main instruments and equipment that are mainly required for metabolomics are Eppendorf tubes (2.0 and 1.5 ml), liquid nitrogen, freezer (–80°C), freeze-dryer, thermo mixer, micro centrifuge, speed vacuum concentrator, ultrasonicator, table-top vortex, pH meter, etc. The requirements may vary based on the methodology one adopts.

7.2.1 ^{1}H NMR-Based Metabolite Profiling of Plant Samples

^{1}H NMR of plant extracts helps in identifying metabolites as well as undertaking metabolomics studies. The chemical shifts of the signals acquired in the ^{1}H NMR spectra can reveal the various metabolites present in the sample such as amino acids, sugars, organic acids, and secondary metabolites. The different regions of the ^{1}H NMR spectra correspond to different characteristic signals of the protons of the metabolites, e.g., 0- to 3-ppm region represents the amino acids; 3 to 5.5 ppm shows the sugar region; and 6 to 10 ppm is known for the aromatic region [19]. The following steps are generally followed for ^{1}H NMR analysis (outlined in Figure 7.2) of plant tissues.

7.2.1.1 Metabolite Extraction

The first critical step is careful plant sample preparation for ^{1}H NMR [20, 21]. The metabolism of the samples needs to be quenched using liquid nitrogen, which also facilitates immediate powdering of samples or long-term storage. The fresh or lyophilized plant tissues (~20–50 mg) are generally subjected to metabolite extraction using suitable solvents (water, methanol, chloroform, etc.) and conditions (temperature, pH, etc.). The extracts obtained are routinely concentrated, dried under vacuum, or lyophilized and then redissolved in deuterated solvents (D_2O, $CDCl_3$, CD_3OD) along with compatible internal standards (DSS [0.01%, w/v], TMS [0.01%, w/v], etc.) followed by ^{1}H NMR spectroscopy.

7.2.1.2 ^{1}H NMR Spectroscopy

In practice, the deuterated plant extract is placed in a 5-mm NMR tube and subjected to ^{1}H NMR (frequency of 500- or 600-MHz instrument) for

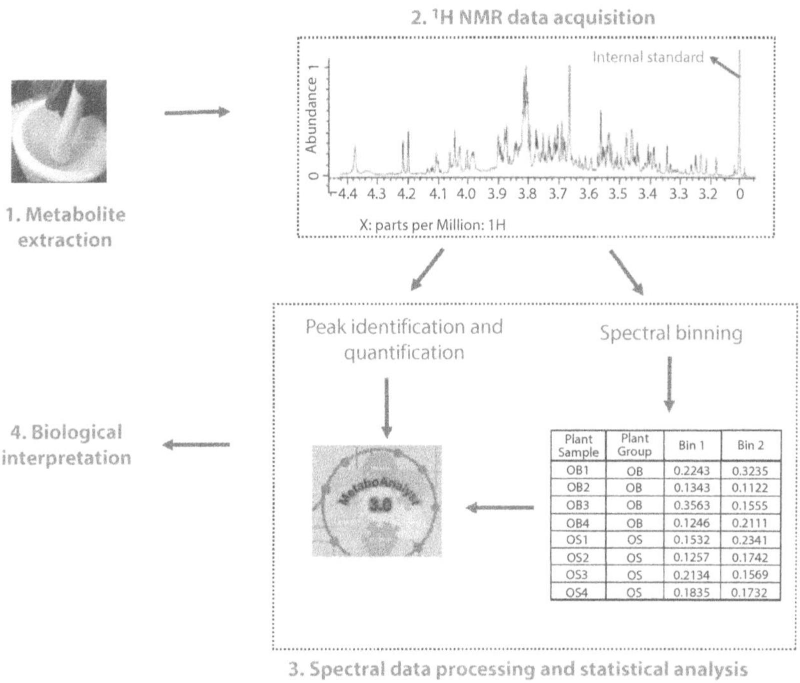

Figure 7.2 Steps for ^{1}H NMR data analysis—1. Metabolite extraction: Sample preparation by homogenizing the intact plant tissue for metabolite extraction in a suitable solvent. Dry and redissolve in suitable deuterated solvent 2. Data acquisition: The extracts are then processed using ^{1}H NMR followed by data acquisition and processing leading to the peak abundance and their chemical shifts 3. Spectral data processing and statistical data analysis: The chromatogram is further processed for peak identification and quantification. Perform spectral binding in order to divide the complete spectra into different bins of size 0.04. Combine all the bin intensities and list them as per different plant groups (example OB, OS) followed by data normalizations and statistical analysis using MetaboAnalyst (http://www.metaboanalyst.ca/). 4. Biological interpretation: The obtained metabolite data can be correlated to different applications leading to its biological interpretation.

a set number of scans (64–128 scans). Other optimal acquisition parameters like temperature, relaxation delay (5 s), water signal pre-saturation, and pulse width (11.6 μs) need to be set in order to obtain a comparable signal-to-noise ratio. The obtained spectra are phased manually and properly normalized as per the internal standard (TMS or DSS at 0.0 ppm), using processing software (JEOL Delta™ or a suitable one). For more detailed step-by-step ^{1}H NMR-based metabolomics analysis of plants, refer to Kruger *et al.*

7.2.1.3 Qualitative and Quantitative Analysis of NMR Signals

The workflow for ^1H NMR data analysis is presented in Figure 7.2. The peak identities are generally assigned based on the ppm and the splits (singlet, doublet, etc.) corresponding to the ^1H NMR signals of known standards. The relative abundances of each peak can be obtained from integrating the intensities. As the ^1H NMR spectra of plant tissues are often crowded, deconvolution of each peak is a major limitation [22]. However, the complete spectra can be used for ^1H NMR fingerprinting [23, 24] to discriminate the effect of perturbations on plant tissues by integrating with multivariate statistical analysis. For that, the raw spectra are generally divided into small bins, and the window size of equal width must be assigned to complete integrated region (e.g., 0.04 ppm). It must be ensured that the spectral regions (bins) that coincide with the signals from the deuterated solvent and internal standard are excluded from the analysis. The retained bins along with their cumulative signal intensities are generally normalized before subjecting to multivariate statistical analysis using free online tools such as MetaboAnalyst [25], which is further elaborated in Section 7.2.3.

7.2.2 Gas Chromatography–Mass Spectroscopy (GC-MS) Based Metabolite Profiling

GC-MS is known to be one of the most popular mass spectrometry-based analytical approaches for metabolite profiling. This technique involves separation as well as identification of the metabolites in one go. Although GC-MS is best applicable to analyze volatile compounds, liquid samples can also be subjected for analysis after derivatization. Derivatization reactions chemically alter the metabolites and make them amenable for efficient volatilization and separation in the GC column. The various steps for metabolomics using GC-MS are described in subsequent details as depicted in Figure 7.3.

7.2.2.1 Sample Preparation

The most crucial step for metabolite analysis using GC-MS is sample preparation, which needs standardization based on the experimental objectives and tissue type. Generally, plant tissues (leaf/root/stem, etc.) of interest are first quenched using liquid N_2, powdered, lyophilized, and stored in a freezer. Metabolite extraction is performed by dissolving either fresh or dried plant tissues (~50–100 mg) in a suitable solvent (1–2 ml of H_2O/methanol/acetic acid, etc.) followed by heating (50°C to 90°C) for a set time (10–20 min), followed by centrifugation at 13,000 × *g* at 4°C for

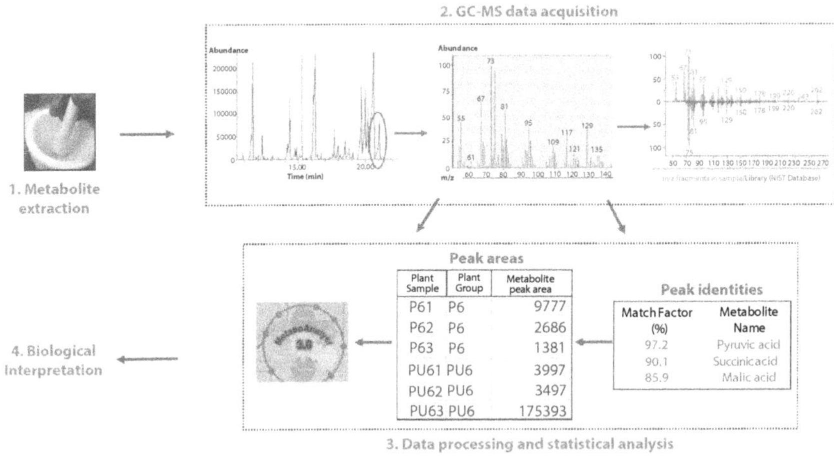

Figure 7.3 Steps for GC-MS data analysis—1. Sample preparation by homogenizing the intact plant leaves/roots/stems using liquid nitrogen. This is followed by metabolite extraction followed by derivatization. 2(a). GC-MS spectra showing the total ion chromatogram. 2(b). Show m/z fragmentation pattern of the selected metabolite peak. 2(c). Library match of the selected fragment with the NIST library. 3(a). List the metabolites as per their matching factor (% probability). 3(b). Peak areas of the metabolites from each spectra of different plant samples are listed. 3(c). The mass spectral data are subjected to statistical analysis using MetaboAnalyst (http://www.metaboanalyst.ca/). 4. Biological interpretation from the multivariate statistical analysis leads to biomarker identification, etc.

10 min to separate the water/methanol (or other solvents) soluble metabolites from insoluble pellets. Aliquots of the supernatant (50–100 μl) are vacuum-dried followed by derivatization. In order to capture the central metabolite information from a plant sample, methoxamine trimethylsilane (MeOx-TMS) derivatization is generally used [26]. The general protocol that one may follow for MeOx-TMS derivatization includes addition of 35 μl of pyridine containing methoxylamine hydrochloride (20 mg/ml) to the dried samples and incubation for 2 h at 37°C, 900 rpm, on a thermomixer. This is followed by the addition of 49 μl of MSTFA (N-methyl-N-(trimethylsilyl)-trifluoroacetamide) and incubation at 37°C for 30 min at 900 rpm on a thermomixer. Centrifuge the mixture at 13,000 × *g* for 10 min and transfer the supernatant to a glass vial insert for GC-MS data acquisition.

7.2.2.2 GC-MS Data Acquisition

The protocol used for metabolite data acquisition in GC-MS generally needs optimization for comprehensive analysis. A range of GC columns used for different types of metabolite fractions (polar, lipophilic, etc.) are available (e.g., RTX-5, DB-5MS, DB-2MS, DB-Wax, etc.) commercially. For data acquisition, typically 1 μl of sample extract (derivatized/underivatized) is injected into an automatic liquid sampler in either split/split fewer modes. Molecules get separated by the respective column using helium as a carrier gas at a constant flow rate of 1 ml/min. In general, the front injection temperature is always set constant at 250°C. A temperature gradient of 1 min at 50°C isothermal, followed by a 10°C min^{-1} oven temperature gradient to a final 200°C, and then holding for 4 min at 200°C, followed by a 10°C min^{-1} to a final 300°C temperature for 10 min is used in many plant metabolomics works; 70-eV electron beam in electron ionization mode generates ion. This fragments the metabolites, which are detected by the mass analyzer in a selected range of m/z (50–600 m/z). The output is GC-MS spectra of total ion chromatograms representing several peaks that correspond to the metabolites.

7.2.2.3 GC-MS Data Pretreatment and Metabolite Profiling

The purpose of data pretreatment is to make the data more comparable for further analysis. It must be noted that a wide range of tools are available, and here we present an outline based on our experience. Different pretreatment methods are available including baseline correction of recorded GC-MS spectra using Metalign software [27]. After complete pretreatment procedure, Automated Mass Spectral Deconvolution & Identification System (AMDIS) [28] is known to be used for peak identification from the spectra. AMDIS software is based on MS search from NIST 11 (NIST, National Institute of Standards and Technology, Maryland) library (Golm database) [29]. After, peak identification using NIST Golm database, chemstation (Agilent) software is used to obtain the peak integrals. All the pieces of information including peak identities and integrals are then combined to prepare a metadata file that includes m/z comparison of different fragments, their elution times, and hits against the library. This metadata file in csv format is used for further statistical analysis.

7.2.2.4 Validation of Identified Metabolites

The identified peaks in the NIST library [30] are further validated by running commercially available standards. Only high-quality standard metabolites should be used for cross-validation of already assigned peaks. Also, to validate the identified metabolites, prepare the mix from aliquots of different ordered standards by mixing them in appropriate solvents followed by GC-MS. Cross-check the retention time of each metabolite in the reaction mix with the already identified and assigned peaks. The identity of only those metabolites should be considered accurate whose retention time as well as the mass ion fragments coincide between samples and standards.

7.2.3 Multivariate Data Analysis

The high-throughput multivariate data generated using different analytical approaches must be statistically analyzed for proper biological interpretation. Different multivariate statistical data analysis approaches can be used including unsupervised (PCA) or supervised (PLS, PLS-DA) methods. In addition, one can generate information about metabolite fold changes and identify key metabolites contributing to the variations, etc., which can be presented as heat maps, Dendrogram, VIP scores, loading plots, etc. The detailed explanation about these is out of scope of the current chapter. The readers are directed to the available different online and standalone software for metabolomics such as MetaboAnalyst [16, 25], XC-MS [31], and SIMCA-P [32]. In PCA, the complex multivariate datasets are reduced in dimensionality and are presented as orthogonal principal components that capture the variations. The variations are presented as PCA score plots across various principal components in different axis (PC1, PC2).

In MetaboAnalyst 3.0, data transformation is done by Pareto scaling (mean-centered and divided by the square root of the standard deviation of each variable), and log transformation (generalized logarithm transformation) method was applied for PCA [16, 25]. For detailed multivariate data analysis in metabolomics, please refer to Worley and Powers [33]. The workflow showing the statistical approaches that can be adopted using MetaboAnalyst [16] is depicted in Figure 7.4.

Currently, MetaboAnalyst 4.0 [34], an updated version, is available (http://www.metaboanalyst.ca/), which features 12 modules with four main divisions that also help in building metabolic pathways along with other features using different tools. Among all the available 12 modules, one module is confined for the pathway analysis from all the identified metabolites.

A specific identity is assigned to each metabolite in terms of compound identification numbers from different databases (HMDB—Human metabolome database, Pubchem, KEGG—Kyoto encyclopedia of genes and genomes). This module then helps in visualization of metabolic pathways

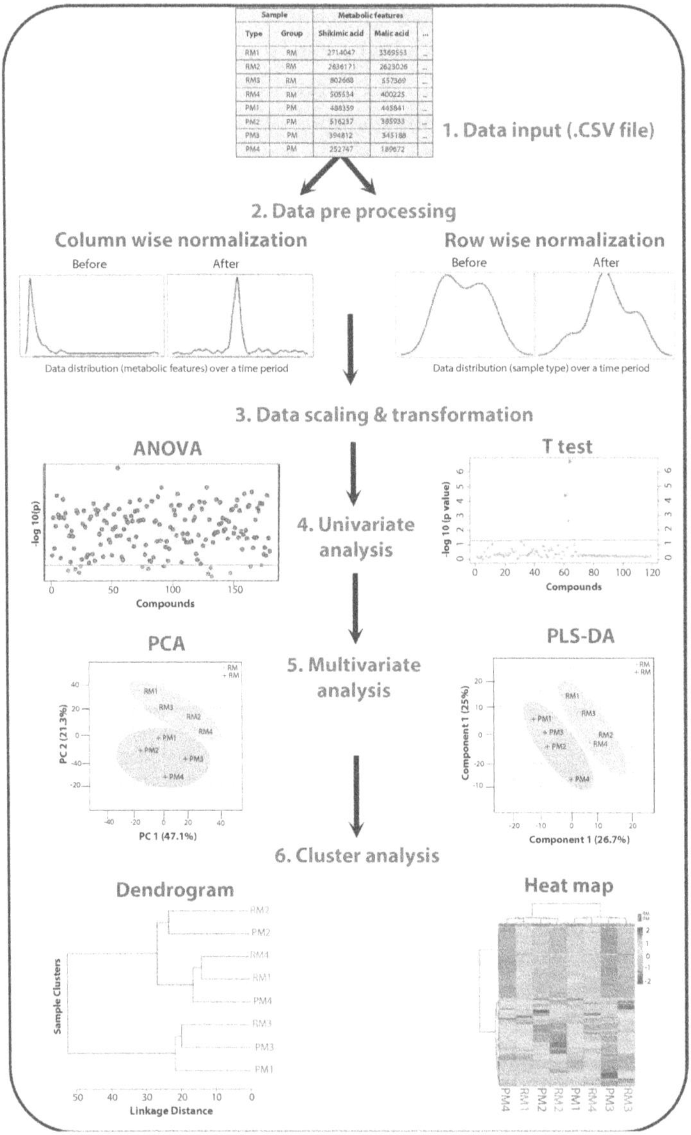

Figure 7.4 Data interpretation using various statistical approaches—different approaches can be utilized for statistical interpretation of metabolomics data. Collected data must be pre-processed in terms of normalization, data scaling, and transformation. Pre-processing is required to make the available data more consistent and comprehensive. The data after complete processing are subjected to univariate and multivariate analysis as per the user's choice. Different univariate approaches include ANOVA (analysis of variance) and t-test, whereas multivariate analysis includes PCA and PLS-DA. Cluster analysis, dendrogram, and heat map can also be performed in order to correlate between the experimental samples based on identified metabolites. The figures are generated using MetaboAnalyst [25].

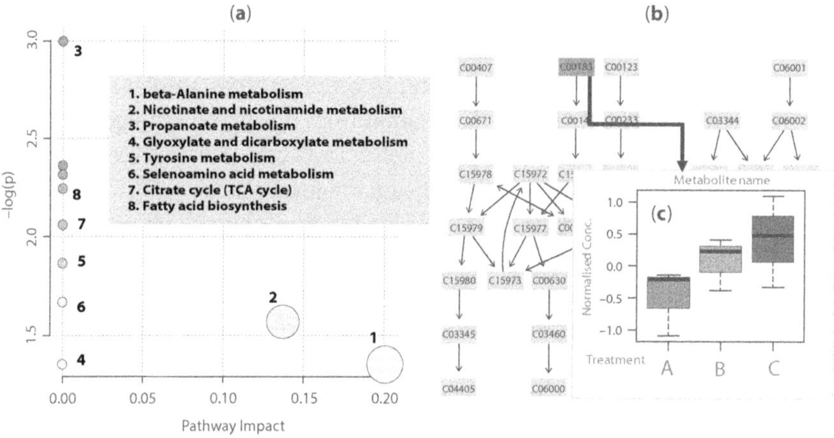

Figure 7.5 Representation of pathway mapping using MetaboAnalyst: the extent of impact of the identified metabolite to each pathway (a) along with the KEGG compound no. through different nodes (b) and the normalized concentration variation of each metabolite under different treatment conditions is mapped (c). Refer to MetaboAnalyst [25] for more details.

displaying comprehensive contribution of the metabolites and the interactions among them as exemplified in Figure 7.5.

7.3 Selected Applications of Metabolomics and Metabolite Profiling

Applications of NMR and GC-MS in plant metabolomics have rapidly grown in the last decade and are widely being applied. The advances of these techniques compounded with better and rapid methods for the qualitative and quantitative analysis of complex metabolic data sets lead to a wide range of applications [35–38]. Metabolite profiling and metabolomics lead to applications such as quality analysis of plant products, studying chemically synthesized compounds, discovery of disease biomarkers, establishing food quality and authenticity, identifying the metabolic variation under stress condition, and evaluating plant-based medicines [39–44]. The key metabolomics applications in broader context include the following:

- **Plant medicine:** Plant metabolomics was earlier used as a technology for studying plant metabolism and their biochemical pathways, but nowadays it can be used to study the macro and micro constituents of medicinal plants. These active constituents are considered to have potential features to fight from several diseases and thus can be used for preparing herbal medicines toward preventing a disease condition [45, 46].

- **Food quality and authenticity:** In order to cross-validate the food quality being served, metabolomics is found to be a good option. The health beneficial metabolites can thus be categorized in lesser time to authenticate the food quality [47].
- **Disease biomarker:** A unique metabolite/phytochemical in any cellular system is considered to be a biomarker for that system. Evaluation of any biological processes or response can be quantitatively measured by the presence of biomarkers. Thus, disease biomarker combining with modern clinical metabolomics approach helps in exploring novel drug discovery, identifying biomarkers for detection and diagnosis, which provides better clinical approach in serious diseases [48, 49].
- **Quality analysis of phytochemicals:** Multiple factors such as geographical regions, biotic and abiotic stress, etc. affect the chemical composition of the plant extracts. Industrial products, particularly herbal medicine need to ensure proper quality control with reproducibility and reliability of the product. Analytical techniques (NMR, HPLC, LC-MS, GC-MS, etc.) based metabolomics can ensure control in quality and better reproducibility [50–52].

Acknowledgments

SKM acknowledges Science and Engineering Research Board (SERB) Early career research funding (File No: ECR/2016/001176), MHRD, for the work. ML and MS acknowledge SERB, MHRD, and IIT Mandi for financial support to carry out relevant work.

Competing Interests

The authors declare no competing financial interests.

References

1. Fiehn, O., Kopka, J., Dörmann, P., Altmann, T., Trethewey, R.N., Willmitzer, L., Metabolite profiling for plant functional genomics. *Nat. Biotechnol.*, 18, 11, 115, 2000.

2. Roessner, U., Willmitzer, L., Fernie, A., Metabolic profiling and biochemical phenotyping of plant systems. *Plant Cell Rep.*, 21, 3, 189–196, 2002.
3. Dixon, R.A., Gang, D.R., Charlton, A.J., Fiehn, O., Kuiper, H.A., Reynolds, T.L., Seiber, J.N., Applications of metabolomics in agriculture. *J. Agricult. Food Chem.*, 54, 24, 8984–8994, 2006.
4. Cascante, M. and Marin, S., Metabolomics and fluxomics approaches. *Essays Biochem.*, 45, 67–82, 2008.
5. Hollywood, K., Brison, D.R., Goodacre, R., Metabolomics: Current technologies and future trends. *Proteomics*, 6, 17, 4716–4723, 2006.
6. Tarpley, L., Duran, A.L., Kebrom, T.H., Sumner, L.W., Biomarker metabolites capturing the metabolite variance present in a rice plant developmental period. *BMC Plant Biol.*, 5, 1, 8, 2005.
7. Roessner, U., Luedemann, A., Brust, D., Fiehn, O., Linke, T., Willmitzer, L., Fernie, A.R., Metabolic profiling allows comprehensive phenotyping of genetically or environmentally modified plant systems. *Plant Cell*, 13, 1, 11–29, 2001.
8. Atanasov, A.G., Waltenberger, B., Pferschy-Wenzig, E.M., Linder, T., Wawrosch, C., Uhrin, P., Rollinger, J.M., Discovery and resupply of pharmacologically active plant-derived natural products: A review. *Biotechnol. Adv.*, 33, 8, 1582–1614, 2015.
9. Cocuron, J.C., Anderson, B., Boyd, A., Alonso, A.P., Targeted metabolomics of *Physaria fendleri*, an industrial crop producing hydroxy fatty acids. *Plant Cell Physiol.*, 55, 3, 620–633, 2014.
10. Fiehn, O., Metabolomics—The link between genotypes and phenotypes, in: *Functional Genomics*, pp. 155–171, Springer Netherlands, 2002.
11. Zhang, A., Sun, H., Wang, P., Han, Y., Wang, X., Modern analytical techniques in metabolomics analysis. *Analyst*, 137, 2, 293–300, 2012.
12. Krishnan, P., Kruger, N.J., Ratcliffe, R.G., Metabolite fingerprinting and profiling in plants using NMR. *J. Exp. Bot.*, 56, 410, 255–265, 2004.
13. Lisec, J., Schauer, N., Kopka, J., Willmitzer, L., Fernie, A.R., Gas chromatography mass spectrometry–based metabolite profiling in plants. *Nat. Protoc.*, 1, 1, 387, 2006.
14. Shulaev, V., Cortes, D., Miller, G., Mittler, R., Metabolomics for plant stress response. *Physiol. Plant.*, 132, 2, 199–208, 2008.
15. Villas-Bôas, S.G., Højer-Pedersen, J., Åkesson, M., Smedsgaard, J., Nielsen, J., Global metabolite analysis of yeast: Evaluation of sample preparation methods. *Yeast*, 22, 14, 1155–1169, 2005.
16. Xia, J., Sinelnikov, I.V., Han, B., Wishart, D.S., MetaboAnalyst 3.0—Making metabolomics more meaningful. *Nucl. Acids Res.*, 43, W1, W251–W257, 2015.
17. Kim, H.K., Choi, Y.H., Verpoorte, R., NMR-based plant metabolomics: Where do we stand, where do we go? *Trends Biotechnol.*, 29, 6, 267–275, 2011.

18. Halket, J.M., Waterman, D., Przyborowska, A.M., Patel, R.K., Fraser, P.D., Bramley, P.M., Chemical derivatization and mass spectral libraries in metabolic profiling by GC/MS and LC/MS/MS. *J. Exp. Bot.*, 56, 410, 219–243, 2004.
19. Markley, J.L., Brüschweiler, R., Edison, A.S., Eghbalnia, H.R., Powers, R., Raftery, D., Wishart, D.S., The future of NMR-based metabolomics. *Curr. Opin. Biotechnol.*, 43, 34–40, 2017.
20. Jahangir, M., Nuringtyas, T.R., Ali, K., Wilson, E.G., Choi, Y.H., Verpoorte, R., NMR-based metabolomics: Understanding plant chemistry and identification of biologically active compounds, in: *NMR-Based Metabolomics*, pp. 246–263, 2018.
21. Halabalaki, M., Bertrand, S., Stefanou, A., Gindro, K., Kostidis, S., Mikros, E., Wolfender, J.L., Sample preparation issues in NMR-based plant metabolomics: Optimisation for vitis wood samples. *Phytochem. Anal.*, 25, 4, 350–356, 2014.
22. Jacobsen, N.E., *NMR Data Interpretation Explained: Understanding 1D and 2D NMR Spectra of Organic Compounds and Natural Products*, John Wiley & Sons, 2016.
23. Kruger, N.J., Troncoso-Ponce, M.A., Ratcliffe, R.G., ^{1}H NMR metabolite fingerprinting and metabolomic analysis of perchloric acid extracts from plant tissues. *Nat. Protoc.*, 3, 6, 1001–1012, 2008.
24. Amberg, A., Riefke, B., Schlotterbeck, G., Ross, A., Senn, H., Dieterle, F., Keck, M., NMR and MS methods for metabolomics, in: *Drug Safety Evaluation*, pp. 229–258, Humana Press, New York, NY, 2017.
25. Xia, J. and Wishart, D.S., Using MetaboAnalyst 3.0 for comprehensive metabolomics data analysis. *Curr. Protoc. Bioinformatics*, 14–10, 2016.
26. Masakapalli, S.K., Kruger, N.J., Ratcliffe, R.G., The metabolic flux phenotype of heterotrophic Arabidopsis cells reveals a complex response to changes in nitrogen supply. *Plant J.*, 74, 4, 569–582, 2013.
27. Lommen, A., MetAlign: Interface-driven, versatile metabolomics tool for hyphenated full-scan mass spectrometry data preprocessing. *Anal. Biochem.*, 81, 8, 3079–3086, 2009.
28. Behrends, V., Tredwell, G.D., Bundy, J.G., A software complement to AMDIS for processing GC-MS metabolomic data. *Anal. Biochem.*, 415, 2, 206–208, 2011.
29. Kopka, J., Schauer, N., Krueger, S., Birkemeyer, C., Usadel, B., Bergmüller, E., Willmitzer, L., GMD@ CSB. DB: The Golm metabolome database. *Bioinformatics*, 21, 8, 1635–1638, 2004.
30. Babushok, V.I., Linstrom, P.J., Reed, J.J., Zenkevich, I.G., Brown, R.L., Mallard, W.G., Stein, S.E., Development of a database of gas chromatographic retention properties of organic compounds. *J. Chromatogr. A*, 1157, 1–2, 414–421, 2007.
31. Smith, C.A., Want, E.J., O'Maille, G., Abagyan, R., Siuzdak, G., XCMS: Processing mass spectrometry data for metabolite profiling using nonlinear peak alignment, matching, and identification. *Anal. Chem.*, 78, 3, 779–787, 2006.

32. Wu, Z., Li, D., Meng, J., Wang, H., Introduction to SIMCA-P and its application, in: *Handbook of Partial Least Squares*, pp. 757–774, Springer, Berlin, Heidelberg, 2010.
33. Worley, B. and Powers, R., Multivariate analysis in metabolomics. *Curr. Metabolomics*, 1, 1, 92–107, 2013.
34. Chong, J., Soufan, O., Li, C., Caraus, I., Li, S., Bourque, G., Wishart, D.S., Xia, J., MetaboAnalyst 4.0: Towards more transparent and integrative metabolomics analysis. *Nucl. Acids Res.*, 2018.
35. Ghosh, D., Statistical analysis of proteomics, metabolomics, and lipidomics data using mass spectrometry. 295 pages. *Int. Stat. Rev.*, 85, 3, 544–554, 2017.
36. Khakimov, B., Rasmussen, M.A., Kannangara, R.M., Jespersen, B.M., Munck, L., Engelsen, S.B., From metabolome to phenotype: GC-MS metabolomics of developing mutant barley seeds reveals effects of growth, temperature and genotype. *Sci. Rep.*, 7, 1, 8195, 2017.
37. Meyer, M.R., Hans-Joachim Hübschmann: Handbook of GC-MS: Fundamentals and applications. *Anal. Bioanal. Chem.*, 408, 6, 1535, 2016.
38. Al-Rubaye, A.F., Hameed, I.H., Kadhim, M.J., A review: Uses of gas chromatography-mass spectrometry (GC-MS) technique for analysis of bioactive natural compounds of some plants. *Int. J. Toxicol. Pharmacol. Res.*, 9, 1, 81–85, 2017.
39. Moros, G., Chatziioannou, A.C., Gika, H.G., Raikos, N., Theodoridis, G., Investigation of the derivatization conditions for GC–MS metabolomics of biological samples. *Bioanalysis*, 9, 1, 53–65, 2017.
40. Mumm, R., Hageman, J.A., Calingacion, M.N., de Vos, R.C.H., Jonker, H.H., Erban, A., Ward, J.L., Multi-platform metabolomics analyses of a broad collection of fragrant and non-fragrant rice varieties reveals the high complexity of grain quality characteristics. *Metabolomics*, 12, 2, 38, 2016.
41. Wishart, D.S., Metabolomics: Applications to food science and nutrition research. *Trends Food Sci. Technol.*, 19, 9, 482–493, 2008.
42. Hall, R.D., Brouwer, I.D., Fitzgerald, M.A., Plant metabolomics and its potential application for human nutrition. *Physiol. Plant.*, 132, 2, 162–175, 2008.
43. Schauer, N. and Fernie, A.R., Plant metabolomics: Towards biological function and mechanism. *Trends Plant Sci.*, 11, 10, 508–516, 2006.
44. Khalil, M.N., Fekry, M.I., Farag, M.A., Metabolome based volatiles profiling in 13 date palm fruit varieties from Egypt via SPME GC–MS and chemometrics. *Food Chem.*, 217, 171–181, 2017.
45. Okada, T., Mochamad Afendi, F., Altaf-Ul-Amin, M., Takahashi, H., Nakamura, K., Kanaya, S., Metabolomics of medicinal plants: The importance of multivariate analysis of analytical chemistry data. *Curr. Comput. Aided Drug Des.*, 6, 3, 179–196, 2010.
46. Shyur, L.F. and Yang, N.S., Metabolomics for phytomedicine research and drug development. *Curr. Opin. Chem. Biol.*, 12, 1, 66–71, 2008.

47. Cevallos-Cevallos, J.M., Reyes-De-Corcuera, J.I., Etxeberria, E., Danyluk, M.D., Rodrick, G.E., Metabolomic analysis in food science: A review. *Trends Food Sci. Technol.*, 20, 11–12, 557–566, 2009.
48. Kohler, I., Hankemeier, T., van der Graaf, P.H., Knibbe, C.A., van Hasselt, J.C., Integrating clinical metabolomics-based biomarker discovery and clinical pharmacology to enable precision medicine. *Eur. J. Pharm. Sci.*, 109, S15–S2, 2017.
49. Yeung, P.K., Metabolomics and biomarkers for drug discovery. *Metabolites*, 8, 1, 11, 2018.
50. Heyman, H.M. and Meyer, J.J.M., NMR-based metabolomics as a quality control tool for herbal products. *S. Afr. J. Bot.*, 82, 21–32, 2012.
51. Lee, K.M., Jeon, J.Y., Lee, B.J., Lee, H., Choi, H.K., Application of metabolomics to quality control of natural product derived medicines. *Biomol. Ther.*, 25, 6, 559, 2017.
52. Albrecht, C.F., Stander, M.A., Grobbelaar, M.C., Colling, J., Kossmann, J., Hills, P.N., Makunga, N.P., LC–MS-based metabolomics assists with quality assessment and traceability of wild and cultivated plants of *Sutherlandia frutescens* (Fabaceae). *S. Afr. J. Bot.*, 82, 33–45, 2012.

8

OMICS-Based Approaches for Elucidation of Picrosides Biosynthesis in *Picrorhiza kurroa*

Varun Kumar

Department of Biotechnology and Bioinformatics, Jaypee University of Information Technology, Waknaghat, Solan (HP), India

Abstract

Picrorhiza kurroa is one of the top 15 medicinal herbs existing in India with respect to revenues generated by the traded material. The *P. kurroa* rhizomes containing Picroside-I (P-I) and Picroside-II (P-II) as active constituents have been utilized in various commercial herbal drug formulations used for the treatment of liver ailments due to their hepatoprotective, antioxidative, anti-allergic, anti-asthmatic, anticancerous, and immunomodulatory activities. Considering its enormous bioactivities, the exhaustive collection, unorganized cultivation, and illegal trading have put *P. kurroa* in the list of critically endangered medicinal herbs. Consequently, the reduced population of *P. kurroa* in the wild has led to its adulteration, which affects its quality in terms of P-I and P-II composition, thereby posing concerns over the efficacy of resulting herbal drugs. Owing to this, there is an urgent need to develop mass production platform for quality herbal raw material of *P. kurroa* with improved levels of P-I and P-II. Therefore, in this chapter, we have discussed how picrosides are biosynthesized, various approaches used to elucidate their biosynthetic routes, and potential strategies to engineer metabolic carbon flux toward picrosides production so as to improve the quality of *P. kurroa*.

Keywords: Metabolic flux, picroside-I and II, MVA/MEP pathway, shikimate/phenylpropanoid pathway, iridoid pathway

Email: varun26g@gmail.com

Rintu Banerjee, Garlapati Vijay Kumar, and S.P. Jeevan Kumar (eds.) OMICS-Based Approaches in Plant Biotechnology, (145–166) © 2019 Scrivener Publishing LLC

8.1 Introduction

The incidences of chronic liver disorders are rising throughout the world and have also been seen in India irrespective of sex, race, and age. Liver diseases are considered to be silent killers as persons would not realize about this problem until it becomes serious. Liver cirrhosis is an end result of various liver diseases, which in worst conditions affects the functioning of liver by increasing resistance to blood flow through the liver. Globally, the cirrhosis of liver is caused by alcohol consumption, non-alcoholic fatty liver, and viral hepatitis. According to the latest data published by the WHO (World Health Organization), 216,865 deaths occur in India due to liver diseases, which is 2.44% of the total deaths caused (http://www.worldlifeexpectancy.com/india-liver-disease).

Over the years, synthetic and recombinant drugs, viz., corticosteroids and interferons, have been used for the treatment of various liver disorders but simultaneously associated with harmful side effects including agitation, depression, retinal ischemia, anxiety, and nephrotic syndrome [1]. Therefore, plant-based remedies came into spotlight for the treatment of liver disorders. The various herbal drugs such as colchicine, glycyrrhizin, silymarin, and curcumin were obtained from medicinal plants and are considered as potent hepatoprotective agents [1]. Among these herbal drugs, high doses of colchicine induced toxic effects and therefore prescribed at very low doses [2]. Further, the majority of reports showed association of curcumin with the treatment of liver cirrhosis, but only a limited number of researchers have suggested its role in the prevention of virus-induced hepatic injury [1]. In contrast, glycyrrhizin has been reported to be effective against chronic hepatitis [1]. However, silymarin has been considered as one of the favored drugs for the treatment of liver disorders; though simultaneous screening of other medicinal herbs revealed *P. kurroa*-based herbal drugs to be highly effective against various liver disorders [3].

Picrorhiza kurroa Royle ex Benth, locally known as Kutki or Karu, is a medicinal herb of family Plantaginaceae (earlier recognized as Scrophulariaceae). It is found in the Himalaya regions of China, Nepal, and India at a high altitude ranging from 3000 to 5000 m [3, 4]. From ancient times, *P. kurroa* has been used as a traditional medicine by local communities for the treatment of respiratory, allergic, stomach, and liver disorders because of its bioactivities such as hepatoprotective [5], antioxidative [6], anti-allergic [7], anti-asthmatic [7], anticancerous [8], and immunomodulatory activity [9]. The herbal drugs prepared from *P. kurroa* are commercially available in the market (Figure 8.1), and the

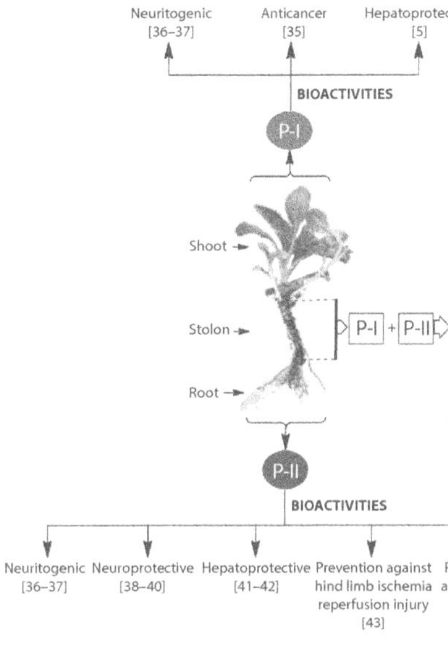

P. kurroa Based Commercial Herbal Drug Formulations

Herbal Drug Formulations	Manufacturer	P-I (%)	P-II (%)	References
Katuki	Zandu Pharma Works Ltd., Mumbai, India	1.29	1.16	[49]
Arogya	Zandu Pharma Works Ltd., Vapi, Gujrat, India	1.01	0.55	[49]
Kutaki	Tansukh Herbal Pvt. Ltd., Lucknow, U.P., India	4.17	3.25	[49]
Livocare	Dindayal Aushdi Pvt. Ltd. Gwalior, M.P., India	0.06	0.14	[49]
Livomap	Maharishi Ayurveda Pvt. Ltd., New Delhi, India	0.12	0.06	[49]
Livomyn	Charka Pharma Pvt. Ltd., Mumbai, India	0.07	0.01	[49]
Livplus	BACFO Pharmaceuticals Ltd., Noida, U.P., India	0.22	0.49	[49]
Pravekliv	Pravek Kalp Herbal products Pvt. Ltd., Noida, U.P., India	0.15	0.12	[49]
Vimliv	OM Pharma Ltd., Bangalore, India	0.001	Traces	[49]

Figure 8.1 Bioactivities, sites of biosynthesis/accumulation in *P. kurroa*, and commercial herbal drug formulations of Picroside-I and Picroside-II. The numbers in square brackets indicate respective references, which are elaborated in the reference list.

demand of raw material has raised to 415 MT/year, which is ~40 times higher as compared to its requirement in 1990 [10]. This figure gives only a sign of its consumption as most of its collection and trade is illegal from wild/natural sources, the major resource base of *P. kurroa*. The extensive and unorganized collection of *P. kurroa* raised serious socioeconomic concerns among farmers, herbal industries, and policy makers due to its reduced population in the natural habitat. *P. kurroa* propagates vegetatively in natural habitat where it arises in the form of a young bud on mature stolons, which subsequently produce a complete plant with roots, shoots, and stolons. Therefore, during illegal procurement and lack of knowledge on harvesting, complete plants get uprooted and then rhizomes are separated and sold in the market while the rest of the plant material is thrown away. This has resulted in the loss of young buds, which arouse on the mature stolons thereby obstructing its propagation in the wild. Owing to a heavy extraction of *P. kurroa* from the wild (the entire Himalayan range), it is listed in the *Negative List of Export (GOI), Appendix II of Cities, Red Data Book of Indian Plants and Prioritized List of Medicinal Plants for Conservation* by the International Union for Conservation of Nature (IUCN) [11].

The imposition of legal restrictions on the collection of raw material from the wild led to the adulteration of *P. kurroa*, which has reduced its quality in terms of active constituents composition, i.e., P-I and P-II, thereby posing concerns over the efficacy of resulting herbal drugs. One such poor quality herbal raw material of *P. kurroa* from China is sold in the markets of India at a cost of Rs 700/kg compared to Rs 350/kg from H.P., with ~4 times lower levels of total P-I and P-II [12].

Therefore, to develop mass production platform for quality herbal raw material of *P. kurroa* with improved levels of P-I and P-II, it is necessary to keep a view of existing knowledge on picrosides biosynthesis so as to provide useful insights into genetic interventions possible to increase the contents of picrosides in *P. kurroa*.

8.2 Cross-Talk of Picrosides Biosynthesis Among Different Tissues of *P. kurroa*

Secondary metabolites are specialized metabolites that mainly synthesized in specific tissues of plants such as leaves, roots, rhizomes, and flower [13, 14]. In *P. kurroa*, it has been observed that P-I is produced in shoots, while stolons/rhizome tissues synthesized both P-I and P-II. In contrast, roots of *P. kurroa* contained only P-II (Figure 8.1) [15]. Moreover, the analyses of *P. kurroa* plants of different age, growth, and development stages revealed significant variations in both P-I and P-II contents, viz., 0.05–0.80% P-I in shoots, 0.15–0.98% P-I and 0.10–0.81% P-II in rhizomes, and 0.07–0.21% P-II in roots [15]. Therefore, it was evident from this analysis that shoots, stolons/rhizomes, and roots are independent biosynthetic organs for P-I and P-II in *P. kurroa*.

8.3 Strategies Used for the Elucidation of Picrosides Biosynthetic Route in *P. kurroa*

The elucidation of biosynthetic routes for secondary metabolites in plants is a challenging task considering that these are produced in specific cells/ tissues at a particular developmental stage and genotype, influenced by the environment and altitude [13]. In the modern era, analysis of differential transcriptomes and metabolomes provided greater insights not only into the biological pathways and molecular mechanisms that regulate the contents of major chemical constituents but also growth and development of organs/tissues involved in metabolites biosynthesis [16, 17]. In *P. kurroa*, various approaches have been used to discern the biosynthesis of picrosides as described in the following sections:

8.3.1 Retro-Biosynthetic Approach

Retro-biosynthesis is an approach that starts with defining a target compound of interest and then traces back the known chemical transformations for its modification to identify potential reactions and precursors [18]. In *P. kurroa*, by looking into the chemical structures of P-I and P-II, it has been proposed that catalpol (CAT) is the basic structural backbone that is esterified with *trans*-cinnamic acid (CA) for the biosynthesis of P-I and vanillic acid (VA) for the production of P-II [19]. Further, CA and VA are the phenolic acids reported to be produced through shikimate/phenylpropanoid pathway, thereby indicated its involvement in the biosynthesis of picrosides [20]. The biosynthetic route for catalpol was established through deuterium-labelled feeding experiments in *Scrophularia*, *Scutellaria*, and *Paulownia* [21, 22]. It is biosynthesized from aucubin, and the latter one is produced from 8-epideoxyloganic acid through a series of chemical intermediates, viz., mussaenosidic acid, deoxygeniposidic acid, geniposidic acid, and bartsioside [21, 22]. The biosynthesis of 8-epideoxyloganic acid proceeds via geranyl pyrophosphate (GPP) through a series of cyclization reactions including geraniol, 10-hydroxygeraniol, 10-oxogeranial, iridodial, epi-iridodial, iridotrial, and boschnaloside [19]. The validation of iridoid pathway was carried out by detecting boschnaloside, mussaenosidic acid, bartsioside, aucubin, and catalpol in *P. kurroa* through LC-ESI-MS/MS followed by their confirmation through fragmentation patterns [19]. Further, the precursor GPP is biosynthesized via both mevalonate (MVA) and non-mevalonate (MEP) pathways [23]. Therefore, the proposed biosynthetic pathways of P-I and P-II involve four different modules, i.e., MVA, MEP, shikimate/phenylpropanoid, and iridoid pathways in *P. kurroa* [19].

8.3.2 *In Vitro* Feeding of Different Precursors and Inhibitors

Precursor feeding is an approach in which various compounds, which act as intermediate metabolic precursors for target secondary metabolites, have been supplied to the culture media in order to observe their effect either on the production of target metabolites or to determine the flux through respective pathways [24–26]. The *P. kurroa* shoots grown in tissue culture conditions were exogenously fed with various precursors and inhibitors not only to determine the relative contribution of different pathway modules but also to validate the proposed pathway of picrosides biosynthesis. The inhibition of MEP pathway with fosmidomycin (an inhibitor of DXPR enzyme) reduced P-I content by 90%, which was much higher as compared to mevinolin (an inhibitor of HMGR enzyme) that reduced P-I by only ~12%; this indicated that MEP pathway contributes more toward P-I biosynthesis [27]. Further, the exogenous addition of immediate precursors for P-I biosynthesis, i.e., CA and CAT (separately and in combination), *in vitro* at different concentrations (1 mg/100 mL, 2.5 mg/100 mL, and 5 mg/100 mL), revealed no significant variations in P-I content among CAT-fed *P. kurroa* shoots compared to control shoots (without treatment). However, *P. kurroa* shoots fed only with CA revealed 1.6-fold elevation in P-I content compared to control shoots, which thereby indicates the possibility of limited CA availability in CAT-fed shoots [26]. In contrast, the combined addition of CA and CAT at 2.5 mg/100 mL concentration showed maximum increase in P-I content with 4.2-fold [26], thus validating the proposed pathway that reported that CAT is esterified with CA for the biosynthesis of P-I in *P. kurroa* [19].

8.3.3 Metabolomics of Natural Variant Chemotypes of *P. kurroa*

The secondary metabolites are mostly synthesized in a tissue-specific manner [13, 14], which adds complications in elucidation of their biosynthetic routes and associated genes information. With the advent of modern high-throughput technologies such as transcriptomics, proteomics, and metabolomics, it became comparatively easier to get insight into molecular mechanisms underlying the biosynthesis of target secondary metabolites [28]. However, these technologies fail to explain the genetic control of secondary metabolic pathways. Therefore, the utilization of genetic diversity existing in a plant species, for the production of secondary metabolites, has become a revolutionary approach to track target secondary metabolites biosynthesis through the use of omics [29]. In *P. kurroa*, the validation of the proposed pathway for P-II biosynthesis faced challenge due to its insignificant production in shoots of *P. kurroa*. Interestingly, the natural variations existing for P-II content among different *P. kurroa* accessions collected from various geographical locations of India provide insight into the intermediates leading to the biosynthesis of P-II [20]. The quantitative analyses of metabolic intermediates, viz., CA, *p*-coumaric acid (*p*-CA), ferulic acid (FA), protocatechuic acid (PRCA), and VA, among stolon tissues of different *P. kurroa* accessions varied for P-II content (1.3–2.6%), revealed flux diversion of shikimate/phenylpropanoid pathway toward VA production in high P-II content accessions [20]. The levels of FA also showed similar patterns as that of VA, thereby indicating that VA biosynthesis might proceed via FA in *P. kurroa*. Further confirmation was provided by adding FA exogenously at 150-μM concentration *in vitro*, which showed a twofold increase in the levels of VA compared to control shoots. Moreover, positive correlation was observed upon integration of transcript levels with intermediate metabolites contents, which establishes the biosynthetic route of P-II in *P. kurroa* [20].

8.4 Strategies Used for Shortlisting Key/Candidate Genes Involved in Picrosides Biosynthesis

The biosynthesis of picrosides involves integration of different pathways including MVA/MEP, shikimate/phenylpropanoid, and iridoid/monoterpene pathways in *P. kurroa* [19]. Therefore, a large number of steps are proposed to be involved in the biosynthesis of P-I and P-II, leading to a complex biosynthetic architecture [3, 27]. To shortlist key/candidate genes/enzymes involved in the biosynthesis of P-I and P-II, combination of approaches was used in the past as outlined in Figure 8.2 and described in the following sections.

8.4.1 Comparative Genomics

Comparative genomics refers to the comparison of genome characteristics of different organisms/species. In *P. kurroa*, genes/enzymes that are involved in the biosynthesis of picrosides and also catalyze similar enzymatic reactions in other plant species were retrieved, and the corresponding genes were cloned through the comparative genomics approach. Pandit *et al.*, [23] cloned two genes of the MVA pathway, i.e., *HMGS* and *PMK*, and three genes of the MEP pathway, viz., *MECPS*, *HDS*, and *ISPD*, in *P. kurroa*. Moreover, the expression analysis of all 15 genes belonging to MVA/MEP pathways vis-à-vis P-I and P-II contents in *P. kurroa* revealed 57- to 166-fold up-regulation of *PMK*, *DXPS*, *ISPE*, *MECPS*, and *ISPD* vis-à-vis the P-I content, while increased expression of *DXPR* and *HDS* with ~9- to 12-fold was observed in congruence with the P-II content [23]. This study showed that the genes up-regulated with respect to P-I and P-II contents belong to the MEP pathway, except *PMK*, which belongs to the MVA pathway, thereby indicating the major role of

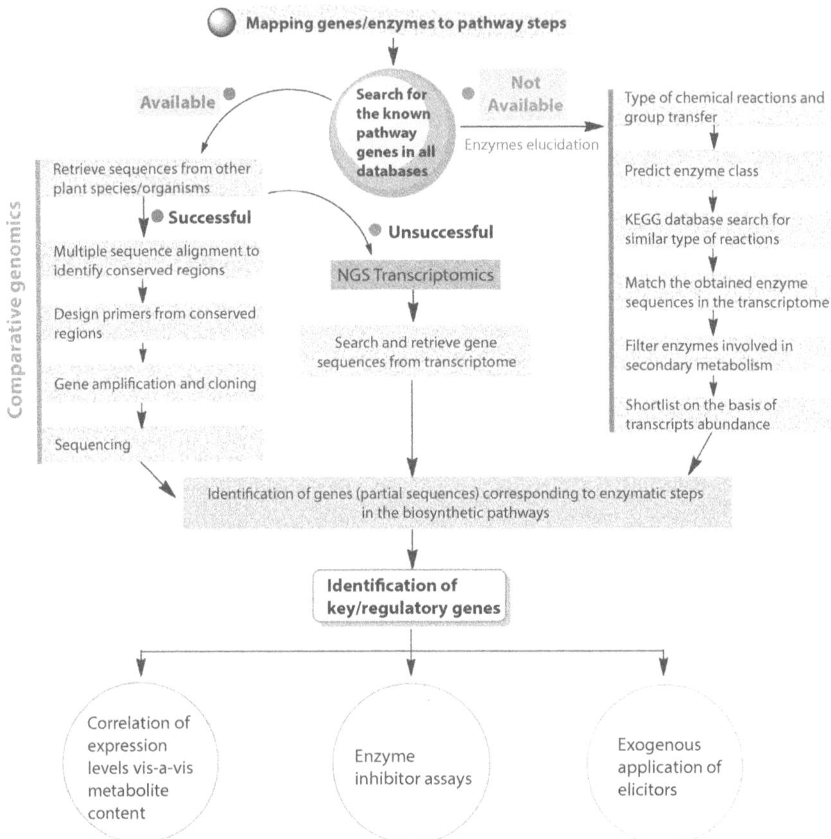

Figure 8.2 Overview of strategies used for the shortlisting of key picrosides biosynthetic genes in *P. kurroa*.

the MEP pathway in picrosides biosynthesis. Nevertheless, comparative genomics approach has been successfully employed to shortlist key genes in MVA/MEP pathways, but it is limited to only conserved and known pathway steps.

8.4.2 Differential Next-Generation Sequencing (NGS) Transcriptomes and Expression Levels of Pathway Genes Vis-à-Vis Picrosides Content

The genome-wide transcriptome profiling through NGS is an effective approach even in nonmodel plant species to discern key genes/transcripts in complex biosynthetic pathways [30]. In *P. kurroa*, different NGS transcriptomes were generated from phenotypes varying in picrosides contents and consequently used for obtaining information for most of the genes encoding enzymes catalyzed steps in the pathway. A total of six transcriptome datasets were generated from differential picrosides content phenotypes of *P. kurroa*, i.e., shoot tissues grown in field (2.7% P-I) as well as *in vitro* at 15°C (0.6% P-I) and 25°C (0.01% P-I); root tissues grown in field (0.4% P-II) and *in vitro* at 25°C (P-II not produced); and stolons of field-grown plants with 1.7% P-I and 0.99% P-II contents [27, 31]. Moreover, four transcriptome datasets were generated from high and low P-I content shoot accessions of *P. kurroa*, i.e., PKS-1 (2.38% P-I), PKS-5 (1.18% P-I), PKS-4 (0.49% P-I), and PKS-21 (0.13% P-I). Similarly, four transcriptome datasets were also generated from high and low P-II content stolon accessions of *P. kurroa*, i.e., PKST-3 (2.65% P-II), PKST-5 (2.19% P-II), PKST-16 (1.52% P-II), and PKST-18 (1.36% P-II) [20]. These transcriptomes obtained from different picrosides content phenotypes provided not only information of genes involved in picrosides biosynthetic pathways but also molecular basis of their regulation in *P. kurroa*. The partial gene sequences for majority of pathway genes obtained from NGS transcriptomes were retrieved, validated, and associated with the contents of picrosides through quantitative real-time polymerase chain reaction (qRT-PCR) on differential content phenotypes. The expression analysis of picrosides biosynthetic pathway genes was carried out under various experimental conditions and resulted in the shortlisting of key candidates as described in Tables 8.1 and 8.2.

8.5 Complete Architecture of Picrosides Biosynthetic Pathway

The shikimate/phenylpropanoid pathway produced CA and VA (precursors of P-I and P-II, respectively) through a series of 10 and 14 enzyme catalyzed reactions [3], respectively, after the initial conversion of PEP and E-4-P to 3-Deoxy-D-arabinoheptulosonate 7-phosphate (DAHP) by an enzyme, DAHPS, 3-Deoxy-D-arabinoheptulosonate 7-phosphate synthase [32]. On the other hand, the MVA pathway occurs in cytosol and produces isoprene unit, i.e., isopentenyl pyrophosphate (IPP), through a series of six enzyme catalyzed reactions, starting with the conversion of acetyl coenzyme A to acetoacetyl coenzyme A by an enzyme, acetoacetyl-CoA thiolase [33].

Table 8.1 Details of strategies used to shortlist key candidate genes/transcripts associated with P-II biosynthesis in *P. kurroa*.

S. No.	Experimental approach	Differential P-II content	Shortlisted genes/ transcripts	Pathway	Fold up-regulated gene expression	Significance/ role	References
1.	Expression analysis of pathway genes among field-grown (natural habitat) vs. *in vitro*-grown (at 15°C and 25°C under tissue culture conditions) *P. kurroa* roots/ stolons	• Roots at 15/25°C—NIL • Field-grown roots—0.4%	*DXPR, HDS*	MEP	9.4–12.12	Formation of GPP	[23]
			F3D, ACT *DHQD, SD, SK, COMT, PAL, EPSPS*	Iridoid S/P	16–96	Proposed enzyme (yet to validate) Formation of VA	[27]
2.	Differential transcriptome analysis of high vs. low P-II content stolon accessions of *P. kurroa*	• PKST-3—2.65% • PKST-5—2.19% • PKST-16—1.52% • PKST-18—1.36%	*DXPS* *HMGR* *PAL*	MEP MVA S/P	Significant positive correlation	Formation of GPP Formation of CA and VA	[20]

S/P: shikimate/phenylpropanoid.

Table 8.2 Details of strategies used to shortlist key candidate genes/transcripts associated with P-I biosynthesis in *P. kurroa*.

S. No.	Experimental approach	Differential P-I content	Shortlisted genes/ transcripts	Pathway	Fold up-regulated gene expression	Significance/ role	References
1.	Expression analysis of pathway genes among field-grown (natural habitat) vs. *in vitro*-grown (at 15°C and 25°C under tissue culture conditions) *P. kurroa* shoots	• Field grown shoots—2.7% • Shoots at 15°C—0.6% • Shoots at 25°C—0.01%	*PMK DXPS, ISPE, MECPS, ISPD*	MVA MEP	107.7 57–166	Formation of GPP	[23]
			HK, PK ICDH, MDH	Glycolysis TCA cycle	8–12.6 1.7–18	Formation of pyruvate	[50]
2.	Expression analysis of pathway genes among high vs. low P-I content shoot accessions of *P. kurroa*	• PKS-1—2.7% • PKS-4—0.3%	*DXPS, ISPD, ISPE* *PMK 2HFD* *EPSPS, SK*	MEP MVA Iridoid S/P	5–16 5 10 20–57	Formation of GPP Proposed enzyme (yet to validate) Formation of CA	[27]

(*Continued*)

Table 8.2 Details of strategies used to shortlist key candidate genes/transcripts associated with P-I biosynthesis in *P. kurroa*. (*Continued*)

S. No.	Experimental approach	Differential P-I content	Shortlisted genes/ transcripts	Pathway	Fold up-regulated gene expression	Significance/ role	References
3.	Expression analysis of pathway genes in *P. kurroa* shoots harvested at different time intervals (0, 10, 20, 30, and 40 days) of plant growth at 15°C *in vitro*	• 0 day—NIL • 10 days—0.02% • 20 days—0.04% • 30 days—0.17% • 40 days—0.16%	*DAHPS* *G-10-H*	S/P Iridoid	4.68 6.28	Formation of CA Formation of CAT	[32]

(*Continued*)

Table 8.2 Details of strategies used to shortlist key candidate genes/transcripts associated with P-I biosynthesis in *P. kurroa*. (*Continued*)

S. No.	Experimental approach	Differential P-I content	Shortlisted genes/ transcripts	Pathway	Fold up-regulated gene expression	Significance/ role	References
4.	Expression analysis of MVA/MEP pathway genes in *P. kurroa* leaves dipped in 5% sucrose solution containing different elicitors (abscisic acid, ABA; methyl jasmonate, MeJA; salicylic acid, SA; and hydrogen peroxide, H_2O_2) for 24–48 h	% increase over control (untreated) at 24 h • ABA—15% • H_2O_2—50% % increase over control (untreated) at 48 h • ABA—45% • H_2O_2—15%	*DXPS, IPPI, HDR/ISPH,*	MEP	ND	Formation of GPP	[51]
			COMT	S/P	ND	Formation of ferulic acid	

(*Continued*)

Table 8.2 Details of strategies used to shortlist key candidate genes/transcripts associated with P-I biosynthesis in *P. kurroa*. (*Continued*)

S. No.	Experimental approach	Differential P-I content	Shortlisted genes/ transcripts	Pathway	Fold up-regulated gene expression	Significance/ role	References
5.	Expression analysis of pathway genes in *P. kurroa* shoots treated with different elicitors (seaweed extract, SWE; abscisic acid, ABA; sodium nitroprusside, SNP; and methyl jasmonate, MeJA) *in vitro* at 15°C	Fold increase over control (untreated) • SWE—2.6-fold • ABA—2.0-fold • SNP—1.35-fold	*HMGS, HMGR, PMK*	MVA	2.71–4.88	Formation of GPP	[52]
			DXPS, ISPD, ISPE, MECPS	MEP	2.04–4.78		

(*Continued*)

Table 8.2 Details of strategies used to shortlist key candidate genes/transcripts associated with P-I biosynthesis in *P. kurroa*. (*Continued*)

S. No.	Experimental approach	Differential P-I content	Shortlisted genes/ transcripts	Pathway	Fold up-regulated gene expression	Significance/ role	References
6.	Expression analysis of pathway genes in *P. kurroa* shoots treated with different precursors (CA, CAT and CA+CAT) *in vitro* at 15°C	Fold increase over control (untreated shoots) • CAT—NIL • CA—1.6 • CA+CAT—4.2	DAHPS, CM *PAL*, *GS*	S/P S/P Iridoid	1.8–2.4 Fold decrease in CA+CAT treated shoots due to feedback inhibition effect—2.4–6.2	Formation of CA Formation of CAT	[26]

S/P: shikimate/phenylpropanoid; ND: not determined.

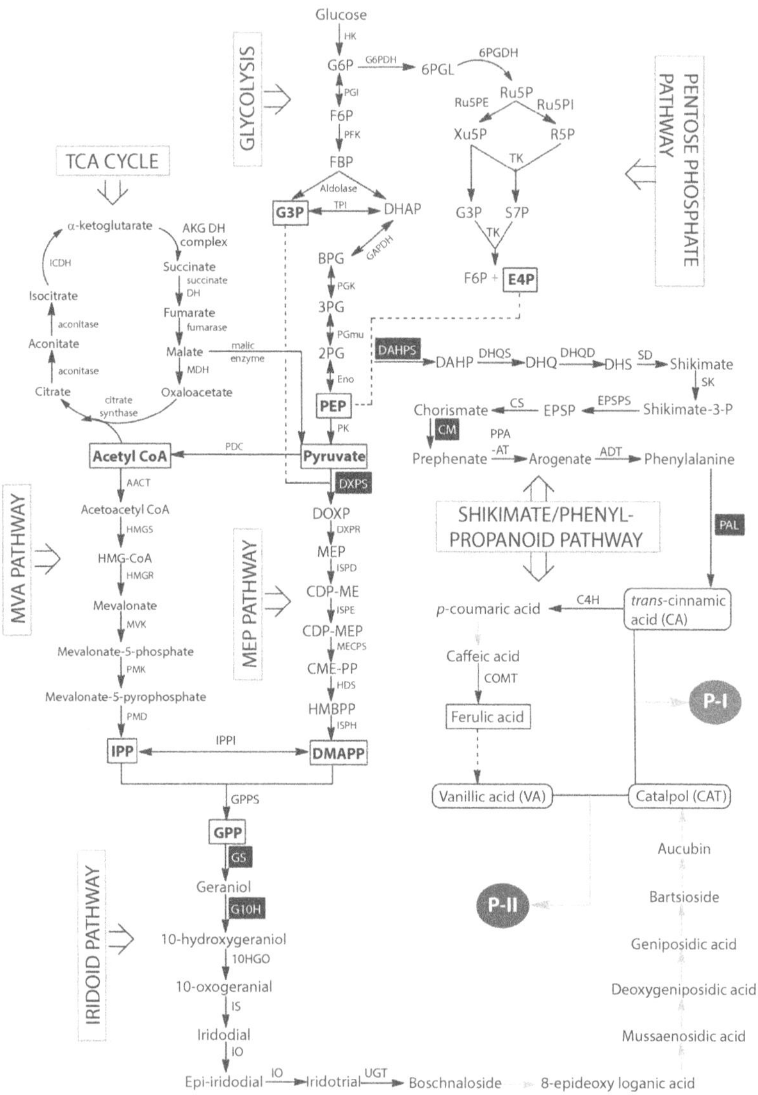

Figure 8.3 Complete metabolic architecture of P-I and P-II biosynthesis in *P. kurroa*. The precursors provided by different integrated pathways are highlighted as boxed. The candidate key genes are in bold and boxed with filled black color. Gray-colored arrows indicate steps with no available enzyme information. The abbreviations are elaborated in the Abbreviation Section of the text.

In contrast, the MEP pathway occurs in plastids and produces isomer of IPP, i.e., dimethylallyl pyrophosphate (DMAPP), through a series of seven enzyme catalyzed reactions, starting with the conversion of pyruvate and G-3-P to 1-deoxy-D-xylulose-5-phosphate by an enzyme, DXPS, 1-deoxy-D-xylulose-5-phosphate synthase [33, 34]. The conversion of IPP to DMAPP is a reversible reaction catalyzed by isopentenyl pyrophosphate isomerase [33, 34]. Further, the fusion of two IPP units produces a monoterpene biosynthetic precursor, i.e., GPP, which leads to the formation of CAT via a series of 14 cyclization reactions through the iridoid pathway [3]. Further, P-I and P-II are biosynthesized via esterification of CAT with CA and VA, respectively (Figure 8.3).

8.6 Challenges and Future Perspectives

Deciphering of biosynthetic pathways for plant secondary metabolites is a highly cumbersome process considering the complexity of their genetic architecture and low levels of metabolic intermediates and corresponding enzymes catalyzing the steps *in planta*. In case of specialized plant metabolites, since multiple pathways contribute to their biosynthesis, it is necessary to decipher metabolic flux through individual pathways, modules, or intermediates to engineer biosynthesis of a target metabolite. In *P. kurroa*, the metabolic flux analysis revealed the role of multiple enzymes in the biosynthesis of P-I across different modules of its biosynthetic pathway. Therefore, this analysis provides a platform to exploit DAHPS, DXPS, chorismate mutase (CM), and phenylalanine ammonia lyase (PAL) for enhanced production of CA through the shikimate/phenylpropanoid pathway, while geraniol synthase (GS) and geraniol-10-hydroxylase (G10H) catalyzed steps might increase the production of CAT via the iridoid pathway. However, as far as P-II is concerned, the major obstacle is its negligence in *P. kurroa* shoots; thus, it renders the establishment of an *in vitro* production platform of both P-I and P-II. Interestingly, OMICS-based approaches utilized to elucidate the complexity of P-II biosynthesis across natural variants enlightened the path for its *in vitro* production. The evidences indicated insignificant levels of last enzyme catalyzing the formation of P-II from VA and CAT in shoots of *P. kurroa*. Hence, the engineering of the last step will not only facilitate the production of P-II but also pose a possibility to establish the sustainable production platform of both P-I and P-II *in vitro*. However, the desired enzyme has yet to be reported and thus paves an interesting challenge for the future.

Abbreviations

P-I: picroside-I; P-II, picroside-II; MVA, mevalonate; MEP, methyl erythritol phosphate; NGS, next-generation sequencing; CAT, catalpol; CA, *trans*-cinnamic acid; VA, vanillic acid; GPP, geranyl pyrophosphate; FA, ferulic acid; PKS, *Picrorhiza kurroa* shoots; PKST, *Picrorhiza kurroa* stolons; qRT-PCR, quantitative real-time polymerase chain reaction; G6P, glucose-6-phosphate; F6P, fructose-6-phosphate; FBP, fructose-1,6-bisphosphate; DHAP, dihydroxyacetone phosphate; G3P, glyceraldehyde-3-phosphate; BPG, 1,3-bisphosphoglycerate; 3PG, 3-phosphoglycerate; 2PG, 2-phosphoglycerate; PEP, phosphoenolpyruvate; HK, hexokinase; PGI, phosphoglucose isomerase; PFK, phosphofructokinase; TPI, triose-phosphate isomerase; GAPDH, glyceraldehyde-3-phosphate dehydrogenase; PGK, phosphoglycerate kinase; PGmu, phosphoglycerate mutase; Eno, enolase; PK, pyruvate kinase; 6PGL, 6-phosphogluconolactone; Ru5P, ribulose-5-phosphate; R5P, ribose-5-phosphate; Xu5P, xylulose-5-phosphate; S7P, sedoheptulose-7-phosphate; E4P, erythrose-4-phosphate; G6PDH, glucose-6-phosphate dehydrogenase; 6PGDH, 6-phosphogluconate dehydrogenase; Ru5PE, ribulose-5-phosphate 3-epimerase; Ru5PI, ribulose-5-phosphate isomerase; TK, transketolase; PDC, pyruvate dehydrogenase complex; ICDH, isocitrate dehydrogenase; AKG DH complex, α-keto glutarate dehydrogenase complex; MDH, malate dehydrogenase; HMG CoA, 3-hydroxy-3-methylglutaryl coenzyme A; DOXP, 1-deoxy-D-xylulose-5-phosphate; CDP-ME, 4-(CDP)-2-methyl-D-erythritol; CDP-MEP, 4-(CDP)-2-methyl-D-erythritol 2-phosphate; CME-PP, 2-C-methyl-D-erythritol 2,4-cyclopyrophosphate; HMBPP, 1-hydroxy-2-methyl 2-(E)-butenyl-4-pyrophosphate; DMAPP, dimethylallyl pyrophosphate; IPP, isopentenyl pyrophosphate; AACT, acetoacetyl-CoA thiolase; HMGS, hydroxymethylglutaryl-CoA synthase; HMGR, hydroxymethylglutaryl-CoA reductase; MVK, mevalonate kinase; PMK, phosphomevalonate kinase; PMD, diphosphomevalonate decarboxylase; DXPS, 1-deoxy-D-xylulose-5-phosphate synthase; DXPR, 1-deoxy-D-xylulose-5-phosphate reductoisomerase; ISPD, 4-diphosphocytidyl-2C-methyl-D-erythritol synthase; ISPE, 4-diphosphocytidyl-2C-methyl-D-erythritol kinase; MECPS, 2-C-methyl-D-erythritol 2,4-cyclopyrophosphate synthase; HDS, 1-hydroxy-2-methyl 2-(E)-butenyl-4-pyrophosphate synthase; ISPH, 1-hydroxy-2-methyl 2-(E)-butenyl-4-pyrophosphate reductase; IPPI, isopentenyl pyrophosphate delta-isomerase; GPPS, geranyl pyrophosphate synthase; GS, geraniol synthase; G10H, geraniol-10-hydroxylase; 10HGO, 10-hydroxygeraniol oxidoreductase; IS, iridoid synthase, IO, iridoid oxidase;

UGT, UDP-glucosyltransferase; DAHP, 3-deoxy-D-arabinoheptulosonate 7-phosphate; DHQ, 3-dehydroquinate; DHS, 3-dehydroshikimate; EPSP, 5-enolpyruvylshikimate-3-phosphate; DAHPS, 3-deoxy-D-arabinoheptulosonate 7-phosphate synthase; DHQS, dehydroquinate synthase; DHQD, dehydroquinate dehydratase; SD, shikimate dehydrogenase; SK, shikimate kinase; EPSPS, 5-enolpyruvylshikimate-3-phosphate synthase; CS, chorismate synthase; CM, chorismate mutase; PPA-AT, prephenate aminotransferase; ADT, arogenate dehydratase; PAL, phenylalanine ammonia lyase; C4H, cinnamic acid-4-hydroxylase; COMT, caffeic acid-3-O-methyltransferase; 2HFD, 2-hydroxyisoflavanone dehydratase; F3D, flavanone 3-dioxygenase; ACT, anthocyanin acyltransferase.

References

1. Muriel, P. and Rivera-Espinoza, Y., Beneficial drugs for liver diseases. *J. Appl. Toxicol.*, 28, 93–103, 2008.
2. Rambaldi, A. and Gluud, C., Colchicine for alcoholic and non-alcoholic liver fibrosis and cirrhosis. *Cochrane Database Syst. Rev.*, 3, CD002148, 2001.
3. Kumar, V., Chauhan, R.S., Tandon, C., Biosynthesis and therapeutic implications of iridoid glycosides from *Picrorhiza* genus: The road ahead. *J. Plant Biochem. Biotechnol.*, 26, 1–13, 2017.
4. Kumar, V., Kumar, V., Chauhan, R.S., Sood, H., Tandon, C., Cost effective quantification of picrosides in *Picrorhiza kurroa* by employing response surface methodology using HPLC-UV. *J. Plant Biochem. Biotechnol.*, 24, 376–384, 2015.
5. Dwivedi, Y., Rastogi, R., Garg, N.K., Dhawan, B.N., Picroliv and its components kutkoside and P-I protect liver against galactosamine induced damage in rats. *Pharmacol. Toxicol.*, 71, 383–387, 1992.
6. Joy, K.L. and Kuttan, R., Anti-oxidant activity of selected plant extract. *Amala Res. Bull.*, 15, 68–71, 1995.
7. Dorsch, W., Stuppher, H., Wagner, H., Gropp, M., Demolin, S., Ring, J., Antiasthmatic effect of *Picrorhiza kurrooa*: Androsin prevents allergen and PAV induced bronchial obstruction in guinea pig. *Int. Arch. Allergy Appl. Immunol.*, 95, 128–133, 1991.
8. Joy, K.L., Rajeshkumar, N.V., Kuttan, G., Kuttan, R., Effect of *Picrorhiza kurroa* extract on transplanted tumours and chemical carcinogenesis in mice. *J. Ethnopharmacol.*, 71, 261–266, 2000.
9. Puri, A., Saxena, R.P., Guru, S.P.Y., Kulshrestha, D.K., Saxena, K.C., Dhawan, B.N., Immunostimulant activity of picroliv, the iridoid glycoside fraction of *Picrorhiza kurrooa* and its protective action against *Leishmania donovani* infection in hamsters. *Planta Med.*, 58, 528–532, 1992.

10. Uniyal, A., Uniyal, S.K., Rawat, G.S., Commercial Extraction of *Picrorhiza kurrooa* Royle ex Benth. in the Western Himalaya. *Mount. Res. Dev.*, 31, 201–208, 2011.
11. Nayar, M.P. and Sastri, A.R.K., *Red data plants of India*, p. 271, CSIR Publication, New Delhi, 1990.
12. Shitiz, K., Pandit, S., Chauhan, R.S., Sood, H., Picrosides content in the rhizomes of *Picrorhiza kurroa* Royle ex Benth. traded for herbal drugs in the markets of North India. *Int. J. Med. Aromatic Plants*, 3, 226–233, 2013.
13. Ramakrishna, A. and Ravishankar, G.A., Influence of abiotic stress signals on secondary metabolites in plants. *Plant Signal. Behav.*, 6, 1720–1731, 2011.
14. Kumar, V. and Chauhan, R.S., Higher amount of steviol detected in the leaves of a non-toxic endangered medicinal herb, *Aconitum heterophyllum. J. Plant Biochem. Biotechnol.*, 25, 442–445, 2016.
15. Pandit, S., Shitiz, K., Sood, H., Chauhan, R.S., Differential biosynthesis and accumulation of picrosides in an endangered medicinal herb *Picrorhiza kurroa. J. Plant Biochem. Biotechnol.*, 22, 335–342, 2013.
16. Garg, R., Shankar, R., Thakkar, B., Kudapa, H., Krishnamurthy, L., Mantri, N., Varshney, R.K., Bhatia, S., Jain, M., Transcriptome analyses reveal |genotype- and developmental stage-specific molecular responses to drought and salinity stresses in chickpea. *Sci. Rep.*, 6, 19228, 2016.
17. Singh, A. and Desgagné-Penix, I., Transcriptome and metabolome profiling of *Narcissus pseudonarcissus* 'King Alfred' reveal components of Amaryllidaceae alkaloid metabolism. *Sci. Rep.*, 7, 17356, 2017.
18. Hadadi, N. and Hatzimanikatis, V., Design of computational retrobiosynthesis tools for the design of *de novo* synthetic pathways. *Curr. Opin. Chem. Biol.*, 28, 99–104, 2015.
19. Kumar, V., Sood, H., Sharma, M., Chauhan, R.S., A proposed biosynthetic pathway of picrosides linked through the detection of biochemical intermediates in the endangered medicinal herb *Picrorhiza kurroa. Phytochem. Anal.*, 24, 598–602, 2013.
20. Kumar, V., Bansal, A., Chauhan, R.S., Modular design of picroside-II biosynthesis deciphered through NGS transcriptomes and metabolic intermediates analysis in naturally variant chemotypes of a medicinal herb, *Picrorhiza kurroa. Front. Plant Sci.*, 8, 564, 2017.
21. Damtoft, S., Jensen, S.R., Jessen, C.U., Knudsen, T.B., Late stages in the biosynthesis of aucubin in Scrophularia. *Phytochemistry*, 33, 1089–1093, 1993.
22. Damtoft, S., Biosynthesis of catalpol. *Phytochemistry*, 35, 1187–1189, 1994.
23. Pandit, S., Shitiz, K., Sood, H., Naik, P.K., Chauhan, R.S., Expression pattern of fifteen genes of non-mevalonate (MEP) and mevalonate (MVA) pathways in different tissues of endangered medicinal herb *Picrorhiza kurroa* with respect to picrosides content. *Mol. Biol. Rep.*, 40, 1053–1063, 2013.
24. Jaisi, A. and Panichayupakaranant, P., Enhanced plumbagin production in *Plumbago indica* root cultures by L-alanine feeding and *in situ* adsorption. *Plant Cell Tiss. Org. Cult.*, 129, 53–60, 2017.

25. Chetri, S.K., Kapoor, H., Agrawal, V., Marked enhancement of sennoside bioactive compounds through precursor feeding in *Cassia angustifolia* Vahl and cloning of isochorismate synthase gene involved in its biosynthesis. *Plant Cell Tiss. Org. Cult.*, 124, 431–446, 2016.
26. Kumar, V., Sharma, N., Sood, H., Chauhan, R.S., Exogenous feeding of immediate precursors reveals synergistic effect on picroside-I biosynthesis in shoot cultures of *Picrorhiza kurroa* Royle ex Benth. *Sci. Rep.*, 6, 29750, 2016.
27. Shitiz, K., Sharma, N., Pal, T., Sood, H., Chauhan, R.S., NGS transcriptomes and enzyme inhibitors unravel complexity of picrosides biosynthesis in *Picrorhiza kurroa* Royle ex. Benth. *PLoS ONE*, 10, e0144546, 2015.
28. Schilmiller, A.L., Pichersky, E., Last, R.L., Taming the hydra of specialized metabolism: How systems biology and comparative approaches are revolutionizing plant biochemistry. *Curr. Opin. Plant Biol.*, 15, 338–344, 2012.
29. Kroymann, J., Natural diversity and adaptation in plant secondary metabolism. *Curr. Opin. Plant Biol.*, 14, 246–251, 2011.
30. Unamba, C.I., Nag, A., Sharma, R.K., Next Generation Sequencing technologies: The doorway to the unexplored genomics of non-model plants. *Front. Plant Sci.*, 6, 1074, 2015.
31. Gahlan, P., Singh, H.R., Shankar, R., Sharma, N., Kumari, A., Chawla, V., Ahuja, P.S., Kumar, S., De novo sequencing and characterization of *Picrorhiza kurrooa* transcriptome at two temperatures showed major transcriptome adjustments. *BMC Genomics*, 13, 126, 2012.
32. Kumar, V., Sharma, N., Shitiz, K., Singh, T.R., Tandon, C., Sood, H., Chauhan, R.S., An insight into conflux of metabolic traffic leading to picroside-I biosynthesis by tracking molecular time course changes in a medicinal herb, *Picrorhiza kurroa. Plant Cell Tiss. Org. Cult.*, 123, 435–441, 2015.
33. Kumar, V., Malhotra, N., Pal, T., Chauhan, R.S., Molecular dissection of pathway components unravel atisine biosynthesis in a non-toxic *Aconitum* species, *A. heterophyllum* Wall. *3 Biotech.*, 6, 1–10, 2016.
34. Zhao, L., Chang, W., Xiao, Y., Liu, H., Liu, P., Methylerythritol phosphate pathway of isoprenoid biosynthesis. *Annu. Rev. Biochem.*, 82, 497–530, 2013.
35. Rathee, D., Thanki, M., Bhuva, S., Anandjiwala, S., Agrawal, R., Iridoid glycosides- kutkin, picroside-I and kutkoside from *Picrorhiza kurroa* Benth inhibits the invasion and migration of MCF-7 breast cancer cells through the down regulation of matrix metalloproteinases. *Arab. J. Chem.*, 6, 49–58, 2013.
36. Li, P., Matsunaga, K., Yamakuni, T., Ohizumi, Y., Potentiation of nerve growth factor-action by picrosides I and II, natural iridoids, in PC12D cells. *Eur. J. Pharmacol.*, 406, 203–208, 2000.
37. Li, P., Matsunaga, K., Yamakuni, T., Ohizumi, Y., Picroside I and II, selective enhancers of the mitogen-activated protein kinase-dependent signaling pathway in the action of neuritogenic substances on PC12D cells. *Life Sci.*, 71, 1821–1835, 2002.
38. Guo, Y., Xu, X., Li, Q., Li, Z., Du, F., Anti-inflammation effects of picroside 2 in cerebral ischemic injury rats. *Behav. Brain Funct.*, 6, 43, 2010.

39. Li, Q., Li, Z., Xu, X.Y., Guo, Y.L., Du, F., Neuroprotective properties of picroside II in a rat model of focal cerebral ischemia. *Int. J. Mol. Sci.*, 11, 4580–4590, 2010.
40. Pei, H., Su, X., Zhao, L., Li, H., Guo, Y., Zhang, M., Xin, H., Primary Study for the therapeutic dose and time window of picroside II in treating cerebral ischemic injury in rats. *Int. J. Mol. Sci.*, 13, 2551–2562, 2012.
41. Gao, H. and Zhou, Y.W., Anti-lipid peroxidation and protection of liver mitochondria against injuries by picroside II. *World J. Gastroenterol.*, 11, 3671–3674, 2005.
42. Gao, H. and Zhou, Y.W., Inhibitory effect of picroside II on hepatocyte apoptosis. *Acta Pharmacol. Sin.*, 26, 729–736, 2005.
43. Comu, F.M., Kilic, Y., Ozer, A., Kirisci, M., Dursun, A.D., Tatar, T., Zor, M.H., Kartal, H., Kucuk, A., Boyunaga, H., Arslan, M., Effect of picroside II on erythrocyte deformability and lipid peroxidation in rats subjected to hind limb ischemia reperfusion injury. *Drug Des. Devel. Ther.*, 10, 927–931, 2016.
44. Choi, J., Choi, B.K., Kim, J.S., Lee, J.W., Park, H.A., Ryu, H.W., Lee, S.U., Hwang, K.W., Yun, W.K., Kim, H.C., Ahn, K.S., Oh, S.R., Lee, H.J., Picroside II attenuates airway inflammation by downregulating the transcription factor GATA3 and Th2-related cytokines in a mouse model of HDM-induced allergic asthma. *PLoS ONE*, 12, e0170832, 2016.
45. Wang, L., Liu, X., Chen, H., Chen, Z., Weng, X., Qiu, T., Liu, L., Effect of picroside II on apoptosis induced by renal ischemia/reperfusion injury in rats. *Exp. Ther. Med.*, 9, 817–822, 2015.
46. Meng, F.J., Hou, Z.W., Li, Y., Yang, Y., Yu, B., The protective effect of picroside II against hypoxia/reoxygenation injury in neonatal rat cardiomyocytes. *Pharm. Biol.*, 50, 1226–1232, 2012.
47. Meng, F.J., Jiao, S.M., Yu, B., Picroside II protects cardiomyocytes from hypoxia/reoxygenation-induced apoptosis by activating the PI3K/Akt and CREB pathways. *Int. J. Mol. Med.*, 30, 263–270, 2012.
48. Wu, N., Li, W., Shu, W., Jia, D., Protective effect of picroside II on myocardial ischemia reperfusion injury in rats. *Drug Des. Dev. Ther.*, 8, 545–554, 2014.
49. Bhandari, P., Kumar, N., Singh, B., Gupta, A.P., Kaul, V.K., Ahuja, P.S., Stability-indicating LC–PDA method for determination of picrosides in hepatoprotective Indian herbal preparations of *Picrorhiza kurroa*. *Chromatographia*, 69, 221–227, 2009.
50. Kumar, V., Shitiz, K., Chauhan, R.S., Sood, H., Tandon, C., Tracking dynamics of enzyme activities and their gene expression in *Picrorhiza kurroa* with respect to picroside accumulation. *J. Plant Biochem. Biotechnol.*, 25, 125–132, 2016.
51. Singh, H., Gahlan, P., Kumar, S., Cloning and expression analysis of ten genes associated with picrosides biosynthesis in *Picrorhiza kurroa*. *Gene*, 515, 320–328, 2013.
52. Sharma, N., Kumar, V., Chauhan, R.S., Sood, H., Modulation of Picroside-I Biosynthesis in Grown Elicited Shoots of *Picrorhiza kurroa In Vitro*. *J. Plant Growth Regul.*, 35, 965–973, 2016.

9

Relevance of Poly-Omics in System Biology Studies of Industrial Crops

Nagendra Nath Das

Department of Biotechnology, Berhampur University, Berhampur, Odisha (India)

Abstract

This chapter emphasizes on the current status, potential benefits, and bottlenecks involved in the role of poly-omics in system biology studies of industrial crops. Pharmaceutical crops, cereal crops, and sugar crops are highly utilized in system biology research, whereas other crops are neglected. Mostly genomics, proteomics, transcriptomics, metabolomics, and phenomics approaches are carried out as omics technologies to study the system biology of industrial crops. Metabolomics approaches are carried out to study metabolite interactions through network modelling. On the other hand, there are scanty research inputs made by the rest of the meta omics technologies (i.e., metagenomics, metatranscriptomics, metaproteomics), and they remain unexplored with regard to industrial crops. In toto, poly-omics approaches in system biology studies are still to be exploited a lot to bring out novel quality traits, which will help to optimize the crop productivity and discovery of new industrial products.

Keywords: System biology, industrial crops, poly-omics, metabolic networks, *de novo* signaling pathway, gene regulatory networks, genotype, evolutionary networks

9.1 Introduction

Omics-related technologies are valuable tools for comprehensive analysis of large amounts of raw data in system biology. Mark Kirschner in 2005 [1]

Email: my.dear.sir@gmail.com

Rintu Banerjee, Garlapati Vijay Kumar, and S.P. Jeevan Kumar (eds.) OMICS-Based Approaches in Plant Biotechnology, (167–182) © 2019 Scrivener Publishing LLC

proposed that "System biology is the study of the behaviour complex biological organizations and processes in terms of the molecular constituents." It rather befits the definition of system biology, but does not imply about involvement of mathematical modelling or computational approaches or addition of latest approaches, e.g., poly-omics technologies. Crop plants are invaluable assets in any market economy standards. Within this context, poly-omics play an important role by supplying high-throughput biological data to improve agronomic quality of crop plants. Poly-omics approaches involved use of functional genomics and proteomics, transcriptomics, metabolomics, proteogenomics, and phenomics tools processes for identifying of unknown gene functions and addressing complex phenotypic and genotypic traits. Close interactions of experimental, computational, and theoretical approaches enable and integrate system biology in a broader perspective (Figure 9.1).

Bottleneck of these approaches infers about sometimes unsuccessful application of novel modelling algorithms and time-cost management in achieving of desired results. Therefore, large-scale biological databases are maintained and updated regularly to reduce the shortcoming of complex approaches in system biology studies [2, 3]. Norman Borlaug in 1945 started the green revolution on implementation of novel genetic variations in plants [4].

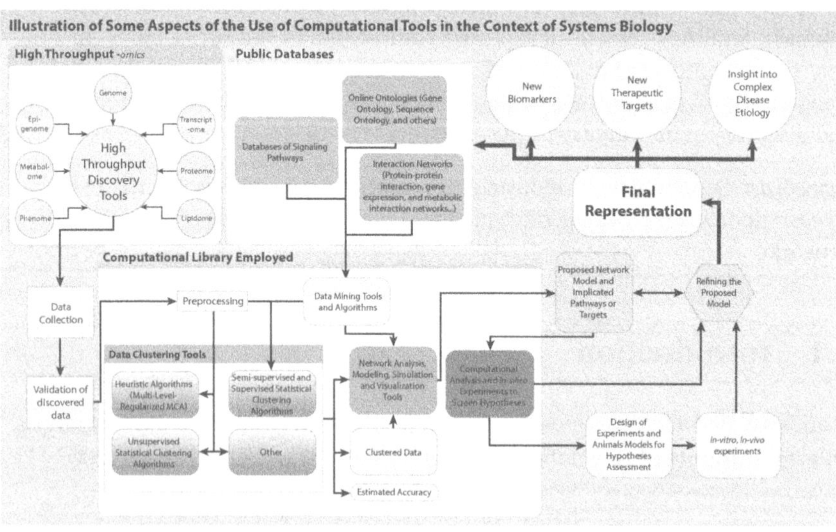

Figure 9.1 Omics data in relation with system biology. (Adapted from Ali *et al.*, (2012). Systems biology, bioinformatics, and biomarkers in neuropsychiatry. *Front. Neurosci.*, https://doi.org/10.3389/fnins.2012.00187.)

Abiotic and biotic stress conditions can affect yield quantity and change at the transcriptome, proteome, and cellular levels of crop plants [5, 6]. Therefore, agricultural production under global warming and climate change needs a call for system biology approach. With new omics technologies application, there is still lack of information about inheritance, ecology, and evolution of plants and animals, which in earlier days enforced scientists like Gregor Mendel or Charles Darwin to reveal the secrets of these key concepts. A total of 16 biological databases and 62 tools are employed to study system biology. Biofortification studies on different crops were enhanced by system biology approach. As crop biology research proceeds, new developments are yet to be achieved in their systemic architectures.

9.2 System Biology of Crops

Integrative and predictive approaches in system biology led to the study of the precise biology of agricultural productivity and the development of smart crops. Crop modelling developments use modern genomics, proteomics, and metabolomics data by parameterizing and redesigning some subroutines of crop systems models. The topology of a network consists of edges and nodes. Motif properties (i.e., feedback and feed forwarded loops) depend on network topology. Network modelling involves two methodologies, (1) dynamic quantitative modelling, which infers about computational simulation and mathematical modelling, and (2) static large-scale biological network modelling, which uses visualization and makes topological models of all omics data [7]. SBML (Systems Biology Markup Language) is used as a more popular programming tool for computational study of models of biological process [8]. At the subcellular scale of plants, two significant approaches were developed, (1) bottom-up (local) and (2) top-down (global). Boolean networks were first used to study the subcellular scale [9], which focused on the dynamics of genes and proteins, and their association with other interacting molecules lies in the bottom-up approach, whereas the top-down approach is dependent on knowledge databases based on omics-related technologies. Stochastic models were based on kinetic parameters of genetical and biochemical networks. Top-down and bottom-up approaches of system biology use network modelling to trace out novel genes or complex traits of superior agronomic qualities present in crop plants, i.e., through signal transduction network, gene regulatory network, and metabolic network. At the tissue-scale study of plants, DS^2 model was used [10]. At the population-scale study of plants, QTL (quantitative trait loci) tool was used [11]. But as the crop model research developed, at the whole plant-scale study, FSPM model was utilized in which the first approach is based on architecture (e.g., Lindenmayer system or popularly known as L-system), and the second approach was based on ecophysiological models [12]. FSPM is still at the nascent level and under continuous stage of development. For example, ethylene gene mutation affected 37% of the total of 869 genes involved in normal fruit maturity and controlled different key metabolic routes for fruit ripening as revealed by a network study [13]. Genome-wide analysis of miRNAs in maize revealed about roots response to nitrogen-limiting conditions [14, 15]. Nutritional qualities on macronutrient and micronutrient levels are improved in crop plants through proteomics [16–19]. *In silico* systems biology approaches with NGS data depict about transgenic crop plant's disease resistance and environment-friendly adaptability against the use of agrochemicals [20]. Interactome studies depict insight into biomolecular interaction networks between different phenotypes and genotypes of rice and other crops [21, 22]. Prototypes of crop systems biology models are made, which may be not complex at the first step in relation to crop improvement strategies.

Crop model designs steadily developed by simulating crop phenology and partitioning approach over the past few decades in system biology. Since 1965, scientists like C.T. De Wit and J.L. Monteith initiated crop modelling [23, 24], and in 1971 to 1974, W.G. Duncan had developed crop models for peanut, cotton, soya bean, and maize [25]. Early mechanistic crop models were generic, e.g., BARCOS [26], SOYMOD [27], SIMCOT [28, 29], and GOSSYM [30, 31]. As crop modelling progressed, many models developed on the basis of portioning parameters, e.g., SUCROS and MACROS family models [32, 33], WOFSOT [34], and hybrid approach models (i.e., co-limited models and source-driven models [35, 36]), for the study of vegetative growth. Later, on the basis of phenology, robust modules were created by use of temperature and day length factors, which affects plant's vegetative as well as reproductive growth phases, e.g., soya bean [37] and maize [38]. For the study of leaf growth of crops, various simulation models were developed, i.e., CERES [39], CropSyst [40], STICS [41], Sirius [42], and IXIM-maize model [43]. Then, the study of 3D growth of plant organs of crops led to various crop models by applying the concept of Lindenmayer systems, also known as L-systems. Stomatal conductance response to CO_2 and water vapor was studied by using the Ball-Berry leaf conductance model [44, 45]. On the basis of nutritional requirement deficiency studies of crops, models were implemented such as CSM-CROPGRO soya bean model [46], DSSAT model [47], CENTURY soil organic matter model [48], and APSIM-N wheat model [49]. Crop models with respect to grain yield were depicted by ECOSYS model [50], MAIZSIM model [51, 52], and ORYZA model [53, 54], because food production depends on the grain yield ratio of crops.

In India, Singh *et al.*, [55] conducted a model sensitive analysis of genetic traits of peanuts under climate change scenarios. Thus, leading researchers and academicians described their state-of-the-art findings and hypothesized from a variety of different approaches. Crop system biology is dynamically evolving with new concepts of models and hypothesis.

9.3 Industrial Crops

Industrial revolution brought raw materials derived from plant products in a capacity to fulfil the needs of human society at the economic level. Once a crop is processed to produce industrial products and possesses economic value, that crop is referred to as industrial crop. In the events of history of industrial crops, one burning example is that earlier human beings produced heat energy by burning wood into charcoal by pyrolysis approach. Similarly, Egyptians used castor oil to burn the lamps, i.e., use of plant oil. As the progress continued in the sector of crops, much more plants cultivated as a crop in earlier days came into purview as industrial crops, which nowadays cover products that entered all facets of our life. Rayon, the synthetic fiber, was derived from cotton plant. Garments made from cotton fabrics with natural dyes have become trendy in fashion boutiques, and this trend enriched the next level for replacement of products of chemical origin by organic insecticides, biofertilizers, biodetergents, and repellents, which are based from plant-origin products. Soya bean, a crop that has multiple uses, can illustrate this point vividly. Similarly, use of sugar and maize for ethanol production and jatropa for biodiesel in Brazil and USA raised these crop commodities' value in the world market [56]. Nowadays, seaweeds are treated as an industrial crop in Japan and other countries in the world [57]. Considering the skin-lotion market, jojoba oil replaces sperm whale oil regarding skin-softening properties, and this helps indirectly in the prevention of killing of sperm whales, which are species that are nearly extinct in the world. Some crops with industrial value are prohibited from large-scale production due to regulatory restrictions, e.g., Castor contains a toxin protein, ricin, which can be deployed in chemical warfare [58]. Since there is no proper classification of industrial crops available, this chapter emphasizes on their classification, which is presented in Table 9.1.

Table 9.1 List of industrial crops.

Industrial crops common names		**Botanical names**
Pharmaceutical crops	1. Stevia 2. Tobacco 3. Alfalfa 4. All Ayurvedic plants 5. Spirulina microalgae (nutritional supplement)	*Stevia rebaudiana* *Nicotiana tabacum* *Medicago sativa* ----- *Arthrospira platensis*
Fiber crops	1. Flax 2. Jute 3. Bamboo 4. Indian mallow 5. Sun hemp 6. Cotton 7. Rubber	*Linum usitatissimum* *Corchorus capsularis* *Bambuseae* *Abutilon theophrasti* *Crotolaria juncia* *Gossipium* spp. *Ficus elastica*
Sugar crops	1. Sugarcane 2. Sugarbeet 3. Dahlia 4. Barley 5. Maple tree 6. Ivory nut palm 7. Sweet sorghum 8. Date 9. Potato 10. Sweet potato	*Saccharum officinarum* *Beta vulgaris* *Dahlia pinnata* *Hordeum vulgare* *Acer saccharum* *Phytelephas macrocarpa* *Sorghum bicolor* *Phoenix dactylifera* *Solanum tuberosum* *Ipomoea batatas*
Dye yielding crops (some selected crops)	1. Green crop (indigo blue) 2. Himalayan rhubarb 3. Abir powder 4. Hemp	*Indigofera tinctoria* Linn. -- *Curcuma zedoaria* Rosc. *Cannabis sativa*

(*Continued*)

Table 9.1 List of industrial crops. (*Continued*)

Industrial crops common names		**Botanical names**
Cereal crops	1. Rice	*Oryza sativa*
	2. Wheat	*Triticum aestivum*
	3. Corn	*Zea mays*
	4. Barley	*Hordeum vulgare*
	5. Millet	*Panicum miliaceum, Eleusine coracana, Setaria italica, Pennisetum glaucum*
	6. Oat	*Avena sativa*
	7. Rye	*Secale cereale*
	8. Fonio	*Digitaria exilis*
	9. Canary grass	*Phalaris* spp.
	10. Sorghum	*Sorghum bicolor*
	11. Buckwheat	*Fagopyrum esculentum*
	12. Spelt	*Triticum spelta*
Oilseed crops	1. Camelina seed oil	*Camelina sativa*
	2. Tug oil tree	*Vernicia fordii*
	3. Jatropa	*Jatropha curcas*
	4. Lavender	*Lavandula*
	5. Rose oil	*Rosa* spp.
	6. Cuphea (a hybrid)	*(Cuphea viscosissima Jacq. x C. lanceolata W.T. Aiton)*
	7. Miscanthus	*Miscanthus sinensis*
	8. Switchgrass	*Panicum virgatum*
	9. Oil palm	*Elaeis guineensis*
	10. Sunflower	*Helianthus annuus* L.
	11. Soybean	*Glycine max*
	12. Canola	*Brassica napus*
	13. Kewra	*Pandanus odorifer*
	14. Lemon oil	*Citrus limonum*
	15. Neem oil	*Azadirachta indica*
	16. Basil oil	*Ocimum basilicum*

(*Continued*)

Table 9.1 List of industrial crops. (*Continued*)

Industrial crops common names		**Botanical names**
Cover crops	1. Bermuda grass 2. Carabao grass 3. Stargrass	*Cynodon dactylon* *Paspalum conjugatum* *Cynodon plectostachyus*
Fruit crops	1. Mango 2. Grape 3. Orange 4. Papaya 5. Banana 6. Pineapple 7. Plantain (dessert banana) 8. Avocado 9. Guava 10. Apple 11. Cashew 12. Citrus 13. Date 14. Jackfruit 15. Lychee 16. Watermelon 17. Strawberry 18. Coconut 19. Peach 20. Blueberry 21. Nut meg	*Mangifera indica* *Vitis vinifera* *Citrus sinensis* *Carica papaya* *Musa acuminata* *Ananas comosus* *Musa paradisiaca* *Persea americana* *Psidium guajava* *Malus domestica* *Anacardium occidentale* *Citrus* spp. *Phoenix dactylifera* *Artocarpus heterophyllus* *Litchi chinensis* *Citrullus lanatus* *Fragaria ananassa* *Cocos nucifera* *Prunus persica* *Cyanococcus* spp. *Myristica fragrans*

(*Continued*)

Noticeably, some of the same crops are categorized into other crops, which are also belonging to industrial crops, because of their multiutility values. Some of the new industrial crops are (a) *Physaria fendleri* (syn. lesquerella) as an alternative to castor oil; (b) Crambe as a replacement of coconut oil and palm oil; (c) Meadowfoam and Milkweed as a replacement for petroleum derivatives; and (d) Kenaf (a new fiber crop) as a replacement of imported newspaper prints [59]. Nowadays, few terms have been associated with industrial crops, e.g., designer crops, horticultural crops, food crops, forage crops, and cash crops.

Table 9.1 List of industrial crops. (*Continued*)

Industrial crops common names		**Botanical names**
Vegetative crops	1. Tomato	*Solanum lycopersicum*
	2. Potato	*Solanum tuberosum*
	3. Garlic	*Allium sativum*
	4. Onion	*Allium cepa*
	5. Broccoli	*Brassica oleracea*
	6. Ginger	*cv.italica*
	7. Mustard	*Zingiber officinale*
	8. Cabbage	*Brassica juncea*
	9. Cauliflower	*Brassica oleracea*
	10. Ladies' finger	*Brassica oleracea var.*
	11. Radish	*Abelmoschus esculentus*
		Raphanus raphanistrum subsp. *sativus*
	12. Tapioca	*Manihot esculenta*
	13. Carrot	*Daucus carota* subsp. *sativus*
	14. Lemon	*Citrus limon*
	15. Chilli	*Capsicum frutescens*
	16. Brinjal/eggplant	*Solanum melongena*
	17. Cucumber	*Cucumis sativus*
	18. Bell pepper	*Capsicum annuum*
	19. Black pepper	*Piper nigrum*
	20. Lettuce	*Lactuca sativa*
	21. Coriander	*Coriandrum sativum*
	22. Parsley	*Petroselinum crispum*
	23. Jalapeno	*Capsicum annuum jalapeno*
	24. Avocado	*Persea americana*
	25. Asparagus	*Asparagus officinalis*
Plantation crops	1. Coffee	*Coffea* spp.
	2. Coconut	*Cocos nucifera*
	3. Tea	*Camellia sinensis*
	4. Rubber	*Ficus elastica*
Phytoremedial crops	1. Indian mustard	*Brassica juncea* L.
	2. White willow	*Salix* spp.
	3. Poplar tree	*Populus deltoids*
	4. Indian grass	*Sorghastrum nutans*
	5. Sunflower	*Helianthus annuus* L.

9.4 Poly-Omics Application in System Biology Studies of Industrial Crops

Poly-omics approaches, i.e., genomics, proteomics, transcriptomics, and phenomics, have been applied in industrial crops to trace out genetic mechanisms, protein–DNA interaction networks, protein–protein interactions, and selection of novel biomarkers. To identify specific genomic changes contributing to advantageous phenotypic performance in varying environmental conditions, poly-omics played a vital role to address system biology approach of industrial crops in abiotic and biotic stress environments. Data obtained as such through these techniques are stored in various databases, so that computational work can be done on the basis of hypothesis and large complex experimental data availability of industrial crops. Then computational data and experimental data are compared together in system biology approach to find novel traits or genes, which can be utilized further to enhance industrial crop productivity. In meta-omics studies, only the metabolomics part is studied in vast amount. Cereal crops and some selected fiber crops, sugar crops, and oilseed crops have been studied at the poly-omics level. RNA-seq of seeds is analyzed together with 662 phenotypic traits of maize to trace out stress adaptation mechanism [60]. It was found out that F-box proteins also play critical roles in seed development, grain filling, and response to abiotic stress in maize [61]. Phenomics studies in maize revealed that a genome-wide association study (GWAS) was performed on 130 divergent traits, with 24,595 SNPs (minor allele frequency [MAF] R5%; only 4.4% of the total), which depicts about three agronomic phenotypic traits of maize at the global scale [62]. Transcriptome regulation changes to the adaptation process in maize were traced out by RNA-seq [63]. In case of *Vernicia fordii* (tung oil tree), comparative transcriptomics analysis reveals different gene expression mechanism against wilt disease resistance in its sister species [64]. An overview of recent gel-based and gel-free proteomics methods adopted as biotechnological tools that could provide knowledge for crop improvement was highlighted by Rohila *et al.*, [65], in which wheat, maize, soya bean, and rice crops depicted physiological traits in relation to stress conditions. Both wild and cultivated barley plants are able to withstand harsh environmental conditions, which is essential for the development of enhanced varieties of barley [66], which promises to be a very important contributor to the continued food security of the human population. Transcriptome and proteomic analysis of mango (*Mangifera indica* Linn) fruits revealed that transcripts mapped to 128 pathways on KEGG database depict many unknown transcripts with 2,754 nonredundant proteins [67]. In comparative genomic analysis of *Carica papaya*, it was revealed that *Phytophthora* (a pathogen) also downregulates homologs of Cp9 and Cp15 in *Glycine max* and *Solanum tuberosum*, respectively [68]. Transcription analysis of softening-related genes during postharvest of papaya fruit depicts that polygalacturonase and endoxylanase are correlated negatively with papaya fruit firmness [69]. From the study of comparative analysis of genomic data and transcriptomics data, it was found out how the host pathogen *Phytophthora* interacts within these three genera and affects their physiological response systems in a certain distinct manner. This highlights poly-omics approach toward system biology studies of industrial crops.

9.5 Concluding Remarks

NGS technologies pave the way for new platforms in system biology approaches by providing global omics-related data that are helpful in innovation or discovery of remedial concepts to address global crop productivity issues. Despite advances in new algorithms, models, and software tools, many major challenges remain unexplored in system biology studies of industrial crops, i.e., both at the conceptual and practical levels. An interesting fact is that system biology is a science of the twentieth century and a new multidisciplinary study by encompassing theoretical, computational, and experimental data altogether in one stream.

Crop productivity is affected by global climate change, and as the human population grows drastically, the need for plant-derived products from crops grows ever increasingly. Surprisingly, due to industrial crops' multifactorial nature, there are very scanty research inputs available about molecular network studies that aim to improve crop productivity in a greater ratio along with poly-omics approaches. In this scenario, system biology modelling provides an unexpurgated view about crop productivity, especially in the sector of industrial crops.

Abbreviations

FSPM: functional–structural plant model
NGS: next-generation sequencing

References

1. Kirschner, M.W., The meaning of systems biology. *Cell*, 121, 4, 503–504, 2005.
2. Locke, J.C., Kozma-Bognar, L., Gould, P.D., Feher, B., Kevei, E., Nagy, F., Turner, M.S., Hall, A., Millar, A.J., Experimental validation of a predicted feedback loop in the multi-oscillator clock of *Arabidopsis thaliana*. *Mol. Syst. Biol.*, 2, 59, 2006.
3. Zeilinger, M.N., Farre, E.M., Taylor, S.R., Kay, S.A., Doyle, F.J., 3rd, A novel computational model of the circadian clock in *Arabidopsis* that incorporates PRR7 and PRR9. *Mol. Syst. Biol.*, 2, 58, 2006.
4. Murphy, D.J. (Ed.), *Plant Breeding and Biotechnology*, Cambridge University Press, U.K., 2007.
5. Suzuki, N., Rivero, R.M., Shulaev, V., Blumwald, E., Mittler, R., Abiotic and biotic stress combinations. *New Phytol.*, 203, 32–43, 2014.
6. Atkinson, N.J. and Urwin, P.E., The interaction of plant biotic and abiotic stresses: From genes to the field. *J. Exp. Bot.*, 63, 3523–3543, 2012.
7. Gilbert, *et al.*, Computational methodologies for modelling, analysis and simulation of signaling networks. *Briefings Bioinform*, 7, 4, 339–353, 2006.
8. Hucka, *et al.*, The systems biology markup language (SBML): A medium for representation and exchange of biochemical network models. *Bioinformatics*, 19, 4, 524–531, 2003.
9. Bruggeman, F.J. and Westerhoff, H.V., The nature of systems biology. *Trends Microbiol.*, 15, 45–50, 2007.
10. Jaeger, J., Irons, D., Monk, N., Regulative feedback in pattern formation: towards a general relativistic theory of positional information. *Development*, 135, 3175–3183, 2008.
11. Xu, *et al.*, Genetic mapping of quantitative trait loci in crops. *Crop J.*, 5, 2, 175–184, 2017.
12. Lindenmayer, A., Mathematical models for cellular interactions in development I. Filaments with one-sided inputs. *Journal of Theor. Biol.*, 18, 280–299, 1968.
13. Gupta, S.M., Srivastava, S., Sane, A.P., Nath, P., Differential expression of genes during banana fruit development, ripening and 1-MCP treatment: Presence of distinct fruit specific, ethylene induced and ethylene repressed expression. *Postharvest Biol. Technol.*, 42, 16–22, 2006.
14. Xu, *et al.*, Genome-wide identification of microRNAs in response to low nitrate availability in maize leaves and roots. *PLoS ONE*, 6, e28009, 2011.
15. Zhao, *et al.*, Cloning and characterization of maize miRNAs involved in responses to nitrogen deficiency. *PLoS ONE*, 7, e29669, 2012.
16. Li, *et al.*, Changes in the transcriptomic profiles of maize roots in response to iron deficiency stress. *Plant Mol. Biol.*, 85, 349–363, 2014.
17. Singh, U.M., Sareen, P., Sengar, R.S., Kumar, A., Plant ionomics: A newer approach to study mineral transport and its regulation. *Acta Physiol. Plant.*, 35, 2641–2653, 2013.

18. Sun, S.M.S. and Liu, Q., Transgenic approaches to improve the nutritional quality of plant proteins. *In Vitro Cell Dev. Biol. Plant.*, 40, 155–162, 2004.
19. Keeler, S.J., Maloney, C.L., Webber, P.Y., Patterson, C., Hirata, L.T., Falco, S.C., Rice, J.A., Expression of *de novo* high lysine a-helical coiled-coil proteins may significantly increase the accumulated levels of lysine in mature seeds of transgenic tobacco plants. *Plant Mol. Biol.*, 34, 15–29, 1997.
20. Yin, X. and Struik, P.C., Applying modelling experiences from the past to shape crop systems biology: The need to converge crop physiology and functional genomics. *New Phytol.*, 179, 629–642, 2008.
21. Seo, *et al.*, Towards establishment of a rice stress response interactome. *PLoS Genet.*, 7, e1002020, 2011.
22. Le, *et al.*, BioModels Database: A free, centralized database of curated, published, quantitative kinetic models of biochemical and cellular systems. *Nucleic Acids Res.*, 34, 689–691, 2006.
23. De Wit, C.T., *Photosynthesis of Leaf Canopies*, vol. 663, p. 57, Centre for Agricultural Publications and Documentation, Wageningen, 1965.
24. Monteith, J.L., Light distribution and photosynthesis in field crops. *Ann. Botany*, 29, 17–37, 1965.
25. Duncan, W.G., Leaf angles, leaf area, and canopy photosynthesis. *Crop Sci.*, 11, 482–485, 1971.
26. De Wit, C.T. (Ed.), *Simulation of Assimilation, Respiration and Transpiration of Crops*, John Wiley & Sons, New York, 1978.
27. Meyer, G.E., Curry, R.B., Streeter, J.G., Mederski, H.J., SOYMOD/OARDC—A dynamic simulator of soybean growth, development and seed yield. 1. Theory, structure and validation. *Res. Bull.*, 1113, 1979.
28. Hesketh, J.D., Baker, D.N., Duncan, W.G., Simulation of growth and yield in cotton—Respiration and carbon balance. *Crop Sci.*, 11, 394–398, 1971.
29. Baker, D.N., Hesketh, J.D., Duncan, W.G., Simulation of growth and yield in cotton.1. Gross photosynthesis, respiration, and growth. *Crop Sci.*, 12, 431–435, 1972.
30. Baker, D.N., Lambert, J.R., McKinion, J.M., GOSSYM: A simulator of cotton crop growth and yield. *South Carolina Agricultural Experiment Station Technical Bulletin*, 1089, 135, 1983.
31. Reddy, V.R., Baker, D.N., Jenkins, J.N., Validation of GOSSYM.2. Mississippi conditions. *Agric. Syst.*, 17, 133–154, 1985.
32. Penning, de Vries, F.W.T., Jansen, D.M., ten Berge, H.F.M., Bakema, A., Simulation of ecophysiological processes of growth in several annual crops. *Simulation Monographs*, 29, 271, 1989.
33. Spitters, C.J.T., Keulen, H.V., Kraalingen, D.W.G.V., A simple and universal crop growth simulator: SUCROS87, in: *Simulation and Systems Management in Crop Production*, R. Rabbinge (Ed.), pp. 141–181, 1989.
34. van Diepen, C.A., Wolf, J., Herman van Keulen Rappoldt, C., WOFOST: A simulation model of crop production. *Soil Use Manage*, 5, 1, 1989.

35. Boote, K.J., Pickering, N.B., Allen, L.H., Jr., Plant modeling: Advances and gaps in our capability to predict future crop growth and yield in response to global climate change, in: *Advances in Carbon Dioxide Effects Research*, L.H. Allen Jr., M.B. Kirkham, D.M. Olszyk, C.E. Whitman (Eds.), pp. 179–228, ASA Special Publication No. 61, ASA-CSSA-SSSA, Madison, WI, 1997.
36. Wilkerson, G.G., Jones, J.W., Boote, K.J., Ingram, K.T., Mishoe, J.W., Modeling soybean growth for crop management. *TransAsabe.*, 26, 63–73, 1983.
37. Boote, K.J., Jones, J.W., Hoogenboom, G., Simulation of crop growth: CROPGRO model, in: *Agricultural Systems Modeling and Simulation*, R.M. Peart and R.B. Curry (Eds.), pp. 651–692, Marcel Dekker Inc., New York, 1998.
38. Stewart, D.W., Dwyer, L.M., Carrigan, L.L., Phenological temperature response of maize. *Agron. J.*, 90, 73–79, 1998.
39. Ritchie, J.T., Singh, U., Godwin, D.C., Bowen, W.T., Cereal growth, development and yield, in: *Understanding Options for Agricultural Production*, G.Y. Tsuji, G. Hoogenboom, P.K. Thornton (Eds.), pp. 79–98, Kluwer Academic Publishers, Dordrecht, 1998.
40. Stockle, C.O., Donatelli, M., Nelson, R., CropSyst, a cropping systems simulation model. *Eur. J. Agron.*, 18, 289–307, 2003.
41. Ripoche, D. *et al.*, STICS: A generic model for the simulation of crops and their water and nitrogen balances. I. Theory and parameterization applied to wheat and corn. *Agronomie*, 18, 311–346, 1998.
42. Jamieson, P.D., Brooking, I.R., Semenov, M.A., MeMaster, G.S., White, J.W., Porter, J.R., Reconciling alternative models of phenological development in winter wheat. *Field Crops Res.*, 103, 36–41, 2007.
43. Lizaso, J.I., Boote, K.J., Jones, J.W., Porter, C.H., Echarte, L., Westgate, M.E., Sonohat, G., CSM-IXIM: A new maize simulation model for DSSAT Version 4.5. *Agron. J.*, 103, 766–779, 2011.
44. Ball, J.T., A model predicting stomatal conductance and its contribution to the control of photosynthesis under different environmental conditions. *Prog. Photosynthesis Res. Proc. Int. Congress 7th*, 4, 1987.
45. Collatz, G.J., Ball, J.T., Grivet, C., Berry, J.A., Physiological and environmental-regulation of stomatal conductance, photosynthesis and transpiration—A model that includes a laminar boundary-layer. *Agric. Forest Meteorol.*, 54, 107–136, 1991.
46. Boote, K.J., Hoogenboom, G., Jones, J.W., Ingram, K.T., Modeling nitrogen fixation and its relationship to nitrogen uptake in the CROPGRO model, in: *Quantifying and Understanding Plant Nitrogen Uptake for Systems Modeling*, L. Ma, L. Ahuja, T. Bruulsema (Eds.), pp. 13–46, Taylor & Francis Group LLC, Boca Raton, 2009.
47. Daroub, S.H., Gerakis, A., Ritchie, J.T., Friesen, D.K., Ryan, J., Development of a soil-plant phosphorus simulation model for calcareous and weathered tropical soils. *Agric. Syst.*, 76, 1157–1181, 2003.

48. Gijsman, A.J., Hoogenboom, G., Parton, W.J., Kerridge, P.C., Modifying DSSAT crop models for low-input agricultural systems using a soil organic matter-residue module from CENTURY. *Agron. J.*, 94, 462–474, 2002.
49. Asseng, S., Keating, B., Huth, N., Eastham, J., Simulation of perched water tables in a duplex soil. *International Conference on Modelling and Simulation*, Hobart, Tasmania, 1997.
50. Grant, R.F. and Baldocchi, D.D., Energy-transfer over crop canopies simulation and experimental-verification. *Agric. Forest Meteorol.*, 61, 129–149, 1992.
51. Yang, Y., Kim, S.H., Timlin, D.J., Fleisher, D.H., Quebedeaux, B., Reddy, V.R., Simulating canopy transpiration and photosynthesis of corn plants under contrasting water regimes using a coupled model. *Trans. Asabe.*, 52, 1011–1024, 2009a.
52. Kim, S.-H., Yang, Y., Timlin, D.J., Fleisher, D.H., Dathe, A., Reddy, V.R., Staver, K., Modeling temperature responses of leaf growth, development, and biomass in maize with MAIZSIM. *Agron. J.*, 104, 1523–1537, 2012.
53. Bouman, *et al.* (Eds.), *ORYZA2000: Modeling Lowland Rice*, vol. 1, International Rice Research Institute, Los Banos, Philippines, 2001.
54. Matthews, R.B., Kropff, M.J., Bachelet, D., Horie, T., Lee, M.H., Centeno, H.G.S., Shin, J.C., Mohandass, S., Singh, S., Defeng, Z., Modeling the impact of climate change on rice production in Asia, in: *Climate Change and Rice*, R.B. Matthews, M.J. Kropff, D. Bachelet, H.H. van Laar (Eds.), pp. 314–325, CAB International, Wallingford, UK, 1995.
55. Singh, P., Boote, K.J., Kumar, U., Srinivas, K., Nigam, S.N., Jones, J.W., Evaluation of genetic traits for improving productivity and adaptation of groundnut to climate change in India. *J. Agron. Crop Sci.*, 198, 399–413, 2012.
56. Chittaranjan, K., Wild Crop Relatives: Genomic and breeding resources, in: *Industrial Crops*, K. Chittaranjann (Ed.), Springer-Verlag, Berlin Heidelberg, 2011.
57. Loureiro, R., Gachon, C.M.M., Rebours, C., Seaweed cultivation: Potential and challenges of crop domestication at an unprecedented pace. *New Phytol.*, 206, 489–492, 2015.
58. Griffiths, G.D. and Phillips, G.J., Jane, Holley, Inhalation toxicology of ricin preparations: Animal models, prophylactic and therapeutic approaches to protection. *Inhal. Toxicol.*, 19, 873–887, 2007.
59. Abdelrhman, H.A., Shahwahid, M., Paridah, M.T., Samad, A.R.A., Habib, MAAEI, Financial and technical assessment of kenaf cultivation for producing fiber utilized in automotive components. *Bus. Eco. J.*, 7, 254, 2016.
60. Bhargava, S. and Sawant, K., Drought stress adaptation: Metabolic adjustment and regulation of gene expression. *Plant Breed.*, 132, 21–32, 2013.
61. Jain, M., Nijhawan, A., Arora, R., Agarwal, P., Ray, S., Sharma, P., Kapoor, S., Tyagi, A.K., Khurana, J.P., F-box proteins in rice. genome-wide analysis, classification, temporal and spatial gene expression during panicle and seed development, and regulation by light and abiotic stress. *Plant Physiol.*, 143, 1467–1483, 2007.

62. Wen, W., Li, D., Li, X., Gao, Y., Li, W., Li, H., Liu, J., Liu, H., Chen, W., Luo, J., Metabolome-based genome-wide association study of maize kernel leads to novel biochemical insights. *Nat. Commun.*, 5, 3438, 2014.
63. Wang, Z., Gerstein, M., Snyder, M., RNA-Seq: A revolutionary tool for transcriptomics. *Nat. Rev. Genet.*, 10, 57–63, 2009.
64. Chen, *et al.*, Comparative transcriptomics atlases reveals different gene expression 65. pattern related to *Fusarium* wilt disease resistance and susceptibility in two *Vernicia* species. *Front Plant Sci.*, 7, 1974, 2016.
65. Rohila, *et al.*, Proteomics: A biotechnology tool for crop improvement. *Front. Plant Sci.*, 4, 35, 2013.
66. Jamshidi, A. and Javanmard., H.R., Evaluation of barley (*Hordeum vulgare* L.) genotypes for salinity tolerance under field conditions using the stress indices. *Ain Shams Eng. J.*, 1–7, 2017.
67. Wu, *et al.*, Transcriptome and proteomic analysis of mango (*Mangifera indica* Linn) fruits. *J. Proteomics.*, 105, 19–30, 2014.
68. Porter, B.W., Zhu, Y.J., Christopher, D.A., *Carica papaya* genes regulated by *Phytophthora palmivora*: A model system for genomic studies of compatible *Phytophthora*-plant interactions. *Tropical Plant Biol.*, 2, 84–97, 2009.
69. Dessiree, *et al.*, Transcription analysis of softening-related genes during postharvest of papaya fruit (*Carica papaya* L. 'Pococí' hybrid). *Postharvest Biol. Technol.*, 125, 42–51, 2017.

Part 3

BIOINFORMATICS

10

Emerging Advances in Computational Omics Tools for Systems Analysis of Gramineae Family Grass Species and Their Abiotic Stress Responsive Functions

Pandiyan Muthuramalingam[1], Rajendran Jeyasri[1], Dhamodharan Kalaiyarasi[1], Subramani Pandian[1], Subramanian Radhesh Krishnan[1,2], Lakkakula Satish[1,3,4], Shunmugiah Karutha Pandian[1] and Manikandan Ramesh[1*]

[1]Department of Biotechnology, Science Campus, Alagappa University, Karaikudi, Tamil Nadu, India
[2]Phytopharma Testing Laboratory, Herbal Division, T. Stanes & Company Ltd, Coimbatore, India
[3]Department of Biotechnology Engineering, Marcus Family Campus, Ben-Gurion University of the Negev, Beer Sheva, Israel
[4]The Jacob Blaustein Institutes for Desert Research, David Bergman Campus, Ben-Gurion University of the Negev, Beer Sheva, Israel

Abstract

C3 and C4 grass species (C3: rice, C4: foxtail millet, sorghum, maize) are simultaneously exposed to one or more abiotic stresses (AbS) either independently or in combination, which leads to loss of nutritional value and yield loss. Throughout the life cycle, these plants face unique and combined AbS (CAbS). A large number of studies have been conducted to understand the molecular mechanisms imparting plant tolerance to diverse unique stresses. Still, the CAbS tolerance mechanism has not been sufficiently studied due to multiple complexities involved. In recent past, several computational omics databases and NGS-based sequencing technologies helped researchers to understand the molecular cross-talks and cellular systems of the plants especially in Gramineae grass species. An innovation in omics and NGS-based research has provided much needed platform to characterize and identify the significance, gain molecular insights, and potential CAbS responsible genes, and promote

*Corresponding author: mrbiotech.alu@gmail.com

Rintu Banerjee, Garlapati Vijay Kumar, and S.P. Jeevan Kumar (eds.) OMICS-Based Approaches in Plant Biotechnology, (185–216) © 2019 Scrivener Publishing LLC

translational research through available integrated multiple omics approaches. In view of these, this chapter elaborates the novel aspects of the plant cellular systems and CAbS responsive genes via various integrative omics approaches. The chapter also delineates the current status, challenges, and future prospects of plant cellular systems and CAbS responsive gene functional discovery for candidate gene selection and also for understanding their role in enhancing crop yield and tolerance level.

Keywords: Omics, Gramineae grass species, abiotic stress, systems biology

10.1 Introduction

Genomics studies in model and important crop plants such as *Oryza sativa*, *Setaria italica*, *Sorghum bicolor*, and *Zea mays* evolved gene mining and provided the comprehensive information about gene function in the first decade of the 21st century. In the current decade, speculating the modern scientific advances, multiple omics technologies paved the way for us to understand the multifaceted wide molecular biological systems that decipher the diverse functions in plants especially in food crops. These emerging approaches are used to drive the improvement of genomic resources in model plant species and accelerate the translational research area by incorporating the information and knowledge across food crops [1]. The emerging advances in next-generation sequencing (DNA and RNA) technologies have revolutionized the biological science field, attributing unparalleled innovations in throughput and sequencing scale and executions of novel applications after genome sequencing. Especially, while promoting genome sequencing projects, NGS approaches provide essential applications such as RNA sequencing (RNA-Seq) for noncoding RNAome and transcriptome, resequencing the whole genome used for variation analysis, chip-seq analysis for DNA–protein interactions, and quantitative detection of epigenomic dynamics [2]. Toting up to these technologies, which speculate on signal transduction and regulation of transcriptional networks, several technologies have emerged, together with molecular interactome analysis used for cross-talks formed by protein–DNA, protein–RNA, and protein–protein interactions [3], hormanomic signatures for understanding the phytohormone mediated cellular signaling [4], analysis of metabolic pathways through metabolomics [5], and phenomic analysis for plant physiological studies [6]. This emerging omics area coupled with genomic possessions treats every molecular cross-talk as an interconnected element of plant physiological and cellular systems. Computational biology plays a pivotal role in all aspects of multiple omics-related research to drive a diverse number of genomic datasets efficiently and extracts fruitful knowledge. Integration of collected omics outcome will get deeper and revise our perceptive and ease of information exchange with other food crops [7]. Hence, repository for multiple omics upshots derived from innovative analytical databases and platforms encompasses the crucial infrastructure for systems analysis. In the current chapter, we provided an overview of the newly developed possessions derived from systems approaches to computational systems biology and bioinformatics, omics investigations in Gramineae family cereal biology, and NGS-related applications and descriptive upshots of plant genomics. We also reviewed the emerging omics scale significance that has recently come into sight in plant omics science especially in Gramineae family crops, together with the nutritional omics, interactome, epigenome, hormonome, metabolome, and phenome. Efforts are made to analyze the incorporation of omics datasets as major concerns for vital analysis and unique and combined abiotic stress (CAbS) responsive gene discovery for candidate gene selection for understanding their role in enhancing crop yield, tolerance level, and understanding of many plant molecular biological mechanisms. Ultimately, we utilize omics-related analyses to understand the plant molecular functions and abiotic stress physiology. For this chapter, we provided a paradigm of current upshot in computational systems biology, plant bioinformatics, and omics, and presented an outline of resources available.

10.2 Gramineae Family Grass Species

10.2.1 *Oryza sativa*

Rice (*Oryza sativa*) is one of the most important staple foods around the world, particularly in Asian countries. The main attributes of rice are small genome size, economic importance, genetic close synteny with many cereals and millets, and its self-fertilizing nature. It has currently emerged as an ideal monocot systemic model for understanding the cereal biology, C3 photosynthesis, and systems analysis [8–10].

10.2.2 *Setaria italica*

Foxtail millet (*Setaria italica*) is the second most cultivated millet around the globe including Asia, America, Australia, Europe, and Africa. Its short life cycle, small genome size, genetically close relatedness to many other millets, inbreed nature, bioenergy potential, and abiotic stress tolerance accentuated foxtail millet as a novel systemic model for evaluating C4 photosynthesis, biofuel traits, and molecular level of stress biology [11].

10.2.3 *Sorghum bicolor*

Sorghum (*Sorghum bicolor*) is a perennial grass grown in tropical, semi-tropical, arid, and semiarid environments and cultivated throughout the world particularly in Africa, Asia, Australia, and Europe. *S. bicolor* is an efficient C4 model crop, source of biofuel, food, animal feed and other agricultural production. Sorghum is another model system to understand the molecular insights of abiotic stress resistance [12, 13].

10.2.4 *Zea mays*

Maize (*Zea mays*) is the world's most important food crop. This unisexual, C4 plant model contains large genome size, distinct chromosome, and genetic collinearity with rice, sorghum, *Setaria*, *Brachypodium*, and other cereals for analyzing and understanding the stress molecular biology, mutant, orthologous genes and regulatory elements [14]. Well-annotated C3 and C4 food crop species and the availability of whole genome information have captured the attention of plant scientists worldwide. Gramineae species genome details and other facts are provided in Table 10.1.

10.3 Abiotic Stress

In nature, an adverse climatic condition usually contains multiple physico-chemical factors, and one stressor is normally accompanied with or followed by one more stressor [15]. To envisage the contribution of unique stress, a guarded variable method was initiated and crop plants were exposed to a single stressor to simplify the system [16]. Morpho-physiological and metabolic changes were observed in unique AbS as has been reviewed previously [17–19]. Though, in natural conditions, plants are simultaneously affected by not only a single stress, once it occurs, it will be followed by multiple stressors [20]. For instance, flooding frequently causes low oxygen stresses, and salinity often leads to osmotic stress [18]. Here we focus the effects of combined abiotic stress (CAbS) on C3 and C4 food grass species, which are similar to the field environment. To analyze the molecular cross-talks, metabolic networks and their regulatory functions in response to developmental changes, differentiation, unique and CAbS, through integrated diverse omics approaches especially transcriptome, metabolome, proteome, and hormanome can be applied. Publicly available integrated omics tools and databases (Table 10.2) were used to identify and annotate the

Table 10.1 Gramene species FAOSTAT report.

Genus/species and facts	*Oryza sativa*	*Setaria italica*	*Sorghum bicolor*	*Zea mays*
Tribe	Oryzeae	Paniceae	Andropogoneae	Maydeae
Subfamily	Ehrhartoideae	Panicoideae	Panicoideae	Panicoideae
Common name	Rice	Foxtail millet	Sorghum	Maize/corn
World production	618,534,989 Mt	27,388,444 Mt	58,620,842 Mt	694,575,552 Mt
World harvest	153,783,818 Ha	35,717,582 Ha	44,703,950 Ha	147,170,849 Ha
Genome size	400–466 Mbp	490 Mbp	750–770 Mbp	3289 Mbp
Chromosome number	12	09	10	10
2n	24	18	20	20
Photosynthetic pathway	C3	C4	C4	C4

Mt: metric ton; Ha: hectares; Mbp: mega base pairs.

Table 10.2 Bioinformatic databases and tools for multiple omics approaches.

Crops	Database	Purpose	URL	Reference
Rice **Maize** **Sorghum**	GRASSIUS	Comprehensive collection of transcription factors	http://grassius.org/grasstfdb.php	[21]
Rice **Maize** **Sorghum**	GRASSIUS	Co-regulation transcription factors	http://grassius.org/grasscoregdb.php	[21]
Rice **Maize** **Sorghum** **Foxtail millet**	Phytozome	Comparative genomics	https://phytozome.jgi.doe.gov/pz/portal.html	[22]

(Continued)

Table 10.2 Bioinformatic databases and tools for multiple omics approaches. (*Continued*)

Crops	Database	Purpose	URL	Reference
Rice **Maize** **Sorghum** **Foxtail millet**	Gramene	Comparative functional genomics	http://www.gramene.org/	[23]
Rice **Maize** **Sorghum** **Foxtail millet**	Plant reactome gramene pathways	Plant metabolic and regulatory pathway analysis	http://plantreactome.gramene.org/	[23]
Rice **Maize** **Sorghum**	Expression atlas	Gene expression and biological conditions	https://www.ebi.ac.uk/gxa/plant/experiments	[24]
Rice **Maize** **Sorghum** **Foxtail millet**	Gramene QTL Database	QTL prediction and analysis	http://archive.gramene.org/qtl/	[25]
Rice	RiceNetDB	Genome-scale multiple level network reconstruction and comprehensive rice genome annotated information	http://bis.zju.edu.cn/ricenetdb/	[26]

(*Continued*)

Table 10.2 Bioinformatic databases and tools for multiple omics approaches. (*Continued*)

Crops	Database	Purpose	URL	Reference
Rice	RiceXPro	Repository of gene expression profiles derived from microarray analysis of tissues and organs	http://ricexpro.dna.affrc.go.jp/	[27]
Rice	RAP-DB	Comprehensive analysis of the genome structure and function of rice on the basis of the annotation	http://rapdb.dna.affrc.go.jp/	[28]
Rice	R-Oryzabase	Comprehensive major DNA sequences and literature-based database	https://shigen.nig.ac.jp/rice/oryzabase/	[29]
Rice	MCDRP	Manually curated database of rice proteins	http://www.genomeindia.org/biocuration/index.php	[30]
Foxtail millet	Foxtail millet database	Genome sequence Gene annotation Genetic markers and population Synteny and collinearity data	http://foxtailmillet.genomics.org.cn/	[31]

(*Continued*)

Table 10.2 Bioinformatic databases and tools for multiple omics approaches. (*Continued*)

Crops	Database	Purpose	URL	Reference
Rice **Maize** **Sorghum** **Foxtail millet**	NCBI RefSeq	Gene annotation	http://www.ncbi.nlm.nih.gov/refseq/	[32]
Rice **Maize** **Sorghum** **Foxtail millet**	Ensembl Plants	Genome sequence Gene annotation	https://plants.ensembl.org/index.html	[33]
Foxtail millet	FmMDB	Molecular marker information Physical maps Comparative maps	http://www.nipgr.res.in/foxtail.html	[34]
Foxtail millet	FmMiRNADB	miRNA data Physical map Comparative maps Target analysismiRNA-based marker information	http://59.163.192.91/FmMiRNADb/index.html	[35]

(*Continued*)

Table 10.2 Bioinformatic databases and tools for multiple omics approaches. (*Continued*)

Crops	Database	Purpose	URL	Reference
Foxtail millet	FmTFDB	Transcription factors Physical map Tissue-specific expression data Functional annotation	http://59.163.192.91/FmTFDb/index.html	[36]
Rice and maize	Yale Plant Genomics	Epigenome analysis	http://plantgenomics.biology.yale.edu/	[37–39]
Rice	PRIN	Interactome analysis	http://bis.zju.edu.cn/prin/	[40]
Rice	Rice SNP-Seek Database	Provides genotype, phenotype, and variety information for rice	http://snp-seek.irri.org/	[41]
Rice	OryGenesDB Data	Display sequence information	http://orygenesdb.cirad.fr/data.html	[42]
Rice	BGI Rice Database	Sequence alignment, small RNA and transcriptome analysis, *de novo* genome assembly, gene structure and function annotation, as well as evolutionary and comparative analysis	http://rise2.genomics.org.cn/page/rice/index.jsp	[43]

(*Continued*)

Table 10.2 Bioinformatic databases and tools for multiple omics approaches. (*Continued*)

Crops	Database	Purpose	URL	Reference
Rice	*Oryza sativa* Genome DB–PlantGDB	Displays high-quality spliced alignments for EST, cDNA, PUT, and proteins	http://www.plantgdb.org/OsGDB/	[44]
Sorghum	MOROKOSHI Database	Transcriptome analysis	http://sorghum.riken.jp/morokoshi/Home.html	[45]
Sorghum	Sorghum functional genomics database	Functional genomics database with multidimensional network analysis	https://omictools.com/sorghum-functional-genomics-database-tool	[46]
Sorghum	PGSB Repeat Database	Comparative analysis and visualization of plant genome data	http://pgsb.helmholtz-muenchen.de/plant/recat/	[47]
Sorghum	Genome data from sweet and grain database	Genome sequence information	http://gigadb.org/dataset/100012	[48]
Sorghum	PlnTFDB	Identify and catalogue all *Plant* genes involved in transcriptional control	http://plntfdb.bio.uni-potsdam.de	[49]

(*Continued*)

Table 10.2 Bioinformatic databases and tools for multiple omics approaches. (*Continued*)

Crops	Database	Purpose	URL	Reference
Foxtail millet	SIFGD	Analyses of gene function or regulatory modules	http://structuralbiology.cau.edu.cn/SIFGD/	[50]
Foxtail millet	New PLACE web server	Presence of *cis*-regulatory elements was identified by Signal Scan Search	https://sogo.dna.affrc.go.jp/cgi-bin/sogo.cgi?page=analysis&lang=en	[51]
Maize	MaizeGDB	Community database for maize genetics and genomics	http://www.maizegdb.org/	[52]
Maize	TIGR maize database	Sequences, assemblies, and annotations	http://maize.tigr.org	[53]
Maize	ZmDB	Sequencing and phenotypic analysis of the gene	http://zmdb.iastate.edu/.	[54]
Maize	MaizeCyc pathway database	Library of evidences collected from the maize genome annotation, gene-based phylogeny trees, and comparison to known genes and pathways from rice and *Arabidopsis* against the PathoLogic module of Pathway Tools	http://maizecyc.maizegdb.org	[55]

(*Continued*)

Table 10.2 Bioinformatic databases and tools for multiple omics approaches. (*Continued*)

Crops	**Database**	**Purpose**	**URL**	**Reference**
Maize	*Zea mays* Plant Structure Ontology database	Structural ontology	http://www.plantontology.org	[56]
Maize	Chromatin Database	Annotated group of maize protein coding transcripts downloaded	www.ChromDB.org	[57]

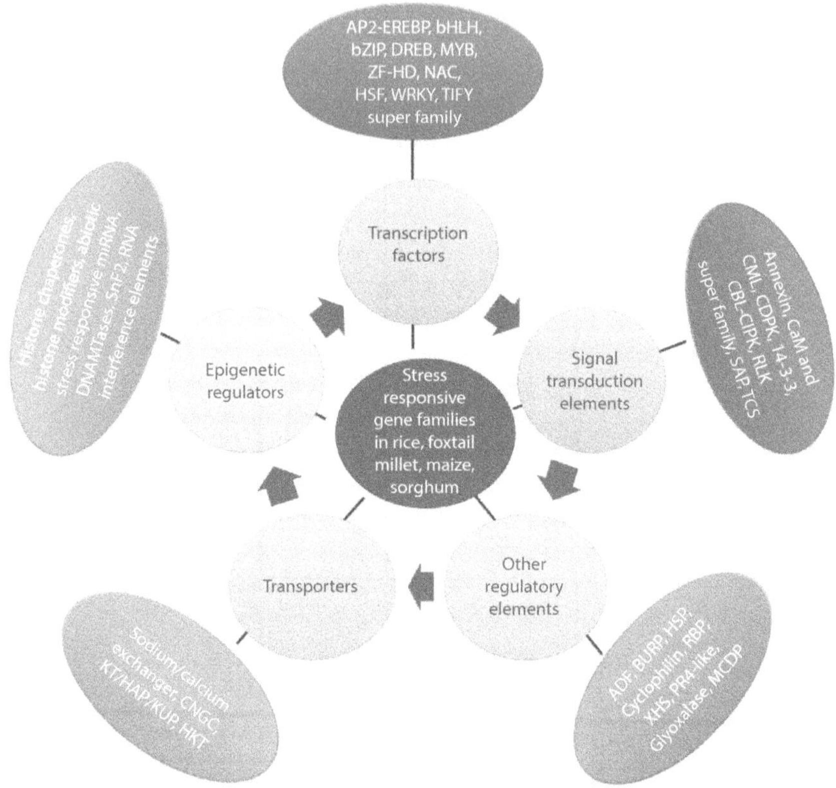

Figure 10.1 Illustration of C3 and C4 stress responsive gene families and their associated regulatory functions. Conceptual model was adapted from Sharma *et al.*, [58].

unique and CAbS genes in C3 and C4 grass species. Based on the integrated multiple omics-related systems analysis and with the help of PubMed search, cellular function and stress responsive gene families and their corresponding molecular mechanism were identified (Figure 10.1).

10.4 Emerging Sequencing Technologies

Recent advances in (DNA and RNA) sequencing technologies that gradually produce large level sequence information have elicited generational changes in plant genomics [2]. Insights of the historical shift in DNA and RNA sequencing were given by Hutchison [59]. A huge number of widespread available NGS platforms [60], as well as the Genome Analyzer or Hiseq (Illumina Solexa) [61, 62], the 454 FLX (Roche) [63], SOLiD (Life

Technologies), and newly emerged platforms such as PacBio RS (Pacific Biosciences) [64] and single molecular sequencing by Heliscope (Helicos) [65], and also available in Ion Torrent (Life Technologies), depend upon the semiconductor chip [66]. In view of instance, a lengthy reader of NGS, the available version of Roche 454 platform, is able to generate approximately 1 million reads in ~400 bp per run, and a small reader, the available Illumina Hiseq2000, is able to generate sequence data up to 600 billion bases (Gb) in 100-bp two paired end reads in a single NGS run. A literature survey on high-throughput NGS-based technologies including computational methods and applications of experimental evidences has been published [2, 67, 68]. This chapter provides a complete view of emerging approaches related to NGS and possible upshots in plants, particularly in C3 and C4 crops on Gramineae family.

10.4.1 NGS-Based Genomic and RNA Sequencing

In this decade, multiple genomes have been sequenced in plants by NGS technology [69–72]. Still the complex plant genomes and their *de novo* assembly remain a major hurdle; a group of combined approaches by using basic Sanger and/or 454 Roche pyrosequencing platforms along with some other NGS sources give more competent techniques of genome assembly, in particular NGS platform. Sequencing of reference genome coupled with resequencing data and their applications carries out the quality of NGS approaches. Gradual achievement of genomic variant datasets facilitates ultrahigh-throughput identification of the alternative genes (alleles) associated with phenotypic assortment and potential gene mutations. A methodological way to distinguish the mutation in *Arabidopsis* that was induced by ethyl methanesulfonate (EMS) is determined with the help of NGS technology [73]. Polymorphisms in DNA such as insertion–deletion (InDels) and SNPs have been identified by NGS-based resequencing, which could enable the identification of those alterations in nucleotides in closely related cultivars and ecotypes [74, 75]. High-throughput polymorphism scrutiny is a pivotal tool for developing a map-based approach [76]. In recent years, genome-wide association study (GWAS) is predominantly used to scrutinize the genetic makeup of numerous complex traits in plants especially of Gramineae family [76]. Whole genome sequencing and resequencing of the entire genome in plants by NGS approaches have been made probable by increased throughput, low cost, and enhancement of the analysis efficacy of the genotype to phenotype relationship [77].

10.4.2 Tanscriptome Analysis Based on NGS

RNA-Seq NGS technologies are a widely popular approach to collect, collate, and quantify the large-scale sequences of noncoding and coding RNAs swiftly [78]. RNA-Seq is widely used in identifying splicing variants by sequenced mapping of DNA fragments to reference genome sequence of rice and *Arabidopsis* [79, 80]. These methods contain more knowledge about splicing and alternative splicing mechanisms in transcriptomes at quantitative patterns of every isoform as well as single base resolution. Because it allows frequent attainment of sequences, polymorphisms, and expression signatures, RNA-Seq, which is based on NGS, has been used for the gradual growth of genomic, epigenomic, and transcriptomic repositories in Gramineae family plant species [81, 82]. While reference genome sequences are unavailable, NGS-based RNA-Seq analysis of *de novo* assembly could enable gradual acquirement of early genomic scale data sequences to afford preliminary depiction of new and crop plant species [83–85]. Plant genomics expanded rapidly by NGS-based sequencing applications (Table 10.3). By searching the Entrez records in the NCBI (https://www.ncbi.nlm.nih.gov/Taxonomy/Browser/wwwtax.cgi), Sequence Read Archive (SRA) in the NCBI (http://www.ncbi.nlm.nih.gov/sra), DDBJ Sequence Read Archive (https://www.ddbj.nig.ac.jp/dra/index-e.html), and European Nucleotide Archive (ENA) (http://www.ebi.ac.uk/ena/home), all of these repositories assemble the datasets to raw sequencing format from NGS platforms; users can choose how

Table 10.3 Status of NGS technology in Gramineae family food crops (https://www.ncbi.nlm.nih.gov/Taxonomy/Browser/wwwtax.cgi, April 2, 2018).

NCBI taxonomy browser	***Oryza sativa***	***Setaria italica***	***Sorghum bicolor***	***Zea mays***
Database name	**No. of experimental information available**			
Nucleotide	177,504	46,098	71,017	642,300
Nucleotide EST	62,441	66,027	209,835	2,019,896
Nucleotide GSS	75,163	96,975	804,473	1,622,820
Protein	21,638	77,753	93,030	272,421
Genome	1	1	1	1

(*Continued*)

Table 10.3 Status of NGS technology in Gramineae family food crops (https://www.ncbi.nlm.nih.gov/Taxonomy/Browser/wwwtax.cgi, April 2, 2018). (*Continued*)

NCBI taxonomy browser	*Oryza sativa*	*Setaria italica*	*Sorghum bicolor*	*Zea mays*
Popset	911	94	527	361
GEO Datasets	11,928	19	673	11,036
PubMed Central	18,980	825	4,350	16,485
Gene	91,385	39,802	41,278	66,600
SRA Experiments	45,452	2,551	5,046	17,140
Bio Project	2,361	190	988	907
Bio Sample	26,699	2,500	3,903	34,594
Bio Systems	751	259	245	268
Assembly	26	3	10	8
Clone DB	172,043	50,086	103,820	1,144,901
Probe	92,681	28,288	10,889	58,036
PubChem BioAssay	436	5	37	415
Taxonomy	1	1	1	1

systematically a crop species whom whole genome has been sequenced and publicly available sequencing data are retrieved for further use.

10.4.3 High-Throughput Omics Layers

With the publicly available C3 and C4 plant models and their corresponding whole genome sequences and the emerging advancement in omics-based research, understanding the computational biology-related tools has become a need to gain more molecular biological and cellular insights. Recently emerged high-throughput omics approaches such as genomic, epigenomic, transcriptomic, proteomic, metabolomic, hormanomic, and interatomic datasets pave the way to diverse budding systematic and/or analytical applications. Furthermore, in rice, sorghum, foxtail millet, and maize, advancement of omics resources has allowed us

Figure 10.2 A simplified omic space with advanced omics layers.

to understand the molecular cross-talks and cellular systems encoding attributes of unique species.

Combinatorial information from multiple omics-based sources is applied to identify the molecular systems insights and biological attributes that accelerate gene discovery and its annotated information. Figure 10.2 represents the thematic flow chart representing the emerging omics layers ranging from "genome" to "phenome"; the theme is called the "omic space."

10.5 Omics Resource in Poaceae Species

The advancement of omics repositories particularly in genomic resources improved in plant species. As a simple example, 64 plant species genome sequence datasets are available at Phytozome database v12.1.6 (https://phytozome.jgi.doe.gov/pz/portal.html). As of now, sequenced genomes in plant species were nominated characteristically for development of genomic resource because (I) they contain specific cellular systems not roofed by model crop plants; (II) which could provide commodity sources like food and energy; or (III) they have evolutionary importance. At this time, we outline the present status of the publicly available omics resources for Gramineae (Poaceae) family.

The Gramineae family includes the most important staple food crops such as *O. sativa*, *S. italica*, *S. bicolor*, and *Z. mays* [8–14]. The *O. sativa* spp., japonica genome project [86], whole genome sequence (WGS) in maize (*Z. mays*), sorghum (*S. bicolor*), and foxtail millet (*S. italica*) have been completed [12, 87, 88]. Rice is one of the most important staple foods around the globe and is a monocot model plant species. To the best of our knowledge, japonica rice genome sequences with quality gene attributes have played pivotal mechanisms in accelerating the huge number of genomic possessions development for mining, isolation, and characterization of potential candidates for genetics, metabolic engineering, and molecular breeding. Maize is also one of the plant models used in basic research of genomic attributes such as epigenetics, evolution, domestication, transposable elements and chromosome structure, complex inheritance, and staple cereals for food and animal feed [89]. Foxtail millet has small diploid (~500 Mb) genome size, has short cycling times (seed to seed, 6 weeks), self-fertilizes, and is accentuated as an ideal plant model for functional genomics studies in cereals, millets, and bioenergy grasses [90, 91]. Sorghum is one of the C4 crop plant models and its whole genome sequence as an envoy species includes biomass resource crops for sugar, starch, and cellulose.

10.6 Role of Functional Omics in Dissecting the Stress Physiology of Gramineae Members

Cereal crops such as rice, sorghum, foxtail millet, and maize in the field are exposed to a multiple number of abiotic stresses (AbS) such as drought, salinity, cold, submergence, temperature, UV, etc. [20, 91, 92]. This urged the plant research society to evaluate the stress mechanisms and their cross-talks in C3 and C4 grass species. Fascinatingly, genome and transcriptome sequencing and annotation projects [31, 86, 89–91] have been identified, and recent literatures paved the way to identify the candidate genes through omics databases, which are predicted to be involved in unique and combined abiotic stress (CAbS) responses [11, 20, 31, 93]. This claim supports to understand the stress tolerance behavior and adaptation to environments and to promote the green revolution of all other food crops.

10.7 Systems Analysis in Gramineae Plant Species

Systems approach mainly depends on various omics technologies. It is a significant tool in classifying and categorizing the wide level of plant cellular systems. Earlier research in plant metabolomics accomplished a huge number of essential progresses in understanding the cellular level gene mechanisms in metabolic organizations by incorporating the metabolome study with specific resources like genome, epigenome, and transcriptome [94–98]. These subsequent achievements, including various omics-related systems analyses, paved the way to improving our knowledge about plant physiology systems at the cellular level. For example, incorporated transcriptome and metabolome analyses were implied to examine high-temperature conditions in developing caryopses of rice plants [99]; also molecular clocks focus on pollination independence and fruit sets induced by pollination [38]. Integrated transcriptome–metabolome analysis is also used to determine the changes in metabolic systems in the field grown plants with mutant lines and transgenic plants [100]. In addition, systems analysis combined with transcriptome, metabolome, and hormonome analyses in transgenics of *Arabidopsis* showed improved growth of leaf and attained molecular insights that control the size of the leaf [101].

An incorporated metabolome and proteome analysis has been implied to distinguish the variations in response to anoxia between wheat and rice coleoptiles [102]. An integrated transcriptome, proteome, and metabolome analysis was applied in maize to characterize and categorize the simultaneous changes in responses to UV-B [103]. These descriptive instances exhibit the essence of systems analysis, which depends on the various omics for framing the crucial mechanisms of plant cellular and physiological systems expected with diverse functions in plants.

Thanks to emerging technological developments of ultrahigh-throughput and genomic level platforms such as a comprehensive study of the association between quantitative instances and genome diversity in multiple omics aspects, whole genome resequencing has facilitated the mining of significant players that may be involved in flexible changes in combinatorial approach through multiple omics scales [104]. Using a large number of variation datasets at the genome scale, all genome accessions and GWAS identified metabolome profile and genetic loci controlling the enzyme functions and biomass [53]. The phytohormonal responses of variations in plants are naturally addressed to identify the interactions between phenomic variations of phytohormonal responses and some other variations in the transcriptome and epigenome [105]. A combinatorial study using plant hormonome signatures for population genomics may be used to detect the relationship between plant hormone abundance and genomic polymorphism traits that may be strongly related to natural adaptation. Earlier research about epigenomic modification, noncoding RNAs, and coding gene transcription has been tied with genome-level datasets related to nucleotide polymorphism in population genomics [106, 107]. In a similar way, epigenomic analysis has also been integrated with genome-level variations that provide the significant indication to the phenotypic diversity associated with epigenetic and genetic regulation.

10.8 Nutritional Omics of Gramineae Species

A Gramineae species grain contains nutritional, therapeutic potential and enormous health benefits. No study has been made to understand the omics and genetics of nutritional traits in C3 and C4 grass species. International and national level efforts need to be mobilized to investigate grain characteristics and study the genetic variations in rice, sorghum, maize, and foxtail millet germplasm using integrated omics and biochemical phenotyping. In addition, the strength of combinatorial omics platform especially on metabolomics and phenomics with GWAS will aid in identification of SNPs, SNP–metabolite pairs, and potential key players controlling grains' nutritional contents. The identified candidates in the germplasm can be introgressed to other Gramineae species through comparative omics (Figure 10.3). The nutritionally improved crops will serve as a healthy source for the global population and evading the nutritional insecurity.

10.9 Future Prospects

The entire elucidation of physicochemical and molecular cross-talks underlying the stress responsive and plant developmental biology highly depends on the comprehensive investigation using omics-based systemic approaches, which is the major base for the application of diverse omics in plant biology. Among them, epigenomics, metabolomics, and hormonomics have a special importance, because molecular elements, metabolites, and signaling molecules are more relevant to the plant physiological phenotype as compared with protein and other biological molecules.

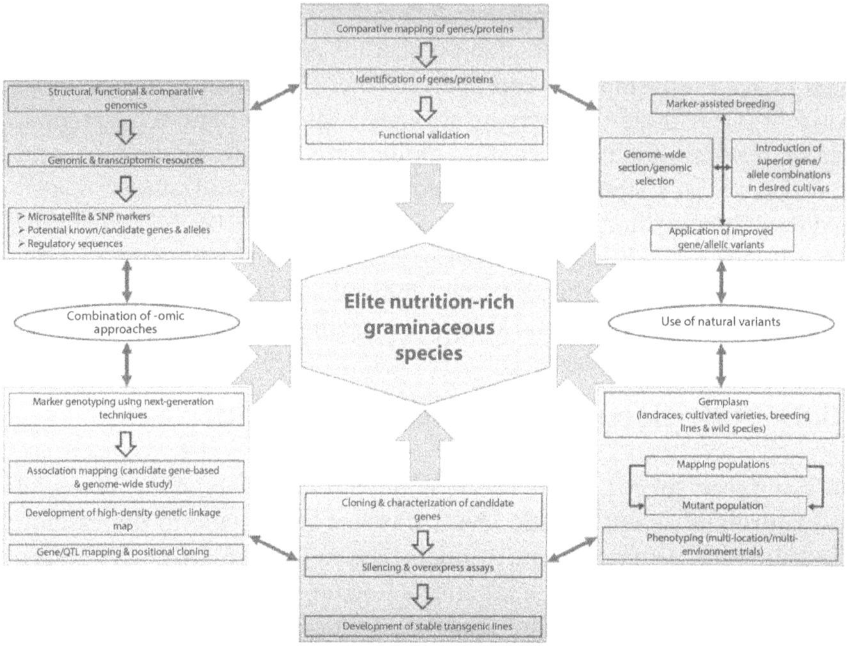

Figure 10.3 Deciphering the nutritional traits in Gramineae species by integrated omics approaches adapted from Muthamilarasan and Prasad [11].

Hence, future research in this thrust area will speculate the direction on epigenome, metabolome, and hormonome and other integrated omics platforms to facilitate the accurate and effective understanding about plant cellular systems physiology, improving the crop yield and quality especially in C3 and C4 crop species.

10.10 Conclusion

Emerging technologies in diverse systematic repositories are giving opportunities to address the genome-level upshots, invigorating researchers' familiarity and eminence of fundamental queries in plant biology. In the meantime, researchers are mainly facing diverse global problems such as water, global warming, energy and food security, and climate changes. Comprehension of the unique plant molecular function and AbS tolerance mechanisms that come into sight, in particular, plant species especially in C3 and C4 crop plants, is also significant to discover potential candidates to enhance the diverse functions. An integration of a broad spectrum of omics datasets from C3 and C4 plant species is done to enhance the research in translational area to contrive plant molecular systems in response to the demands of plant physiologists and stress biologists, and also the rising demands of mankind.

Acknowledgments

Pandiyan Muthuramalingam (Rc.SO (P)/DBT-BIF/15207/2017 dated February 02, 2018) thanks the DBT-Bioinformatics Infrastructure Facility Scheme, New Delhi, India for the financial support in the form of fellowship. The authors gratefully acknowledge the use of Bioinformatics Infrastructure Facility, Alagappa University, funded by the Department of Biotechnology, Ministry of Science and Technology, Government of India grant (No. BT/BI/25/015/2012).

References

1. Mochida, K. and Shinozaki, K., Genomics and bioinformatics resources for crop improvement. *Plant Cell Physiol.*, 51, 497–523, 2010.
2. Lister, R., Gregory, B.D., Ecker, J.R., Next is now: New technologies for sequencing of genomes, transcriptomes, and beyond. *Curr. Opin. Plant Biol.*, 12, 107–118, 2009.
3. Arabidopsis Interactome Mapping Consortium, Evidence for network evolution in an *Arabidopsis* interactome map. *Science*, 333, 601–607, 2011.
4. Kojima, M., Kamada-Nobusada, T., Komatsu, H., Takei, K., Kuroha, T., Mizutani, M., Highly sensitive and high-throughput analysis of plant hormones using MS-probe modification and liquid chromatography–tandem mass spectrometry: An application for hormone profiling in *Oryza sativa. Plant Cell Physiol.*, 50, 1201–1214, 2009.
5. Saito, K. and Matsuda, F., Metabolomics for functional genomics, systems biology, and biotechnology. *Annu. Rev. Plant Biol.*, 61, 463–489, 2010.
6. Yang, W., Duan, L., Chen, G., Xiong, L., Liu, Q., Plant phenomics and high-throughput phenotyping: Accelerating rice functional genomics using multidisciplinary technologies. *Curr. Opin. Plant Biol.*, 16, 180–187, 2013.
7. Shinozaki, K. and Sakakibara, H., Omics and bioinformatics: An essential toolbox for systems analyses of plant functions beyond 2010. *Plant Cell Physiol.*, 50, 1177–1180, 2009.
8. Wu, J., Maehara, T., Shimokawa, T., A comprehensive rice transcript map containing 6591 expressed sequence tag sites. *Plant Cell*, 14, 525–535, 2002.
9. Chen, M., Presting, G., Barbazuk, W.B., An integrated physical and genetic map of the rice genome. *Plant Cell*, 14, 537–545, 2002.

10. Sasaki, T., Matsumoto, T., Antonio, B.A., Nagamura, Y., From mapping to sequencing, post sequencing and beyond. *Plant Cell Physiol.*, 46, 3–13, 2005.
11. Muthamilarasan, M. and Prasad, M., Advances in Setaria genomics for genetic improvement of cereals and bioenergy grasses. *Theor. Appl. Genet.*, 128, 1–14, 2015.
12. Paterson, A.H., Bowers, J.E., Bruggmann, R., Dubchak, I., Grimwood, J., Gundlach, H., Haberer, G., Hellsten, U., Mitros, T., Poliakov, A., Schmutz, J., Spannagl, M., Tang, H., Wang, X., Wicker, T., Bharti, A.K., Chapman, J., Feltus, F.A., Gowik, U., Grigoriev, I.V., Lyons, E., Maher, C.A., Martis, M., Narechania, A., Otillar, R.P., Penning, B.W., Salamov, A.A., Wang, Y., Zhang, L., Carpita, N.C., Freeling, M., Gingle, A.R., Hash, C.T., Keller, B., Klein, P., Kresovich, S., McCann, M.C., Ming, R., Peterson, D.G., Mehboob-ur-Rahman, Ware, D., Westhoff, P., Mayer, K.F., Messing, J., Rokhsar, D.S., The sorghum bicolor genome and the diversification of grasses. *Nature*, 457, 551–556, 2009.
13. Wang, X., Gowik, U., Tang, H., Bowers, J.E., Westhoff, P., Paterson, A.H., Comparative genomic analysis of C4 photosynthetic pathway evolution in grasses. *Genome Biol.*, 10, R68, 2009.
14. Lin, Z., Li, X., Shannon, L.M., Yeh, C.T., Wang, M.L., Bai, G., Peng, Z., Li, J., Trick, H.N., Clemente, T.E., Doebley, J., Schnable, P.S., Tuinstra, M.R., Tesso, T.T., White, F., Yu, J., Parallel domestication of the Shattering1 genes in cereals. *Nat. Genet.*, 44, 720–724, 2012.
15. Kralova, K., Jampilek, J., Ostrovsky, I., Metabolomics—Useful tool for study of plant responses to abiotic stresses. *Ecol. Chem. Eng. S.*, 19, 133–161, 2012.
16. Chaves, M.M., Flexas, J., Pinheiro, C., Photosynthesis under drought and salt stress: Regulation mechanisms from whole plant to cell. *Ann. Bot.*, 103, 551–560, 2009.
17. Shulaev, V., Cortes, D., Miller, G., Mittler, R., Metabolomics for plant stress response. *Physiol. Plant.*, 132, 199–208, 2008.
18. Obata, T. and Fernie, A.R., The use of metabolomics to dissect plant responses to abiotic stresses. *Cell Mol. Life Sci.*, 69, 3225–3243, 2012.
19. Arbona, V., Manzi, M., Ollas, C., Gomez-Cadenas, A., Metabolomics as a tool to investigate abiotic stress tolerance in plants. *Int. J. Mol. Sci.*, 14, 4885–4911, 2013.
20. Muthuramalingam, P., Krishnan, S.R., Pothiraj, R., Ramesh, M., Global transcriptome analysis of combined abiotic stress signaling genes unravels key players in *Oryza sativa* L. An *in silico* approach. *Front. Plant Sci.*, 8, 759, 2017a.
21. Yilmaz, A., Nishiyama, M.Y., Fuentes, B.G., Souza, G.M., Janies, D., Gray, J., Grotewold, E., GRASSIUS: A platform for comparative regulatory genomics across the grasses. *Plant Physiol.*, 149, 171–180, 2009.
22. Goodstein, D.M., Shu, S., Howson, R., Neupane, R., Hayes, R.D., Fazo, J., Mitros, T., Dirks, W., Hellsten, U., Putnam, N., Rokhsar, D.S., Phytozome:

A comparative platform for green plant genomics. *Nucleic Acids Res.*, 40, D1178–D1186, 2012.
23. Tello-Ruiz, M.K., Stein, J., Wei, S., Preece, J., Olson, A., Naithani, S., Gramene 2016: Comparative plant genomics and pathway resources. *Nucleic Acids Res.*, 44, 1133–1140, 2016.
24. Petryszak, R., Keays, M., Tang, Y.A., Fonseca, N.A., Barrera, E., Burdett, T., Expression Atlas update—An integrated database of gene and protein expression in humans, animals and plants. *Nucleic Acids Res.*, 44, 746–752, 2016.
25. Ni, J., Pujar, A., Youens-Clark, K., Yap, I., Jaiswal, P., Tecle, I., Tung, C., Ren, L., Spooner, W., Wei, X., Avraham, S., Ware, D., Stein, L., McCouch, S., Gramene QTL database: Development, content and applications. *Database*, Article ID bap005, 2009, https://doi.org/10.1093/database/bap005.
26. Liu, L., Mei, Q., Yu, Z., Sun, T., Zhang, Z., Chen, M., An integrative bioinformatics framework for genome-scale multiple level network reconstruction of rice. *J. Integrative Bioinform.*, 10, 94–102, 2013.
27. Sato, Y., Antonio, B.A., Namiki, N., Takehisa, H., Minami, H., Kamatsuki, K., Sugimoto, K., Shimizu, Y., Hirochika, H., Nagamura, Y., RiceXPro: A platform for monitoring gene expression in japonica rice grown under natural field conditions. *Nucleic Acids Res.*, 39, D1141–D1148, 2010.
28. Sakai, H., Shin Lee, S., Tanaka, T., Numa, H., Kim, J., Kawahara, Y., Wakimoto, H., Yang, C., Iwamoto, M., Abe, T., Yamada, Y., Muto, A., Inokuchi, H., Ikemura, T., Matsumoto, T., Sasak, T., Itoh, T., Rice Annotation Project Database (RAP-DB): An integrative and interactive database for rice genomics. *Plant Cell Physiol.*, 54, 1–11, 2012.
29. Kurata, N. and Yamazaki, Y., Oryzabase. An integrated biological and genome information database for rice. *Plant Physiol.*, 140, 2006.
30. Raghuvanshi, S., Gour, P., Joseph, S.V., Manually Curated Database of Rice Proteins (MCDRP), a database of digitized experimental data on rice. *Curr. Plant Biol.*, 8, 26–30, 2016.
31. Zhang, G., Liu, X., Quan, Z., Cheng, S., Xu, X., Pan, S., Xie, M., Zeng, P., Yue, Z., Wang, W., Tao, Y., Bian, C., Han, C., Xia, Q., Peng, X., Cao, R., Yang, X., Zhan, D., Hu, J., Zhang, Y., Li, H., Li, H., Li, N., Wang, J., Wang, C., Wang, R., Guo, T., Cai, Y., Liu, C., Xiang, H., Shi, Q., Huang, P., Chen, Q., Li, Y., Wang, J., Zhao, Z., Wang, J., Genome sequence of foxtail millet (*Setaria italica*) provides insights into grass evolution and biofuel potential. *Nat. Biotechnol.*, 30, 549–554, 2012.
32. Pruitt, K.D., Brown, G.R., Hiatt, S.M., Thibaud-Nissen, F., Astashyn, A., Ermolaeva, O., Farrell, C.M., Hart, J., Landrum, M.J., McGarvey, K.M., Murphy, M.R., O'Leary, N.A., Pujar, S., Rajput, B., Rangwala, S.H., Riddick, L.D., Shkeda, A., Sun, H., Tamez, P., Tully, R.E., Wallin, C., Webb, D., Weber, J., Wu, W., Dicuccio, M., Kitts, P., Maglott, D.R., Murphy, T.D., Ostell, J.M., RefSeq: An update on mammalian reference sequences. *Nucleic Acids Res.*, 42, D756–D763, 2014.

33. Kersey, P.J., Allen, J.E., Christensen, M., Davis, P., Falin, L.J., Grabmueller, C., Hughes, D.S., Humphrey, J., Kerhornou, A., Khobova, J., Langridge, N., McDowall, M.D., Maheswari, U., Maslen, G., Nuhn, M., Ong, C.K., Paulini, M., Pedro, H., Toneva, I., Tuli, M.A., Walts, B., Williams, G., Wilson, D., Youens-Clark, K., Monaco, M.K., Stein, J., Wei, X., Ware, D., Bolser, D.M., Howe, K.L., Kulesha, E., Lawson, D., Staines, D.M., Ensembl genomes 2013: Scaling up access to genome-wide data. *Nucleic Acids Res.*, 42, D546–D552, 2014.
34. Suresh, B.V., Muthamilarasan, M., Mishra, G., Prasad, M., FmMDb: A versatile database of foxtail millet markers for millets and bioenergy grasses research. *PLoS ONE*, 8, e71418, 2013.
35. Khan, Y., Yadav, A., Suresh, B.V., Muthamilarasan, M., Yadav, C.B., Prasad, M., Comprehensive genome-wide identification and expression profiling of foxtail millet [*Setaria italica* (L.)] miRNAs in response to abiotic stress and development of miRNA database. *Plant Cell. Tiss. Organ Cult.*, 118, 279–292, 2014.
36. Bonthala, V.S., Muthamilarasan, M., Roy, R., Prasad, M., FmTFDb: A foxtail millet transcription factors database for expediting functional genomics in millets. *Mol. Biol. Rep.*, 41, 6343–6348, 2014.
37. Li, X., Wang, X., He, K., Ma, Y., Su, N., He, H., High-resolution mapping of epigenetic modifications of the rice genome uncovers interplay between DNA methylation, histone methylation, and gene expression. *Plant Cell*, 20, 259–276, 2008.
38. Wang, H., Schauer, N., Usadel, B., Frasse, P., Zouine, M., Hernould, M., Regulatory features underlying pollination-dependent and -independent tomato fruit set revealed by transcript and primary metabolite profiling. *Plant Cell*, 21, 1428–1452, 2009.
39. He, G., Zhu, X., Elling, A.A., Chen, L., Wang, X., Guo, L., Global epigenetic and transcriptional trends among two rice subspecies and their reciprocal hybrids. *Plant Cell*, 22, 17–33, 2010.
40. Gu, H., Zhu, P., Jiao, Y., Meng, Y., Chen, M., PRIN: A predicted rice interactome network. *BMC Bioinform.*, 12, 161, 2011.
41. Alexandrov, N., Tai, S., Wang, W., Mansueto, L., Palis, K., Fuentes, R., Ulat, V., Chebotarov, D., Zhang, G., Li, Z., Mauleon, R., Hamilton, R., McNally, K., SNP-Seek database of SNPs derived from 3000 rice genomes. *Nucleic Acids Res.*, 43, D1023–D1027, 2014.
42. Droc, G., Ruiz, M., Larmande, P., Pereiral, A., Piffanelli, P., Morel, J.B., Dievart, A., Courtois, B., Guiderdoni, E., Perin, C., OryGenesDB: A database for rice reverse genetics. *Nucleic Acids Res.*, 34, D737, 2005.
43. Zhao, W., Wang, J., He, X., Huang, X., Jiao, Y., Dai, M., Wei, S., Fu, J., Chen, Y., Ren, X., Zhang, Y., Ni, P., Zhang, J., Li, S., Wang, J., Wong, G., Zhao, H., Yu, J., Yang, H., Wang, J., BGI-RIS: An integrated information resource and comparative analysis workbench for rice genomics. *Nucleic Acids Res.*, 32, D377–D382, 2003.

44. Duvick, J., Fu, A., Muppirala, U., Sabharwal, M., Wilkerso, M., Lawrence, C., Lushbough, C., Brendel, V., PlantGDB: A resource for comparative plant genomics. *Nucleic Acids Res.*, 36, D959–D965, 2008.
45. Makita, Y., Shimada, S., Kawashima, M., Kondou-Kuriyama, T., Toyoda, T., Matsui, M., MOROKOSHI: Transcriptome database in *Sorghum bicolor. Plant Cell Physiol.*, 56, 1–8, 2014.
46. Tian, T., You, Q., Zhang, L., Yi, X., Yan, H., Xu, W., Su, Z., SorghumFDB: Sorghum functional genomics database with multidimensional network analysis. *Database*, 1–16, 2016.
47. Spannagl, M., Nussbaumer, T., Bader, K., Martis, M., Seidel, M., Kugler, K., Gundlach, H., Mayer, K., PGSB PlantsDB: Updates to the database framework for comparative plant genome research. *Nucleic Acids Res.*, 44, D1141–D1147, 2016.
48. Dong, S., Guo, X., He, B., Genome data from sweet and grain sorghum (*Sorghum bicolor*). *Genome Biol.*, 12, R114, 2011.
49. Riaño-Pachón, D., Ruzicic, S., Dreyer, I., Mueller-Roeber, B., PlnTFDB: An integrative plant transcription factor database. *BMC Bioinformatics*, 8, 42, 2007.
50. You, Q., Zhang, L., Yi, X., Zhang, Z., Xu, W., Su, Z., SIFGD: *Setaria italica* Functional Genomics Database. *Molecular Plant*, 8, 967–970, 2015.
51. Muthamilarasan, M., Khan, Y., Jaishankar, J., Shweta, S., Lata, C., Prasad, M., Integrative analysis and expression profiling of secondary cell wall genes in C4 biofuel model *Setaria italica* reveals targets for lignocelluloses bioengineering. 6, 965, 2015.
52. Lawrence, C., Dong, Q., Polacco, M., Seigfried, T., Brendel, V., MaizeGDB, the community database for maize genetics and genomics. *Nucleic Acids Res.*, 32, D393–D397, 2004.
53. Chan, E.K., Rowe, H.C., Kliebenstein, D.J., Understanding the evolution of defense metabolites in Arabidopsis thaliana using genome-wide association mapping. *Genetics*, 185, 991–1007, 2010.
54. Dong, Q., Roy, L., Freeling, M., Walbot, V., Brendel, V., ZmDB, an integrated database for maize genome research. *Nucleic Acids Res.*, 31, D244–D247, 2003.
55. Monaco, M., Sen, T., Dharmawardhana, P., Ren, L., Schaeffer, M., Naithani, S., Amarasinghe, V., Thomason, J., Harper, L., Gardiner, J., Cannon, E., Doreen Ware, C., Jaiswal, P., Maize metabolic network construction and transcriptome analysis. *Plant Genome*, 6, 1, 2013.
56. Vincent, P., Coe, E., Jr., Polacco, M., *Zea mays* ontology—A database of international terms. *Trends in Plant Sci.*, 8, 11, 2003.
57. Boerner, S. and McGinnis, K., Computational identification and functional predictions of long noncoding RNA in *Zea mays. PLoS ONE*, 7, e43047, 2012.
58. Sharma, G., Giri, J., Tyagi, A.K., Sub-functionalization in rice gene families with regulatory roles in abiotic stress responses. *Critical Rev. Plant Sci.*, 1–55, 2017.

59. Hutchison, C.A., 3rd DNA sequencing: Bench to bedside and beyond. *Nucleic Acids Res.*, 35, 6227–6237, 2007.
60. Gupta, P.K., Single-molecule DNA sequencing technologies for future genomics research. *Trends Biotechnol.*, 26, 602–611, 2008.
61. Bennett, S., Solexa Ltd. *Pharmacogenomics*, 5, 433–438, 2004.
62. Bennett, S.T., Barnes, C., Cox, A., Davies, L., Brown, C., Toward the 1,000 dollars human genome. *Pharmacogenomics*, 6, 373–382, 2005.
63. Margulies, M., Egholm, M., Altman, W.E., Attiya, S., Bader, J.S., Bemben, L.A., Genome sequencing in microfabricated high-density picolitre reactors. *Nature*, 437, 376–380, 2005.
64. Eid, J., Fehr, A., Gray, J., Luong, K., Lyle, J., Otto, G., Real-time DNA sequencing from single polymerase molecules. *Science*, 323, 133–138, 2009.
65. Milos, P., Helicos BioSciences. *Pharmacogenomics*, 9, 477–480, 2008.
66. Rothberg, J.M., Hinz, W., Rearick, T.M., Schultz, J., Mileski, W., Davey, M., An integrated semiconductor device enabling non-optical genome sequencing. *Nature*, 475, 348–352, 2011.
67. Horner, D.S., Pavesi, G., Castrignano, T., De Meo, P.D., Liuni, S., Sammeth, M., Bioinformatics approaches for genomics and post genomics applications of next-generation sequencing. *Brief. Bioinform.*, 11, 181–197, 2010.
68. Metzker, M.L., Sequencing technologies—The next generation. *Nat. Rev. Genet.*, 11, 31–46, 2010.
69. Sato, S., Hirakawa, H., Isobe, S., Fukai, E., Watanabe, A., Kato, M., Sequence analysis of the genome of an oil-bearing tree, *Jatropha curcas* L. *DNA Res.*, 18, 65–76, 2011.
70. Shulaev, V., Sargent, D.J., Crowhurst, R.N., Mockler, T.C., Folkerts, O., Delcher, A.L., The genome of woodland strawberry (*Fragaria vesca*). *Nat. Genet.*, 43, 109–116, 2011.
71. Wang, X., Wang, H., Wang, J., Sun, R., Wu, J., Liu, S., The genome of the mesopolyploid crop species Brassica rapa. *Nat. Genet.*, 43, 1035–1039, 2011.
72. Xu, X., Pan, S., Cheng, S., Zhang, B., Mu, D., Ni, P., Genome sequence and analysis of the tuber crop potato. *Nature*, 475, 189–195, 2011.
73. Uchida, N., Sakamoto, T., Kurata, T., Tasaka, M., Identification of EMS-induced causal mutations in a non-reference *Arabidopsis thaliana* accession by whole genome sequencing. *Plant Cell Physiol.*, 52, 716–722, 2011.
74. Yamamoto, T., Nagasaki, H., Yonemaru, J., Ebana, K., Nakajima, M., Shibaya, T., Fine definition of the pedigree haplotypes of closely related rice cultivars by means of genome-wide discovery of single-nucleotide polymorphisms. *BMC Genomics*, 11, 267, 2010.
75. Arai-Kichise, Y., Shiwa, Y., Nagasaki, H., Ebana, K., Yoshikawa, H., Yano, M., Discovery of genome-wide DNA polymorphisms in a landrace cultivar of Japonica rice by whole-genome sequencing. *Plant Cell Physiol.*, 52, 274–282, 2011.
76. Atwell, S., Huang, Y.S., Vilhjalmsson, B.J., Willems, G., Horton, M., Li, Y., Genome-wide association study of 107 phenotypes in *Arabidopsis thaliana* inbred lines. *Nature*, 465, 627–631, 2010.

77. Rounsley, S.D. and Last, R.L., Shotguns and SNPs: How fast and cheap sequencing is revolutionizing plant biology. *Plant J.*, 61, 922–927, 2010.
78. Garber, M., Grabherr, M.G., Guttman, M., Trapnell, C., Computational methods for transcriptome annotation and quantification using RNA-seq. *Nat. Methods.*, 8, 469–477, 2011.
79. Filichkin, S.A., Priest, H.D., Givan, S.A., Shen, R., Bryant, D.W., Fox, S.E., Genome-wide mapping of alternative splicing in *Arabidopsis thaliana. Genome Res.*, 20, 45–58, 2010.
80. Lu, T., Lu, G., Fan, D., Zhu, C., Li, W., Zhao, Q., Function annotation of the rice transcriptome at single-nucleotide resolution by RNA-seq. *Genome Res.*, 20, 1238–1249, 2010.
81. Castruita, M., Casero, D., Karpowicz, S.J., Kropat, J., Vieler, A., Hsieh, S.I., Systems biology approach in *Chlamydomonas* reveals connections between copper nutrition and multiple metabolic steps. *Plant Cell*, 23, 1273–1292, 2011.
82. Gowik, U., Brautigam, A., Weber, K.L., Weber, A.P., Westhoff, P., Evolution of C4 photosynthesis in the genus flaveria: How many and which genes does it take to make C4? *Plant Cell*, 23, 2087–2105, 2011.
83. Fu, C.H., Chen, Y.W., Hsiao, Y.Y., Pan, Z.J., Liu, Z.J., Huang, Y.M., OrchidBase: A collection of sequences of the transcriptome derived from orchids. *Plant Cell Physiol.*, 52, 238–243, 2011.
84. Su, C.L., Chao, Y.T., Alex Chang, Y.C., Chen, W.C., Chen, C.Y., Lee, A.Y., *De novo* assembly of expressed transcripts and global analysis of the *Phalaenopsis aphrodite* transcriptome. *Plant Cell Physiol.*, 52, 1501–1514, 2011.
85. Wong, M.M., Cannon, C.H., Wickneswari, R., Identification of lignin genes and regulatory sequences involved in secondary cell wall formation in *Acacia auriculiformis* and *Acacia mangium* via *de novo* transcriptome sequencing. *BMC Genomics*, 12, 342, 2011.
86. International Rice Genome Sequencing Project, The map-based sequence of the rice genome. *Nature*, 436, 793–800, 2005.
87. Schnable, P.S., Ware, D., Fulton, R.S., Stein, J.C., Wei, F., Pasternak, S., The B73 maize genome: Complexity, diversity, and dynamics. *Science*, 326, 1112–1115, 2009.
88. Bennetzen, J.L., Schmutz, J., Wang, H., Percifeld, R., Hawkins, J., Pontaroli, A.C., Estep, M., Feng, L., Vaughn, J.N., Grimwood, J., Jenkins, J., Barry, K., Lindquist, E., Hellsten, U., Deshpande, S., Wang, X., Wu, X., Mitros, T., Triplett, J., Yang, X., Ye, C.Y., Mauro-Herrera, M., Wang, L., Li, P., Sharma, M., Sharma, R., Ronald, P.C., Panaud, O., Kellogg, E.A., Brutnell, T.P., Doust, A.N., Tuskan, G.A., Rokhsar, D., Devos, K.M., Reference genome sequence of the model plant. *Setaria. Nat. Biotechnol.*, 30, 555–561, 2012.
89. Walbot, V., 10 reasons to be tantalized by the B73 maize genome. *PLoS Genet.*, 5, e1000723, 2009.
90. Li, P. and Brutnell, T.P., *Setaria viridis* and *Setaria italica*, model genetic systems for Panicoid grasses. *J. Exp. Bot.*, 62, 3031–3037, 2011.

91. Lata, C., Gupta, S., Prasad, M., Foxtail millet: A model crop for genetic and genomic studies in bioenergy grasses. *Crit. Rev. Biotechnol.*, 33, 328–343, 2013.
92. Muthuramalingam, P., Krishnan, S.R., Pandian, S., Ramesh, M., Emerging trends on abiotic stress tolerance investigation in crop plants. *Adv. Biotechnol Microbiol.*, 6, 1, 2474–7637, 2017b.
93. Priya, P. and Jain, M., RiceSRTFDB: A database of rice transcription factors containing comprehensive expression, cis-regulatory element and mutant information to facilitate gene function analysis. *Database*, article ID bat027, 2013, https://doi.org/10.1093/database/bat027.
94. Tohge, T., Nishiyama, Y., Hirai, M.Y., Yano, M., Nakajima, J., Awazuhara, M., Functional genomics by integrated analysis of metabolome and transcriptome of Arabidopsis plants over-expressing an MYB transcription factor. *Plant J.*, 42, 218–235, 2005.
95. Hirai, M.Y. and Saito, K., Analysis of systemic sulfur metabolism in plants using integrated '-omics' strategies. *Mol. Biosyst.*, 4, 967–973, 2008.
96. Watanabe, M., Mochida, K., Kato, T., Tabata, S., Yoshimoto, N., Noji, M., Comparative genomics and reverse genetics analysis reveal indispensable functions of the serine acetyltransferase gene family in *Arabidopsis. Plant Cell*, 20, 2484–2496, 2008.
97. Yonekura-Sakakibara, K., Tohge, T., Matsuda, F., Nakabayashi, R., Takayama, H., Niida, R., Comprehensive flavonol profiling and transcriptome co-expression analysis leading to decoding gene–metabolite correlations in *Arabidopsis. Plant Cell*, 20, 2160–2176, 2008.
98. Okazaki, Y., Shimojima, M., Sawada, Y., Toyooka, K., Narisawa, T., Mochida, K., A chloroplastic UDP-glucose pyrophosphorylase from Arabidopsis is the committed enzyme for the first step of sulfolipid biosynthesis. *Plant Cell*, 21, 892–909, 2009.
99. Yamakawa, H. and Hakata, M., Atlas of rice grain filling-related metabolism under high temperature: Joint analysis of metabolome and transcriptome demonstrated inhibition of starch accumulation and induction of amino acid accumulation. *Plant Cell Physiol.*, 51, 795–809, 2010.
100. Izawa, T., Mihara, M., Suzuki, Y., Gupta, M., Itoh, H., Nagano, A.J., Os-GIGANTEA confers robust diurnal rhythms on the global transcriptome of rice in the field. *Plant Cell*, 23, 1741–1755, 2011.
101. Gonzalez, N., De Bodt, S., Sulpice, R., Jikumaru, Y., Chae, E., Dhont, S., Increased leaf size: Different means to an end. *Plant Physiol.*, 153, 1261–1279, 2010.
102. Shingaki-Wells, R.N., Huang, S., Taylor, N.L., Carroll, A.J., Zhou, W., Millar, A.H., Differential molecular responses of rice and wheat coleoptiles to anoxia reveal novel metabolic adaptations in amino acid metabolism for tissue tolerance. *Plant Physiol.*, 156, 1706–1724, 2011.
103. Casati, P., Campi, M., Morrow, D.J., Fernandes, J.F., Walbot, V., Transcriptomic, proteomic and metabolomic analysis of UV-B signaling in maize. *BMC Genomics*, 12, 321, 2011.

104. Kroymann, J., Natural diversity and adaptation in plant secondary metabolism. *Curr. Opin. Plant Biol.*, 14, 246–251, 2011.
105. Delker, C., Poschl, Y., Raschke, A., Ullrich, K., Ettingshausen, S., Hauptmann, V., Natural variation of transcriptional auxin response networks in *Arabidopsis thaliana. Plant Cell*, 22, 2184–2200, 2010.
106. Shoemaker, R., Deng, J., Wang, W., Zhang, K., Allele-specific methylation is prevalent and is contributed by CpG-SNPs in the human genome. *Genome Res.*, 20, 883–889, 2010.
107. Mochida, K. and Shinozaki, K., Advances in omics and bioinformatics tools for systems analyses of plant functions. *Plant Cell Physiol.*, 52, 2017–2038, 2011.

11

OMIC Technologies in Bioethanol Production: An Indian Context

Pulkit A. Srivastava and Ragothaman M. Yennamalli*

Department of Biotechnology and Bioinformatics, Jaypee University of Information Technology, Waknaghat, Himachal Pradesh, India

Abstract
Swift depletion coupled with hazardous effects of nonrenewable energy sources (such as petrol, coal, and diesel) on the environment has inspired all governments around the globe to participate in identification, development, and commercialization of the technology to produce environment-friendly renewable fuels. In this context, recent studies on bioethanol have reflected its viability to be one such energy source due to its production by renewable and sustainable biomass with comparatively less emissions than fossil fuels. Meanwhile, advancement in next-generation sequencing technology and genetic engineering has provided an opportunity to not only inspect into omics data but also to produce hyper-productive strains for efficient degradation of lignocelluloses material at the industry level. This chapter talks about the availability of omics data for microorganism strains and biomass required for bioethanol production in India (third largest energy consumer, second most populated, and counted among the fastest developing countries).

Keywords: Agricultural bioinformatics, biofuel, genome and proteome data, future trends

11.1 Introduction

Natural fossil fuels like natural gas, coal, petroleum, and diesel have been a primary source for the world's economy. However, the over consumption of fossil fuels has not only resulted in depletion of these nonrenewable

**Corresponding author*: ym.ragothaman@juit.ac.in; ragothaman@gmail.com

Rintu Banerjee, Garlapati Vijay Kumar, and S.P. Jeevan Kumar (eds.) OMICS-Based Approaches in Plant Biotechnology, (217–244) © 2019 Scrivener Publishing LLC

resources but also played a major role in the increase of greenhouse gases (GHGs), global warming, and climate change. These limitations accompanied by an increase in the world's energy requirement have made an emphasis to search for alternate energy sources with the following characteristics: renewable, efficient, sustainable, and cost effective with negative or less emission of GHGs [1]. Fuels generated from plant biomass can be considered as substitute of fossil fuels, termed as biofuel. Although, biofuel is similar to hydrocarbons with respect to emission, it can be more environment-friendly if the production and distribution of biofuel are taken care of.

The very first advantage that biofuels hold over conventional fuels is that they are biological molecules and hence biodegradable. This means that they will not persist in the environment for long durations and would not make an area uninhabitable. Secondly, biofuel production can be manipulated to eliminate sulfur to reduce the net impact of acid rain. Further, the limitations imposed by GHG on use of fossil fuels can be easily overcome by the use of biofuel. Lastly, biofuel production would decrease the energy dependence, i.e., countries having land resources to grow biofuel feedstock could produce their own energy and would no longer be dependent on fossil fuels that are limited to few geographical locations in the world.

The production of biofuel has changed over the course of time and biofuel can be broadly classified into three generations, namely, the first, second, and third generations. First-generation biofuel is made using feedstock that is also consumed as human food like sugar, vegetable oil, and/or starch. However, the food vs. fuel debate paved the path for second-generation biofuel, where only sustainable feedstock is considered for biofuel productions, which are either a non-food crop or plant products that are no longer useful for consumption. Since the second-generation biofuel feedstock covers a wider range from forestry to agriculture and waste material enriched in lignocellulosic content that is difficult to extract, it involves advanced conversion technologies. Therefore, the quality of second-generation biofuel is majorly driven by the production routes opted by the industry generally having the following sequence: pretreatment, hydrolysis, fermentation, distillation, and purification of bioethanol.

Moreover, the outcomes of recent advancement and experimentation on algae-based biofuel production in terms of diversity and quantity with lower resource inputs as compared to that of second-generation biofuel production have given more impetus toward the usage of algae for biofuel production. Besides the diversity of carbon sources they can use and quantity they can produce, the simplified path to genetically manipulate them to produce any kind of biofuel is most remarkable. Hence, algae-based biofuel production marks the third generation. However, they require more understanding and further scientific research.

Bioethanol is a biofuel that can be easily produced by renewable and sustainable plant resources that it can be extensively used for transportation with a possibility of minimized particulate production in compression–ignition engines [47]. In Europe and in the United States, bioethanol has been used since the 1900s but was hampered due to its high production cost. However, after the oil crisis in the 1970s, production of bioethanol was resumed with renewed interest. Although bioethanol has lower energy content than petrol (68%), its cleaner combustion and reduction of CO_2 emission level (by 80%) enable it to be a viable alternate source of energy. Also, studies conducted in the past couple of decades have also shown that replacement of fossil fuels by ethanol-based fuels has resulted a drastic reduction in net average of GHGs by 71% for cellulosic ethanol and 31% for bioethanol [2].

Therefore, the scenario presented by second- and third-generation biofuel (mainly bioethanol) production provides an ample area where even a small improvement or discovery could give a huge impetus toward the efficient production of bioethanol. The major areas that need scientific attention are raw materials, organisms that can produce enzymes being used in pretreatment process, and availability of omics data for further research. Hence, the chapter focuses on the availability of omics data and biomass found specifically in India for bioethanol production and the bacterial organism indigenous to India having cellulolytic enzymes.

11.2 Indian Scenario

Bioethanol in India is generally derived from molasses (by-product of sugar production process). However, with the increase in the energy demand of India and its population, it is impossible that molasses alone would be sustainable for bioethanol production. Besides, the debate on first-generation biofuels as they are derived from food crops has raised questions about net energy, water utilization, and net GHG balance. These criticisms have led to increased interest in second-generation biofuels [3]. In India, more than half of its land is used for agriculture. Also, it is the world's first largest producer of rice (https://www.worldatlas.com/articles/the-countries-producing-the-most-rice-in-the-world.html) and the second largest producer of sugarcane (https://www.worldatlas.com/articles/top-sugarcane-producing-countries.html), wheat (https://www.worldatlas.com/articles/top-wheat-producing-countries.html), and sorghum (https://www.worldatlas.com/articles/top-sorghum-producing-countries-in-the-world.html). This leads to production of massive amounts of crop residues that could be used for second-generation biofuel production. Most of these crop residues are burned by farmers that leads to air pollution. Specifically, this event has been seen in the northern states of India. Agricultural waste has not been given much attention and usually ends being composted or grazed by animals. Agricultural waste, such as rice husk, has been traditionally being burned in the states of Haryana, Punjab, UP, and Rajasthan. This burning is not only an inefficient method of eliminating biomass; it also leads to increased levels of particulate matter in the northern parts of India. In November 2015, the state of Delhi reported the highest particulate matter between the ranges 368–122, much higher than the safe limits (http://www.dailypioneer.com/todays-newspaper/delhi-gasps-for-fresh-air-as-particulate-matter-soars-by-350.html and https://www.nature.com/news/new-delhi-car-ban-yields-trove-of-pollution-data-1.19385). The lack of a supply-chain mechanism of utilizing the biomass left after harvesting is a plausible reason as to why the farmers resort to burning. Thus, industries that can utilize the agricultural waste are a potential area of development that needs attention. Moreover, sustainably produced second-generation biofuel can potentially promote rural development and improve economic conditions in developing regions.

There are two essential requirements for successful production of bioethanol in a developing country like India: availability of second-generation plants and crop residues that are sustainable, and identification of native strains of microorganisms (preferably bacteria) that produce cellulases. The latter is most important as characterization and further manipulation of catalysts via protein engineering could lead to preparing and commercialization of enzymatic cocktails. While enzymatic cocktails are available from companies such as Novozymes, there is a lack of enzymatic cocktail of Indian origin that in our opinion has the ability to revolutionize bioethanol production in India.

11.3 Cellulolytic Enzymes Producing Bacterial Strains Isolated from India

An impediment toward the conversion of plant biomass into bioethanol is the enzymatic hydrolysis of lignocellulosic material. It is a process whereby lignocellulosic content from plant biomass is converted into fermentable sugars (Figure 11.1) through the use of enzymes like cellulases and xylanases. However, production of these enzymes at industrial scale by bacterial and fungal strains involves relatively high cost, and hence

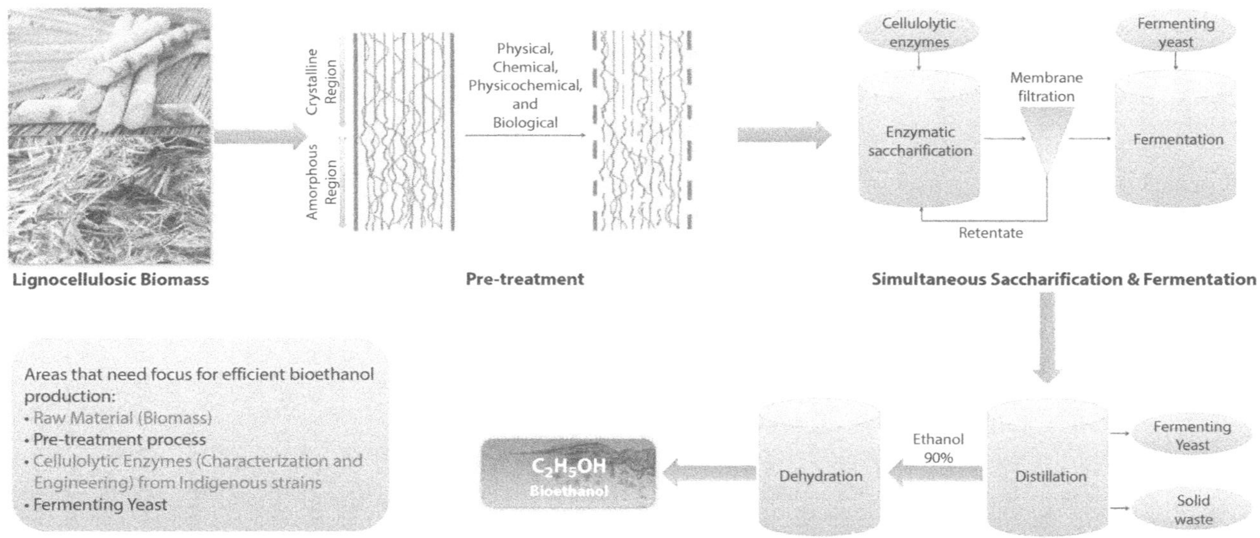

Figure 11.1 Schematic diagrams explaining the bioethanol production from second-generation biomass. Conversion of second-generation sustainable biomass (described in the text) to bioethanol involves multiple steps, where pretreatment renders the cellulose fibers to be exposed and to detach from other recalcitrant polymers (hemicelluloses and lignin). Further saccharification and fermentation followed by distillation delivers the bioethanol that can be blended with gasoline/petrol. However, certain areas require focused attention, with respect to India, such as procurement of biomass and preparing an enzymatic cocktail, which involves complete characterization and protein engineering of cellulases identified from strains isolated in India.

identification of new and active enzymes besides improvement in the efficiency of already known enzymes is one of the major areas of research in bioethanol production.

Cellulases working synergistically with xylanases are the most common enzymes produced by fungal and bacterial strains. But enzymes like amylase, holocellulase, and pectinase not only increase the range of plant biomass that can be degraded but also reduce the cost if they can be produced by only one organism. The greater interest lies in the enzymes produced from thermophiles ranging from 50°C to 90°C as it can be used for harsh industrial process. Moreover, at the industrial scale, it is important to produce copious amount of cellulases, and hence omics study of organisms that can be easily engineered genetically besides having higher growth rate plays a crucial role (bacteria). Here, we give an overview of various promising microorganism strains indigenous to India for bioethanol production. Table 11.1 provides a list of bacterial organisms that can produce the plant biomass-degrading cellulolytic enzymes (such as endo/exoglucanases, cellobiohydrolases, β-glucosidases, and others) with its genome availability.

11.3.1 *Bacillus* Genus of Lignocellulolytic Degrading Enzymes

Bacillus, a genus of gram-positive and rod-shaped bacteria, has around 20 different species (Table 11.1) capable of producing lignocellulolytic degrading enzymes isolated from India. Many of these have the ability to grow at varying pH and temperature making them attractive for industrial purpose of cellulase production.

Among the 20 strains listed, two strains of *Bacillus cereus* and *Bacillus amyloliquefaciens*, respectively, can produce a variety of enzymes like β-glucosidase, cellulase, pectinase, xylanase, and β-1,4-endoglucanase for hydrolysis of plant biomass and have been extensively studied.

11.3.2 *Bhargavaea cecembensis*

Gram-positive, nonmotile, nonspore forming, and rod-shaped bacteria isolated from the Indian Ocean are reported about their cellulase-producing capability [4]. Named after the late Indian scientist, PM Bhargava, the strain of *Bhargavaea cecembensis* was isolated from the hot spring of Manikaran (India). This thermophillic organism is potentially a source of cellulase that can be scaled up for industrial-scale production.

Table 11.1 List of microorganism strains and enzymes isolated reported to be indigenous to India and are proposed to be industrially important for bioethanol production.

Enzyme	Name	Genome	Strain	Reference
Cellulase	*Agrobacterium larrymoorei*	Present	A1	[24]
Amylase, cellulase, xylanse	*Alkalibacterium* sp.	Absent	JN645863	[11]
Endoglucanase, exoglucanase, xylanase	*Amycolatopsis* sp.	Absent	GDS	[25]
Cellulase, xylanase	*Aneurinibacillus aneurinilyticus*	Present	DBT87, DBT15	[26]
Cellulase, β-glucosidase, pectinase, xylanase, β-1,4-endoglucanase	*Bacillus amyloliquefaciens*	Present	IARI-SP-2 KF240848, IARI-R-25	[27, 28]
Cellulase	*Bacillus aquimaris*	Present	NT1	[29]
Cellulase, β-glucosidase, xylanase	*Bacillus baekryungensis*	Absent	IARI-ABL-36	[28]
Cellulase	*Bacillus brevis*	Present	VS-1	[30]
Cellulase	*Bacillus cellulasensis*	Present	NIO-1130T HQ858013	[31]
Amylase, cellulase, pectinase, xylanase, β-glucosidase, β-1,4-endoglucanase	*Bacillus cereus*	Present	JX312575, JN411376, IARI-SP-7 KF240853, DBT14, IARI-L-73	[26–28, 32]

(*Continued*)

Table 11.1 List of microorganism strains and enzymes isolated reported to be indigenous to India and are proposed to be industrially important for bioethanol production. (*Continued*)

Enzyme	Name	Genome	Strain	Reference
Amylase, β-glucosidase, pectinase	*Bacillus firmus*	Present	JN411422, IARI-L-21	[28, 32]
Amylase, cellulase, pectinase, xylanase, β-glucosidase	*Bacillus flexus*	Present	JN645852, IARI-ABL-37	[11, 28]
Amylase, β-glucosidase, pectinase, β-1,4-endoglucanase	*Bacillus licheniformis*	Present	JX312587, IARI-SP-3 KF240849, IARI-ABL-38	[27, 28, 32]
Amylase, cellulase, pectinase, xylanase, β-glucosidase	*Bacillus marisflavi*	Present	IARI-ABL-39	[28]
β-1,4-Endoglucanase	*Bacillus megaterium*	Present	JF343146, IARI-SP-9 KF240855	[27]
Amylase, β-glucosidase, pectinase	*Bacillus mojavensis*	Present	IARI-ABL-40	[28]
β-Glucosidase, pectinase	*Bacillus muralis*	Present	IARI-L-74	[28]
β-1,4-Endoglucanase	*Bacillus mycoides*	Present	JN411283, IARI-SP-8 KF240854	[27, 32]
β-1,4-Endoglucanase	*Bacillus oceanisediminis*	Present	IARI-SP-10 KF240856	[32]

(*Continued*)

Table 11.1 List of microorganism strains and enzymes isolated reported to be indigenous to India and are proposed to be industrially important for bioethanol production. (*Continued*)

Enzyme	Name	Genome	Strain	Reference
Cellulase, β-glucosidase	*Bacillus pumilus*	Present	JN411426, NAIMCC-B01415, IARI-SP-5 KF240851, IARI-L-54	[28, 32, 33]
Amylase, cellulase, xylanase	*Bacillus selenatarsenatis*	Present	JN645858	[11]
β-1,4-Endoglucanase	*Bacillus* sp.	Absent	SV1, 313SI, IARI-SP-4 KF240850	[27, 34, 35]
Amylase, xylanase, β-glucosidase, pectinase	*Bacillus subtilis*	Present	JF343233, JN411353, LFS3, IARI-SP-1 KF240847, IARI-L-69	[27, 28, 32, 36]
β-1,4-Endoglucanase	*Bacillus thuringiensis*	Present	IARI-SP-6 KF240852	[27]
Cellulase	*Bhargavaea cecembensis*	Present	JX312640	[4]
Cellulase	*Cellulomonas cellulans*	Present		[37]
Xylanase	*Cellulosimicrobium cellulans*	Present	CKMX1	[38]
Cellulase	*Citrobacter freundii*	Present		[39]

(*Continued*)

Table 11.1 List of microorganism strains and enzymes isolated reported to be indigenous to India and are proposed to be industrially important for bioethanol production. (*Continued*)

Enzyme	Name	Genome	Strain	Reference
Amylase, cellulase, β-glucosidase, pectinase	*Desemzia incerta*	Present	IARI-L-46	[28]
Amylase, β-glucosidase, pectinase, xylanase	*Exiguobacterium antarcticum*	Present	IARI-L-70	[28]
Amylase, cellulase, β-glucosidase, pectinase, xylanase	*Exiguobacterium indicum*	Present	IARI-R-137	[28]
Amylase, β-glucosidase, pectinase	*Exiguobacterium undae*	Present	IARI-L-116	[28]
Amylase, cellulase, xylanase	*Halobacillus trueperi*	Present	JN645845	[11]
Amylase, cellulase, xylanase	*Halomonas caseinilytica*	Present	JN645870	[11]
Amylase, cellulase, xylanase	*Halomonas muralis*	Present	JN645872	[11]
Amylase, β-glucosidase	*Jeotgalicoccus halotolerans*	Absent	IARI-ABR-5	[28]
Cellulase	*Marinobacter* sp.	Absent	MSI032	[40]
Cellulase	*Paenibacillus barcinonensis*	Absent	MG7	[41]
Cellulase, endoglucanase	*Paenibacillus pabuli*	Present	JF343230	[32]

(*Continued*)

Table 11.1 List of microorganism strains and enzymes isolated reported to be indigenous to India and are proposed to be industrially important for bioethanol production. (*Continued*)

Enzyme	Name	Genome	Strain	Reference
Amylase, cellulase, β-glucosidase	*Paenibacillus* sp.	Present	IHBB 10380, IHB B 3415	[42, 43]
Amylase, cellulase, β-glucosidase, pectinase, xylanase	*Paenibacillus terrae*	Present	JF343212, IARI-L-127	[28, 32]
Amylase, β-glucosidase, pectinase	*Paenibacillus xylanexedens*	Present	IARI-L-76	[28]
Amylase, cellulase, xylanase	*Planococcus antarcticus*	Present	IARI-ABL-9	[28]
β-Glucosidase, pectinase	*Planococcus donghaensis*	Present	IARI-L-39	[28]
Amylase, β-glucosidase, pectinase	*Planococcus kocurii*	Present	IARI-ABL-3	[28]
Amylase, cellulase, β-glucosidase, pectinase, xylanase	*Pontibacillus* sp.	Absent	IARI-ABL-41	[28]
Cellulase	*Pseudomonas fluorescens*	Present		[44]
Amylase, cellulase, xylanase	*Salicola marasensis*	Absent	JN645869	[11]
Amylase, cellulase, xylanase	*Sediminibacillus halophilus*	Present	JN645856	[11]
Cellulase, xylanase	*Serratia rubidaea*	Present	DBT4	[26]
Amylase, β-glucosidase, pectinase	*Sinobaca beijingensis*	Absent	IARI-ABL-18	[28]

(*Continued*)

Table 11.1 List of microorganism strains and enzymes isolated reported to be indigenous to India and are proposed to be industrially important for bioethanol production. (*Continued*)

Enzyme	Name	Genome	Strain	Reference
Amylase, β-glucosidase, pectinase	*Sporosarcina aquimarina*	Absent	IARI-L-77	[28]
Amylase, cellulase, β-glucosidase, pectinase	*Sporosarcina globispora*	Present	IARI-R-111	[28]
Amylase, cellulase, pectinase	*Sporosarcina psychrophila*	Present	IARI-R-110	[28]
Amylase, β-glucosidase, pectinase	*Staphylococcus arlettae*	Present	IARI-ABL-33	[28]
Amylase, β-glucosidase, pectinase	*Staphylococcus xylosus*	Present	IARI-ABR-1	[28]
Xylanase	*Stenotrophomonas maltophilia*	Present	X6 JX220726	[45]
Endoglucanase, cellulase	*Streptomyces albogriseoulus*	Present	C2-221	[7]
Cellulase	*Streptomyces albospinus*	Absent	MTCC 8768	[8]
Cellulase	*Streptomyces griseorubens*	Present	AB184139	[9]
Cellulase	*Streptomyces matensis*	Absent	KF553639	[10]
Endoglucanase, cellulase	*Streptomyces nitrosporeus*	Absent	N2-14	[7]

(*Continued*)

Table 11.1 List of microorganism strains and enzymes isolated reported to be indigenous to India and are proposed to be industrially important for bioethanol production. (*Continued*)

Enzyme	Name	Genome	Strain	Reference
Holocellulase, cellulase, endoglucanase	*Streptomyces* sp.	Absent	ssr-198, S7, M7a, M7b	[5–7]
Cellulase, xylanase	*Streptomyces viridiviolaceus*	Absent	JN645842	[11]
Amylase, cellulase	*Thalassobacillus* sp.	Absent	JN645857	[11]
Amylase, cellulase, β-glucosidase, xylanase	*Virgibacillus halodenitrificans*	Present	IARI-ABR-18	[28]

11.3.3 *Streptomyces* Genus for Hydrolytic Enzymes

Besides being the largest genus for antibiotic production, several species of *Streptomyces* isolated from India have shown to have cellulase-producing ability [5–11]. *S. albogriseoulus*, *S. albospinus*, *S. griseorubens*, *S. matensis*, *S. nitrosporeus*, and *S. viridiviolaceus* have been reported in various reports exhibiting cellulase-producing capability.

Although a significant effort has been made to reduce the production cost of cellulase at industrial scale, it still equals twice the cost of producing ethanol from starch. Hence, realizing the paramount importance of cellulolytic enzymes producing organism, it is necessary to explore new organism and cocktails of enzymes that can be used for production at industrial scale. Besides, modulating transcriptional regulators to produce the desired cocktail of lignocellulolytic degrading enzymes with high yield and productivity can also be an alternative for cost reduction. However, to attain this, one must have in-depth knowledge of the organism's genome, and hence its public availability plays a crucial role.

11.4 Biomass Sources Native to India

In other countries, where bioethanol blending has been successfully implemented (such as the USA and Brazil), second-generation biomass has been actively researched upon and new technologies are developed and simultaneously commercialized for a sustainable bioethanol production. One of the successful crops that have garnered massive attention is switchgrass, a perennial prairie plant mostly found in Midwestern United States. In the case of India, multiple labs have focused on other perennial plants that are not cultivated for food. Here, we list some of those grasses/plants that are found geographically in India and highlight the advances as well.

11.4.1 *Albizia lucida* (Moj)

Albizia lucida, known as Moj, is native to the northeastern part of India. It is a fast-growing tree that is majorly used as fuel by local tribal people. Physiochemical analysis of the plant is carried out to check its use in biofuel production, and it was found that it can be a good alternate for bioethanol production. Specifically, its high cellulose with lower lignin makes it an attractive biomass. Further, the crystallinity content of Moj analyzed using the Ash test was found to be 46.43% and degrades at temperature of 355°C [12].

11.4.2 *Areca catechu* (Betel Nut)

Areca catechu is a species of palm that majorly grows in Asia, parts of east Africa, and tropical Pacific. In India, it is used extensively for cooking and heating. Interestingly, the husk of the betel nut palm is proposed to be used as biomass source. Compared to Moj, the betel nut husk contains higher crystallinity of cellulose and hemicelluloses of 63.84% [12].

11.4.3 *Arundo donax* (Giant Reed)

Arundo donax is a perennial dry energy grass native to Middle East Asia, Indian subcontinent, and Mediterranean Basin. Similar to *Miscanthus*, it has a rhizoidal root system that is tough and forms an extensive knotty network through which it propagates further [13]. It has an ability to grow in erosive and marginal, degraded soils with remarkable increase in biomass in response to nitrogen fertilizers. Additionally, it can improve soil fertility and has higher yield of bioethanol, making it a prominent candidate for bioethanol production [14].

11.4.4 *Pennisetum purpureum* (Napier Grass)

The hybrid variety of Napier grass developed by Tamilnadu Agricultural University, Coimbatore, has a higher biomass yield compared to other perennial grasses. While it has been bred for feedstock purposes, it is highly tolerant to salinity and reduces the pH of the saline soil [15].

11.4.5 *Brassica* Family of Biomass Crops

Brassica carinata (Ethiopian mustard) is a newly introduced semiarid crop in India that is agriculturally significant as it is used as a source for jet engine aviation biofuel [16, 17]. The environmental performance of the biofuel in comparison to conventional gasoline was estimated to be better for ethanol-based application in passenger cars [17].

Another member of *Brassica*, *Brassica napus* (rapeseed), a popular crop for edible oil production, has also been proposed to be a sustainable biomass source for bioethanol production. The investigation of the various factors that influence the saccharification of rapeseed stalks concluded that the lower the quantity of pectin and arabinogalactans, the better the saccharification [18].

Multiple studies have been reported on the common mustard plant, a widely grown agricultural crop in India and South Asia. Specifically, the structural features of dilute acid, steam exploded, and alkali-pretreated mustard stalks and their impacts on enzymatic hydrolysis are investigated [19].

11.4.6 *Cajanus cajan* (Pigeon Pea)/*Cenchrus americanus* (Pearl Millet)/*Corchorus capsularis* (Jute)/*Lens culinaris* (Lentil)/*Saccharum officinarum* (Sugarcane)/*Triticum* sp. (Wheat)/*Zea mays* (Maize)

These crops are grown in huge quantity in India covering almost all parts of the country. Various parts of plants like straw, stalk, waste, and husk are utilized for the production of bioethanol. The roadmap for biofuel production in India [3] shows the total net ethanol availability in 2010/2011 as 27.9 billion liters and estimates an increase in the availability (37.3% by 2020/2021) with the increase in production of these plants. As stated earlier, among the world's largest producer of rice, pigeon pea, pearl millet, jute, sugarcane, wheat, and maize, there are high rates of having biomass resource from various parts that are not edible or form waste.

11.4.7 *Medicago sativa* (Alfalfa)

Alfalfa is a perennial flowering plant of pea family grown almost all around the world. Due to its rich lignocellulosic content and being herbaceous non-edible energy crop, it is considered as a potential source of biomass for bioethanol production. Though not produced in large quantity, West and South-Central India is an ideal location for the production of alfalfa.

11.4.8 *Manihot esculenta* (Cassava)/*Salix viminalis* (Basket Willow)/*Setaria italica* (Foxtail Millet)/*Setaria viridis* (Green Foxtail)

With the efficient and cost-effective lignocellulosic bioethanol production, bioethanol industry would need a more reliable plant biomass that can be produced using minimum water, fertilizer, and agricultural land. To address this issue, various C4 energy plants could be used for bioethanol production as they use an efficient system when compared to C3 plants, hence yielding high amount of biomass. In a recent article [20], the authors investigated the optimization techniques for biomass production and feedstock quality with respect to bioethanol production in C4 plants.

11.4.9 *Vetiveria zizanioides* (Vetiver or Khas)

In many parts of India, Khas grass grows wild and is an important plant for industrial and medicinal uses. Khas/Vetiver's roots are used for cooling purposes, and oil is extracted as well, which is prized for its cooling effect (http://www.vetiver.org/).

Harvesting of Khas/Vetiver's roots is done by uprooting the entire plant and removing the roots after cleaning. Interestingly, systematically cultivated Khas/Vetiver is harvested by removing the roots after 10–12 months, mostly for medicinal purposes. However, the pulp of Khas/Vetiver goes into paper and straw board production.

11.4.10 Millets and *Sorghum bicolor* (Sorghum)

Although there are plants and crop wastes that are being utilized for biofuel production in India (like wheat bran, wheat straw, sugarcane bagasse, rice straw, and rice husk), there are plants that have been investigated by few labs for their utilization in biofuel production. However, the lack of omics (genome, transcriptome, and proteome) data for these plants is acting as a roadblock in the efficient production of bioethanol. Table 11.1 lists a total of 39 Indian plants that have been investigated for their bioethanol production capability and availability of the omics data on the National Center for Biotechnology Information (NCBI) portal. As discussed in Ref. [3], wastes from sugarcane, wheat, maize, jute, pearl millet, pigeon pea, and lentil are being extensively used for bioethanol production. However, plants like Moj, Bonbogori and German grass, foxtail millet, basket willow, and green foxtail can be investigated in depth to evaluate their efficiency in the production of bioethanol. Specifically, sorghum has received a lot of attention recently [21, 22]. Therefore, availability of genome, transcriptome, and proteome of plants not only would pave a better way to understand and overcome the complexity involved in bioethanol production at the industrial level but also would provide various reliable resources from all over the country to meet the challenge of fuels. Moreover, in a country like India with vast land resources to grow biofuel feedstock, it has a potential to not only produce the energy for itself but also distribute the bioethanol to other countries.

11.5 Omics Data and Its Application to Bioethanol Production

"Omics" is a general term applied to biological data that result after sequencing the genome of a particular organism. Derived from the suffix "ome," omics has multiple flavors, such as transcriptomics, metabolomics, etc. With the advent of next-generation sequencing techniques, it has been a matter of hours for any organism's genome to be elucidated. The implication of genome sequence availability is quite long reaching. For example, once the genome of an organism is known, the functional genomics (process of identifying all the genes and annotating their functions) step enables to study specific genes that influence in a specific pathway.

In the case of bioethanol production, the availability of omics data is a major limitation. Among the ~65 strains isolated from India, there are only 49 species' genome available (73% of strains) in the NCBI. While published reports emphasize on the specific enzyme that was isolated and characterized from the strains, they do not report the genome availability or sequencing efforts. The genome data available for most of the ~65 strains of Indian origin can be analyzed with standard organism's genome available in the NCBI using BLAST and other genome comparative tools. As a general rule of thumb, these genome data can be used as the strain's genome, with the caveat that there might be some unique genes present in the Indian strain that are missing in the standard genome's data. To overcome this limitation, it would be ideal to have the strain's genome sequenced by NGS and deposited in the NCBI. Also, many of the metagenome efforts have led to identification of indigenous strains that have possible implications in bioethanol production in India.

The other way around would be a metagenomic study of microorganisms with characteristics like low pH and sustainability at higher temperature. The strains that produce cellulases and possess any one of the above characteristic can be further explored, i.e., what regulatory factors play a critical role in production of cellulase and other plant-biomass degradation. However, all research to manipulate the production of cellulase at industrial scale requires the presence of omics data. Additionally, availability of genomic information would also increase the computational studies so that it can be well integrated with experimental studies to not only increase the production of cellulase but also at the same time decrease the experimental cost.

Similarly, among the biomass sources that are native to India, there are only 15 genome data available [such as *Brassica juncea* (mustard), *Brassica napus* (rapeseed), *Brassica rapa* (field mustard), *Cajanus cajan* (pigeon pea), *Cenchrus americanus* (pearl millet), *Corchorus capsularis* (jute), *Gossypium barbadense* (Brazilian cotton), *Gossypium hirsutum* (upland cotton), *Manihot esculenta* (cassava), *Populus euphratica*, *Populus trichocarpa* (black cottonwood), *Setaria italica* (foxtail millet), *Sorghum bicolor* (sorghum), *Triticum* sp. (wheat), and *Zea mays* (maize)]. Among these 15, only 10 organisms' proteome data are available. However, there are 21 transcriptome data available for the biomass sources. For strain engineering and development, it is essential that the organism's genome, transcriptome, and proteome are all available. Thus, the availability of all three data for sorghum, maize, upland cotton, and Brassica species will enable for further research (Table 11.2).

Table 11.2 Availability of genome, proteome, and transcriptome data for biomass sourced from India.

S. No.	Name (common name)	Genome	Transcriptome	Proteome	References
1	*Albizia lucidia* (Moj)	Absent	Absent	Absent	[12]
2	*Areca catechu* (betel nut)	Absent	Present	Absent	[12]
3	*Arundo donax* (giant reed)	Absent	Absent	Absent	[13, 46]
4	*Bambusoideae* (bamboo)	Absent	Present	Absent	–
5	*Brassica carinata*	Absent	Present	Absent	[17]
6	*Brassica juncea* (Mustard)	Present	Present	Absent	[18]
7	*Brassica napus* (rapeseed)	Present	Present	Present	[18]
8	*Brassica rapa* (field mustard)	Present	Present	Present	[19]
9	*Cajanus cajan* (pigeon pea)	Present	Absent	Absent	[3]
10	*Cenchrus americanus* (pearl millet)	Present	Present	Absent	[3]
11	*Chrysopogon zizanioides* (Vetiver)	Absent	Absent	Absent	–
12	*Corchorus capsularis* (jute)	Present	Absent	Present	[3]
13	*Cynodon dactylon* (Bermuda grass)	Absent	Absent	Absent	[46]

(Continued)

Table 11.2 Availability of genome, proteome, and transcriptome data for biomass sourced from India. (*Continued*)

S. No.	Name (common name)	Genome	Transcriptome	Proteome	References
14	*Echinochloa polystachya* (German grass)	Absent	Absent	Absent	[3]
15	*Gossypium barbadense* (Brazilian cotton)	Present	Present	Absent	–
16	*Gossypium hirsutum* (upland cotton)	Present	Present	Present	[48, 49]
17	*Helianthus tuberosus* (sun root)	Absent	Absent	Absent	[50]
18	*Hibiscus cannabinus* (Kenaf)	Absent	Present	Absent	[51]
19	*Lantana camara* (red sage)	Absent	Absent	Absent	[52]
20	*Lens culinaris* (lentil)	Absent	Present	Absent	[3]
21	*Linum perenne* (blue flax)	Absent	Absent	Absent	–
22	*Manihot esculenta* (cassava)	Present	Present	Present	[20]
23	*Medicago sativa* (alfalfa)	Absent	Present	Absent	[20]

(*Continued*)

Table 11.2 Availability of genome, proteome, and transcriptome data for biomass sourced from India. (*Continued*)

S. No.	Name (common name)	Genome	Transcriptome	Proteome	References
24	*Miscanthus fuscus* (elephant grass)	Absent	Absent	Absent	[20]
25	*Paspalum notatum* (Bahia grass)	Absent	Absent	Absent	[46]
26	*Pennisetum purpureum* (Napier grass)	Absent	Absent	Absent	[46]
27	*Phalaris arundinacea* (reed canary grass)	Absent	Absent	Absent	[13, 53]
28	*Phragmites australis* (common reed)	Absent	Absent	Absent	[54]
29	*Populus euphratica*	Present	Present	Present	[3]
30	*Populus trichocarpa* (black cottonwood)	Present	Present	Present	[3]
31	*Saccharum officinarum* (sugarcane)	Absent	Present	Absent	[3]
32	*Salix viminalis* (basket willow)	Absent	Present	Absent	[3]
33	*Sesbania bispinosa* (prickly Sesban)	Absent	Absent	Absent	–
34	*Setaria italica* (foxtail millet)	Present	Present	Present	[3]

(*Continued*)

Table 11.2 Availability of genome, proteome, and transcriptome data for biomass sourced from India. (*Continued*)

S. No.	Name (common name)	Genome	Transcriptome	Proteome	References
35	*Setaria viridis* (green foxtail)	Absent	Absent	Absent	[3]
36	*Sorghum bicolor* (sorghum)	Present	Present	Present	[21, 22]
37	*Triticum* sp. (wheat)	Present	Present	Absent	[3]
38	*Zea mays* (maize)	Present	Present	Present	[3]
39	*Ziziphus rugosa* (Bonbogori)	Absent	Absent	Absent	[12]

Besides, the higher content of lignin hinders the production of bioethanol and the reason can be imparted to complexity involved in separating lignin from other polymers for efficient enzymatic hydrolysis [23]. Hence, one can think of genetically engineering fuel plants to generate less amount of lignin as compared to other biopolymers and would require information of regulatory factors involved in the synthesis of lignin. Though the biosynthetic pathway of lignin has been deciphered in great depth, regulating its production in various plants would surely involve the knowledge of genome of that particular plant. In addition, one can go for transcriptomics study to know what all genes and regulatory factors are involved in higher production of cellulose and xylose in a given plant so as to increase the content of bioethanol in the long term.

11.6 Conclusion

The success of bioethanol blending as proposed by the Government of India is dependent on two things: (1) availability of biomass sources that are sustainable and (2) characterization of catalysts. For the former, the major limitation is incentive to the farmers to provide the industry enough biomass (such as second-generation biomass) to make it sustainable. In this regard, this chapter summarizes not only agricultural crop residues but also plants that are used majorly as feed or for other purposes. With regards to catalysts, it has been shown that for a developing country like India, there are abundant strains available that produce cellulases and other associated enzymes. Some of these enzymes are attractive for industry in terms of thermostability and other properties that are deemed desirable. At the same time, some of these organisms' genome has been sequenced, and thus there is availability of omics data that can be tapped for further characterization and development of strains that are highly suitable for industrial purposes. In the future, there is a need to generate more omics data (genome, transcriptome, and proteome) to be generated for both the biomass source and strains native to India.

References

1. Nigam P.S., Singh A., Production of liquid biofuels from renewable resources, *Prog. Energy Combust. Sci.*, 37, 52–68, 2011.
2. Koh, L., Ghazoul, J., Biofuels, biodiversity, and people: Understanding the conflicts and finding opportunities. *Biol. Conserv.*, 141, 10, 2450–2460, 2008.
3. Dhar, S., Shukla, P., Low carbon scenarios for transport in India: Co-benefits analysis. *Energy Policy*, 81, 186–198, 2015.

4. Manorama,R., Pindi,P., Reddy,G., Shivaji,S., *Bhargavaea cecembensis* gen. nov., sp. nov., isolated from the Chagos-Laccadive ridge system in the Indian Ocean. *Int. J. Syst. Evol. Microbiol*, 59, 10, 2618–2623, 2009.
5. Singh, S., Tiwari, R., Renuse, S., Pranaw, K., Nain, L., Proteomic analysis of Streptomyces sp. ssr-198 grown on paddy straw. *J. Basic Microbiol.*, 55, 6, 790–797, 2015.
6. Rathnan, R., and Ambili, M., Cellulase Enzyme Production by Streptomyces Sp Using Fruit Waste as Substrate. *Aust. J. Basic & Appl. Sci.*, 5, 12, 1114–1118, 2011.
7. Chellapandi P., Jani H.M., Production of endoglucanase by the native strains of Streptomyces isolates in submerged fermentation., *Braz. J. Microbiol.*, 39, 122–127, 2008.
8. Prasad, P., Bedi, S., Singh, T., *In vitro* Cellulose Rich Organic Material Degradation by Cellulolytic *Streptomyces albospinus* (MTCC 8768). *MJM*, 2012.
9. Prasad, P., Bedi, S., Characterization of a novel thermophilic cellulase producing strain *Streptomyces matensis* strain St-5. *IJCMAS*, 3, 3, 74–88, 2014.
10. Prasad P., Singh T., Bedi S., Characterization of the cellulolytic enzyme produced by *Streptomyces griseorubens* (Accession No. AB184139) isolated from Indian soil, *J. King Saud Univ. - Sci.*, 25, 245–250, 2013.
11. Sahay, H., Mahfooz, S., Singh, A., Singh, S., Kaushik, R., Saxena, A., Arora, D., Exploration and characterization of agriculturally and industrially important haloalkaliphilic bacteria from environmental samples of hypersaline Sambhar lake, India. *World J. Microbio.l Biotechnol.*, 28, 11, 3207–3217, 2012.
12. Sasmal, S., Goud, V., Mohanty, K., Characterization of biomasses available in the region of North-East India for production of biofuels. *Biomass and Bioenergy*, 45, 212–220, 2012.
13. Odero, D., Gilbert, R., Ferrell, J., Helsel, Z., *Production of Giant Reed for Biofuel.* University of Florida, Florida, 2008.
14. Aliberti, A., Ventorino, V., Robertiello, A., Galasso, M., Blaiotta, G., Comite, E., Faraco, V., Pepe, O., Effect of Cellulase, Substrate Concentrations, and Configuration Processes on Cellulosic Ethanol Production from Pretreated *Arundo donax*. *BioRes*, 12, 3, 2017.
15. Ma, C., Naidu, R., Liu, F., Lin, C., Ming, H., Influence of hybrid giant Napier grass on salt and nutrient distributions with depth in a saline soil. *Biodegradation*, 23, 6, 907–916, 2012.
16. Lal, B., Rana, K., Rana, D., Shivay, Y., Sharma, D., Meena, B., Gautam, P., Biomass, yield, quality and moisture use of *Brassica carinata* as influenced by intercropping with chickpea under semiarid tropics. *JSSAS*, 2017.
17. González-García, S., Gasol, C., Gabarrell, X., Rieradevall, J., Moreira, M., Feijoo, G., Environmental aspects of ethanol-based fuels from *Brassica carinata*: A case study of second generation ethanol. *Renew. Sust. Energ. Rev.*, 13, 9, 2613–2620, 2009.
18. Wood, I., Wellner, N., Elliston, A., Wilson, D., Bancroft, I., Waldron, K., Effect of *Brassica napus* cultivar on cellulosic ethanol yield. *Biotechnol. Biofuels*, 8, 1, 2015.

19. Kapoor, M., Raj, T. , Vijayaraj, M., Chopra, A., Gupta, R., Tuli, D., Kumar, R., Structural features of dilute acid, steam exploded, and alkali pretreated mustard stalk and their impact on enzymatic hydrolysis. *Carbohydr. Polym.*, 124, 265–273, 2015.
20. Byrt, C., Grof, C., Furbank, R., C4 Plants as Biofuel Feedstocks: Optimising Biomass Production and Feedstock Quality from a Lignocellulosic Perspective Free Access. *J. Integr. Plant Biol.*, 53, 2, 120–135, 2011.
21. Lingle, S., Opportunities and Challenges of Sweet Sorghum as a Feedstock for Biofuel. *ACS Symp. Ser.*, 177–188, 2010.
22. Qi, C., Yadama, V., Guo, K., Wolcott, M., Thermal stability evaluation of sweet sorghum fiber and degradation simulation during hot pressing of sweet sorghum–thermoplastic composite panels. *Ind. Crops Prod.*, 69, 335–343, 2015.
23. Welker, C., Balasubramanian, V., Petti, C., Rai, K., DeBolt, S., Mendu, V., Engineering Plant Biomass Lignin Content and Composition for Biofuels and Bioproducts. *Energies*, 8, 8, 7654–7676, 2015.
24. Bera, A., Ghosh, A., Mukhopadhyay, A., Chattopadhyay, D., Chakrabarti, K., Improvement of degummed ramie fiber properties upon treatment with cellulase secreting immobilized *A. larrymoorei* A1. *Bioprocess Biosyst. Eng.*, 38, 2, 341–351, 2014.
25. Kshirsagar, S., Saratale, G., Saratale, R., Govindwar, S., Oh, M., An isolated Amycolatopsis sp. GDS for cellulase and xylanase production using agricultural waste biomass. *J. Appl. Microbiol.*, 120, 1, 112–125, 2015.
26. Asem, D., Leo, V., Passari, A., Tonsing, M., Joshi, J., Uthandi, S., Hashem, A., Abd_Allah, E., Singh, B., Evaluation of gastrointestinal bacterial population for the production of holocellulose enzymes for biomass deconstruction. *PLoS One*, 12, 10, p. e0186355, 2017.
27. Pandey, S., Tiwari, R., Singh, S., Nain, L., Saxena, A., Evaluation of beta-1,4- Endoglucanases Produced by Bacilli Isolated from Paper and Pulp Mill Effluents Irrigated Soil. *J. Microbiol. Biotechnol.*, 24, 8, 1073–1080, 2014.
28. Yadav, A., Sachan, S., Verma, P., Kaushik, R., Saxena, A., Cold active hydrolytic enzymes production by psychrotrophic Bacilli isolated from three sub-glacial lakes of NW Indian Himalayas. *J. Basic Microbiol.*, 56, 3, 294–307, 2015.
29. Trivedi, N., Gupta, V., Kumar, M., Kumari, P., Reddy, C., Jha, B., Solvent tolerant marine bacterium *Bacillus aquimaris* secreting organic solvent stable alkaline cellulose. *Chemosphere*, 83, 5, 706–712, 2011.
30. Singh, V., Kumar, A., Production and purification of an extracellular cellulase from *Bacillus brevis* vs-1. *IUBMB Life*, 45, 3, 443–452, 1998.
31. Mawlankar, R., Thorat, M., Krishnamurthi, S., Dastager, S., *Bacillus cellulasensis* sp. nov., isolated from marine sediment. *Arch. Microbiol*, 198, 1, 83–89, 2015.
32. Pandey, S., Singh, S., Yadav, A., Nain, L., Saxena, A., Phylogenetic Diversity and Characterization of Novel and Efficient Cellulase Producing Bacterial Isolates from Various Extreme Environments. *Biosci. Biotechnol. Biochem.*, 77, 7, 1474–1480, 2013.

33. Padaria, J., Sarkar, K., Lone, S., Srivastava, S., Molecular characterization of cellulose- degrading *Bacillus pumilus* from the soil of tea garden, Darjeeling hills, India. *J. Environ. Biol.*, 35, 3, p. 551, 2014.
34. Thomas, L., Ram, H., Singh, V., Inducible cellulase production from an organic solvent tolerant Bacillus sp. SV1 and evolutionary divergence of endoglucanase in different species of the genus Bacillus. *Braz. J. Microbiol.*, 49, 2, 429–442, 2018.
35. Goyal, V., Mittal, A., Bhuwal, A., Singh, G., Yadav, A., Aggarwal, N., Parametric Optimization of Cultural Conditions for Carboxymethyl Cellulase Production Using Pretreated Rice Straw by Bacillus sp. 313SI under Stationary and Shaking Conditions. *Biotechnol. Res. Int.*, 2014, 1–7, 2014.
36. Rawat, R., Tewari, L., Purification and characterization of an acidothermophilic cellulase enzyme produced by Bacillus subtilis strain LFS3. *Extremophiles*, 16, 4, 637–644, 2012.
37. Premkumar, J., Sudhakar, T., Srikiran, K., Isolation and identification of *Cellulomonas cellulans* from silver fish and characterization of cellulase enzyme. *J. Chem. Pharm. Res.*, 7, 1, 346–349, 2015.
38. Walia, A., Mehta, P., Chauhan, A., Kulshrestha, S., Shirkot, C., Purification and characterization of cellulase-free low molecular weight endo β-1,4 xylanase from an alkalophilic *Cellulosimicrobium cellulans* CKMX1 isolated from mushroom compost. *World J. Microbio.l Biotechnol.*, 30, 10, 2597–2608, 2014.
39. Pawar, K., Dar, M., Rajput, B., Kulkarni, G., Enrichment and Identification of Cellulolytic Bacteria from the Gastrointestinal Tract of Giant African Snail, Achatina fulica. *Appl. Biochem. Biotechnol.*, 175, 4, 1971–1980, 2014.
40. Shanmughapriya, S., Kiran, G., Selvin, J., Thomas, T., Rani, C., Optimization, Purification, and Characterization of Extracellular Mesophilic Alkaline Cellulase from Sponge-Associated Marinobacter sp. MSI032. *Appl. Biochem. Biotechnol.*, 162, 3, 625–640, 2009.
41. Asha, B., Revathi, M., Yadav, A., Sakthivel, N., Purification and Characterization of a Thermophilic Cellulase from a Novel Cellulolytic Strain, *Paenibacillus barcinonensis. J. Microbio.l Biotechnol.*, 22, 11, 1501–1509, 2012.
42. Pal, M., Swarnkar, M., Thakur, R., Kiran, S., Chhibber, S., Singh, A., Gulati, A., Complete Genome Sequence of Paenibacillus sp. Strain IHBB 10380 Using PacBio Single-Molecule Real-Time Sequencing Technology. *Genome Announc.*, 3, 2, 2015.
43. Dhar, H., Swarnkar, M., Gulati, A., Singh, A., Kasana, R., Draft Genome Sequence of a Cellulase-Producing Psychrotrophic Paenibacillus Strain, IHB B 3415, Isolated from the Cold Environment of the Western Himalayas, India. *Genome Announc.*, 3, 1, 2015.
44. Sethi, S., Datta, A., Gupta, B., Gupta, S., Optimization of Cellulase Production from Bacteria Isolated from Soil. *ISRN Biotechnol.*, 2013, 1–7, 2013.
45. Raj, A., Kumar, S., Singh, S., A Highly Thermostable Xylanase from *Stenotrophomonas maltophilia*: Purification and Partial Characterization. *Enzyme Res.*, 2013, 1–8, 2013.

46. Hattori, T., Morita, S., Energy Crops for Sustainable Bioethanol Production; Which, Where and How? *Plant Prod. Sci.*, 13, 3, 221–234, 2010.
47. Balat M., Balat H., Recent trends in global production and utilization of bio-ethanol fuel. *Applied Energy*, 86, 11, 2273–2282, 2009.
48. Baig, M., Dharmadhikari, S., Bioethanol Production from Enzymatically Hydrolysed Cotton Stalk: One Approach Towards Sustainable Energy Development. *Curr. World Environ.*, 9, 3, 940–946, 2014.
49. Wang, M., Zhou, D., Wang, Y., Wei, S., Yang, W., Kuang, M., Ma, L., Fang, D., Xu, S., Du, S., Bioethanol production from cotton stalk: A comparative study of various pretreatments. *Fuel*, 184, 527–532, 2016.
50. Song, Y., Wi, S., Kim, H., Bae, H., Cellulosic bioethanol production from Jerusalem artichoke (Helianthus tuberosus L.) using hydrogen peroxide-acetic acid (HPAC) pretreatment. *Bioresour. Technol.*, 214, 30–36, 2016.
51. Saba, N., Jawaid, M., Hakeem, K., Paridah, M., Khalina, A., Alothman, O., Potential of bioenergy production from industrial kenaf (*Hibiscus cannabinus* L.) based on Malaysian perspective. *Renew. Sust. Energ. Rev.*, 42, 446–459, 2015.
52. Kuhad, R., Gupta, R., Khasa, Y., Singh, A., Bioethanol production from Lantana camara (red sage): Pretreatment, saccharification and fermentation. *Bioresour. Technol.*, 101, 21, 8348–8354, 2010.
53. Riffaldi, R., Saviozzi, A., Cardelli, R., Bulleri, F., Angelini, L., Comparison of Soil Organic-Matter Characteristics under the Energy Crop Giant Reed, Cropping Sequence and Natural Grass. *Commun. Soil Sci. Plant Anal.*, 41, 2, 173–180, 2010.
54. Cotana, F., Cavalaglio, G., Pisello, A., Gelosia, M., Ingles, D., Pompili, E., Sustainable Ethanol Production from Common Reed (*Phragmites australis*) through Simultaneous Saccharification and Fermentation. *Sustainability*, 7, 9, 12149–12163, 2015.

Part 4

ADVANCES IN CROP IMPROVEMENT: EMERGING TECHNOLOGIES

12

Genome Editing: New Breeding Technologies in Plants

Kalyani M. Barbadikar[1*†], Supriya B. Aglawe[1†], Satendra K. Mangrauthia[1], M. Sheshu Madhav[1] and S.P. Jeevan Kumar[2]

[1]Biotechnology Section, ICAR–Indian Institute of Rice Research (IIRR), Rajendranagar, Hyderabad, Telangana, India
[2]Biotechnology Section, ICAR–Indian Institute of Seed Science (IISS), Mau, Uttar Pradesh, India

Abstract
The changing climate, increasing population, malnutrition, and extreme and erratic weather conditions present several challenges to many crops. At the same time, a plateau in yield increment has been witnessed in major crops, which call for urgent need of quick, sustainable and eco-friendly technologies. In this respect, genome editing (GE), a new breeding technology (NBT), has shown potential to transform not only in fundamental research of plant biology but more importantly also for addressing growing challenges of food security. GE exploits the homologous recombination (HR) or nonhomologous end-joining (NHEJ) based repair mechanism already existing in organisms. The meganucleases (MgN), zinc-finger nucleases (ZFNs), transcription activator-like effector nucleases (TALENs), and cluster regularly interspaced short palindromic repeats (CRISPR/Cas) have been used for GE. It can be used for desired modification of the native endogenous gene or targeted insertion of cis/trans gene. Applications of this technique are continuously increasing since its discovery, and several developments in original technologies have been brought for precision and accuracy. In this chapter, we present overall view of GE tools, their advantages, shortcomings, practical considerations, and applications in crop improvement.

**Corresponding author*: kalyaniaau@gmail.com
[†]*Equal contribution as first author*

Rintu Banerjee, Garlapati Vijay Kumar, and S.P. Jeevan Kumar (eds.) OMICS-Based Approaches in Plant Biotechnology, (247–286) © 2019 Scrivener Publishing LLC

Keywords: Genome editing, breeding technique, crop improvement, transcriptional reprogramming, epigenetic modification, disease resistance, biotic/abiotic stress tolerance, biosafety

12.1 Introduction: Genome Editing

The global climate change with increased temperature, water scarcity, and proliferating population poses a serious problem for food security and nutrition. Landmark achievements have been recorded using conventional and molecular plant breeding for crop improvement. Withal, new, targeted, and precise techniques called new breeding techniques (NBTs) are needed for expedited improvement of crops for various traits of interest to overcome present challenges. Although through genetic engineering, transgenic crops for various economic traits have been developed, only few products are accepted for commercial cultivation in certain parts of the world. With NBTs, the limitations of conventional plant breeding and biosafety regulations related to transgenic technology can be effectively met. NBTs include sites directed nucleases, *viz.*, meganucleases, zinc-finger nucleases (ZFNs), transcription activator-like effector nuclease (TALEN) and cluster regularly interspaced short palindromic repeats (CRISPR), cisgenics, oligonucleotide directed mutagenesis, grafting, reverse breeding, RNA-dependent DNA methylation, and agroinfiltration. Genome editing (GE) is a novel mode of genetic engineering in which DNA is inserted, replaced, or removed from a genome using artificially engineered nucleases, or molecular scissors. These artificially engineered nucleases are able to cut genome at a specific site, which produces site-specific GE. The nucleases create specific double-stranded breaks (DSBs) at desired locations in the genome and harness the cell's endogenous mechanisms to repair the induced break by natural processes of homologous recombination (HR) and nonhomologous end-joining (NHEJ). Paszkowski and co-workers achieved GE for the first time in *Nicotiana tabacum* by direct gene transfer to tobacco protoplasts [1]. It took another 25 years to develop the information and tools necessary to overcome the major challenges associated with GE, at least for some plant species [2].

GE offers advantage over traditional mutation breeding as it creates the variation/mutations in the particular target site and facilitates site-specific integration of gene of interest in the genome. From the last few years, genetically modified (GM) crops have given rise to great hue and cry. The major public concern is about inserting foreign genes in the plants. Today, so many GM crops are ready for field and bio-safety trials and many are in pipeline, but due to great public concern, many countries made regulatory consideration more stringent while some countries banned field trials of GM crops. In GE, "cisgenesis" is one of the good alternatives for genetic engineering. Identification of many plant endogenous genes that are generally hijacked by pathogens/insects and used for their survival and replication will facilitate development of disease/insect-resistant plant through GE. Likewise, many positive and negative regulators of important agronomic traits are known; one can target such genes through GE for development of better phenotypes such as high yield, dwarfness, good nutritional value, and biotic and abiotic tolerance. Single nucleotide polymorphism (SNP) is a change at single nucleotide in the genome and many SNPs are identified in important crops such as rice, maize, soybean, wheat, barley, sorghum, pigeon pea, chick pea, *etc.* and can also be exploited for development of better crops through GE. One can employ GE techniques for transcriptional reprogramming of endogenous genes, which is one of the most important and useful potential applications of GE in crop improvement. One more potential application of GE attracts plant scientists across the globe for its use in development of non-GM plants with desired traits (nontransgenic genome modification). This may reduce public concern, anxiety, and regulatory costs of GM crops. In other words, one can say that the next generation of transgenic plants developed through GE may truly be described as genetically engineered organisms (GEOs) as opposed to genetically modified organisms (GMOs).

The possibility of making targeted modifications of the DNA in a living cell using sequence-specific nucleases was first described in 2002 [3]. Since then, several of these nucleases have been discovered, such as meganucleases (MgN), ZFNs, and TALENs. Most recently, an efficient GE tool has been developed based on CRISPR associated nucleases system. CRISPR/Cas system has once again caught the attention of researchers across the world on GE. The ease, low cost, and speed of designing of GE techniques make GE the most appropriate and feasible for plant improvement. This chapter presents an overview of different GE techniques and their advantages, disadvantages, and practical considerations while designing an experiment and application in some of the model plants and important crops for GE.

12.2 GE: The Basics

DSB repair mechanics is basic to the use of nucleases in GE. GE is made possible by creation of DSB and its subsequent repair by exploiting cells' natural repair mechanism. So it is important to understand the cell's DNA DSB repair mechanisms to understand the techniques of GE. The NHEJ and homology directed repair (HR) are the known DSB repair pathways that occur in all organisms. For nuclease-based GE, both repair pathways, *i.e.*, NHEJ and HR, can be exploited.

12.2.1 Nonhomologous End-Joining (NHEJ)

NHEJ functions in all kind of cells, from bacteria to man, where a homologous template is not needed. It involves excision of damage or mismatched nucleotides by nucleases and DNA resynthesis by polymerases. The DNA ligase IV complex performs the ligation step of repair. There are two repair systems that have two sub-pathways: the classical NHEJ pathway and the microhomology-based alternative NHEJ pathway. In classical NHEJ (Figure 12.1a), proteins like Ku70 and Ku80 bind to broken DNA ends and are joined together by a ligase resulting in insertion or deletion (indel) of nucleotides. In microhomology-based alternative NHEJ, 5′ ends are cut until 3′ overhangs with homology are created. At their complementary sequence, DNA strands are bind, and flaps of nonhomologous DNA are excised. This results in deletions because the DNA between homologous sections is removed. This pathway mostly results

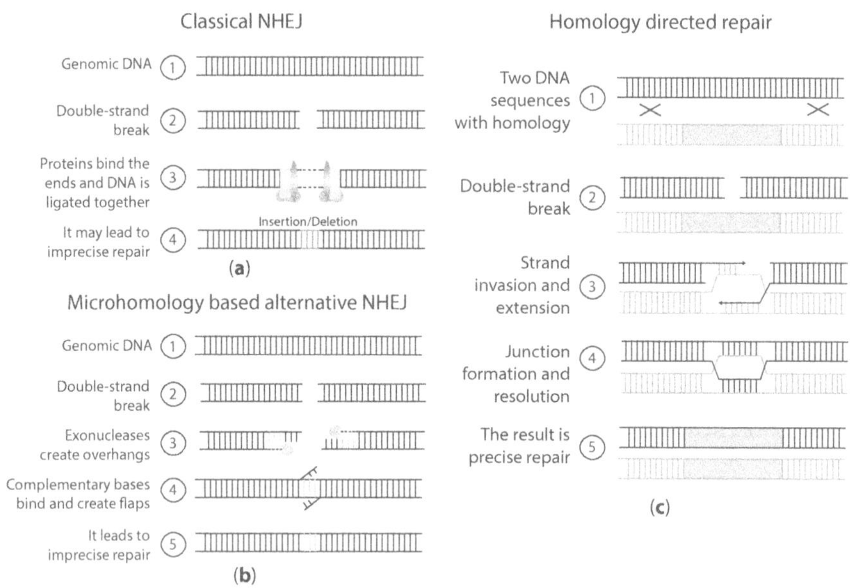

Figure 12.1 Major DNA repair pathways in plants. (a) Classical NHEJ; (b) microhomology-based alternative NHEJ; (c) homology directed repair (HR). Classical NHEJ may lead to insertions or deletions, while microhomology-based alternative NHEJ always results in deletions [129].

in small DNA insertions or deletions (InDels) at the break site. InDels located in a gene's open-reading frame can cause frame shift mutations ultimately resulting in loss of gene function (Figure 12.1b).

12.2.2 Homology Directed Repair (HR)

HR is a DSB mechanism which cause exchange of genetic material between parental chromosomes in meiotically dividing cells. HR is generally an error-free DSB repair pathway, because DNA templates guide accurate replacement of the DNA sequence at the break point (Figure 12.1c). Presence of homologue piece of DNA activates this pathway mostly in G2 and S phases of the cell cycle. In the absence of homologue DNA template, NHEJ can activate instead [4]. Template DNA (in the form of a sister chromatid or homologous chromosome) is not always present, which may obstruct HR. Exogenously provided synthetic template DNA can be used for gene insertion, replacement, deletions, *etc.* HR is mainly restricted to late S/G2 phases in the cell cycle, whereas NHEJ functions throughout the cell cycle, which makes NHEJ the major DSB repair pathway [5].

12.3 Engineered Nucleases: The Key Players in GE

The meganucleases, ZFNs, TALENs, and CRISPR associated nucleases are the GE tools (Figure 12.2). These GE tools differ in mechanism of action, and Table 12.1 presents a comparative account between the four nucleases. As with the potential applications and versatility of these nucleases for crop improvement, the off-target effects, *i.e.*, induced mutations at unintended sites in genomes, also need to be evaluated.

12.3.1 Meganucleases

These are naturally occurring gene-targeting proteins encoded by mobile introns (mobile sequence inside a self-splicing RNA intron or self-splicing protein intein) [6]. These unique enzymes generally have high activity and long recognition sequences (>14 bp); they generate targeted DSBs resulting in site-specific digestion of target DNA [7, 8] (Figure 12.2). The sequence that MgN recognizes is referred to as "homing site"; hence, MgNs are also known as homing endonucleases and LAGLIDADG homing endonucleases (LHEs) on the basis of amino acid sequence [9, 10]. There are about hundreds of MgNs identified in several organisms including bacteria, fungi, and some plants. Most commonly utilized LAGLIDADG family

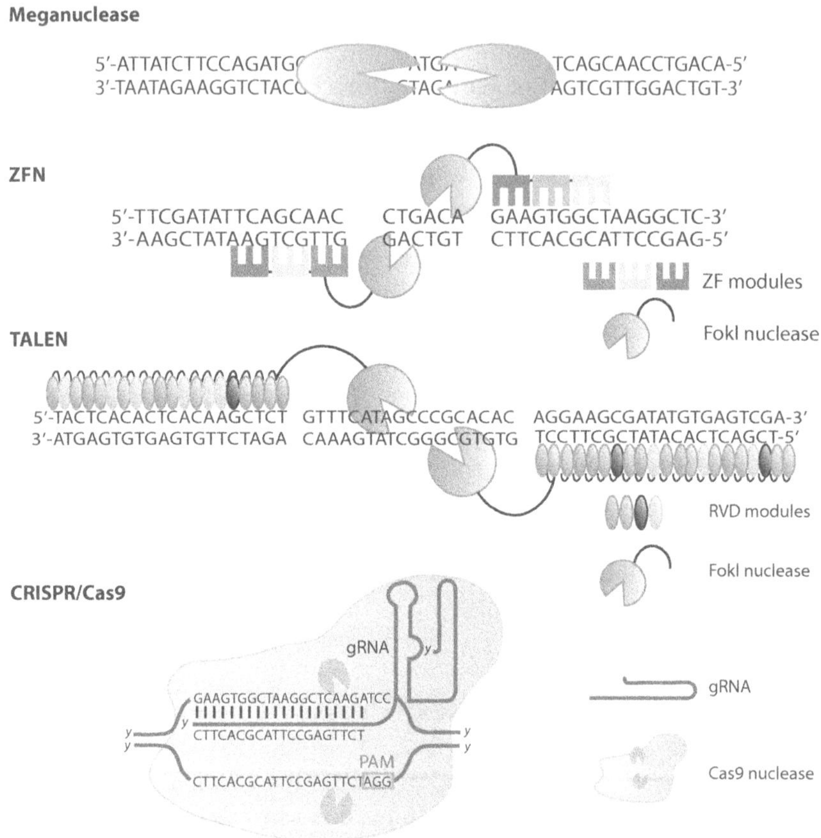

Figure 12.2 Types of different engineered nucleases used for GE, namely, ZFN, TALEN, and CRISPR/Cas [130].

members include I-*Sce*I and I-*Cre*I, which includes proteins with one or two copies of the LAGLIDADG motifs. The family members, *viz.*, I-*Cre*I and I-*Ceu*I enzymes, possess only one copy of motif, which acts as homodimers that recognize palindromic homing site. Enzymes with two copies include family members such as I-DMO1 and I-*Sce*I of the LAGLIDADG motif separated by about 100 residues and act as monomers. It is of interest to note that MgNs hold the promise of even greater specificity than ZFNs, TALENs, or CRISPR/Cas system due to large recognition sites, thus giving a low level of off-target cleavage within a genome and low toxicity. In spite of their greater specificity, this system is not widely adopted in crop improvement. A major limitation of the use of naturally occurring MgNs is that due to the large recognition sites, each MgN may have a homing site

Table 12.1 Comparative summary of GE techniques.

Conventional GE techniques	**MgNs**	**ZFNs**	**TALENs**	**CRISPR/Cas**
Mode of action: Integrate construct at target site through site-specific recombination	Integrate construct at target site through HR	Protein pair assembled adjacent to target site and introduce DSB	Protein pair assembled adjacent to target site and introduce DSB	Hybridization of gRNA recruits Cas nuclease at hybridization site and introduce DSB (presence of PAM sequence facilitates cleavage)
Components: Cre-Lox, Flp-recombinase, Phic3-instegrase, transposons, *etc.*	DNA recognition domain and nuclease domain	Zinc-finger DNA binding domain and *Fok*I nuclease domain	TAL DNA binding domain and *Fok*I nuclease domain	gRNA, Cas9 protein
Nature: Natural	Synthetic	Synthetic	Synthetic	Synthetic
Length of recognition site: Various size	>14 bp	18–24 bp	24–59 bp	20–22 bp
Specificity: High	High	High	High	Moderate

(*Continued*)

Table 12.1 Comparative summary of GE techniques. (*Continued*)

Conventional GE techniques	**MgNs**	**ZFNs**	**TALENs**	**CRISPR/Cas**
Off target cleavage: Very low	Very low	Low	Low	Low (quite higher than ZFNs and TALENs)
Advantages: DSB is not required Allow various genome modifications	DSB is not required Allow various genome modifications	High efficiency Drug/herbicide selection not required DSB repaired by NHEJ or HR Can target any sequence Lower chance of off-target cleavage than CRISPR Successfully used in different plant species	High efficiency Drug/herbicide selection not required DSB repaired by NHEJ or HR Can target any sequence Lower chance of off-target cleavage than CRISPR Successfully used in different plant species	Very high efficiency Drug/herbicide selection not required DSB repaired by NHEJ or HR Can target any sequence Rapid construction easy designing and easy delivery Multiplexing possible Successfully used in different plant species

(*Continued*)

Table 12.1 Comparative summary of GE techniques. (*Continued*)

Conventional GE techniques	**MgNs**	**ZFNs**	**TALENs**	**CRISPR/Cas**
Disadvantages: Time consuming and laborious Low efficiency Require long homology stretch of DNA Drug/herbicide selection required	Difficult to construct novel MgN Large recognition sites make it so that it may have a cut site in genome just once	More difficult to assemble than TALENs and CRISPR	More difficult to assemble than CRISPR	Requirement of PAM sequence may limit target site selection Off-target cleavage more than ZFNs and TALENs

in an organism's genome just once, or possibly not at all. Second limitation is that the cleavage and DNA binding domains of MgNs are not clearly demarcated. The attempts to reengineer DNA contact points of the endonuclease can be challenging and often compromise nuclease activity [9]. Another limitation associated with MgNs is lack of clear correspondence between MgN protein residues and their target DNA sequence specificity. Because of these challenges, very few academic groups and companies routinely engineer MgNs that target novel DNA sites. To circumvent the difficulty of programming the nuclease to recognize a desired target, TALE DNA binding module to target a MgN to the desired sequence in the genome has generated a MgN–TALE chimera (megaTAL) [11]. Similar efforts to combine the precision of MgNs with the flexibility and ease of targeting of CRISPR may further revolutionize genome engineering.

12.3.2 Zinc-Finger Nucleases

To overcome the limitations of MgNs, Kim *et al.*, [12] created novel site-specific endonucleases by linking zinc-finger proteins to the cleavage domain of *Fok*I endonuclease and termed as ZFNs. Lloyd *et al.*, [13] used ZFNs for the first time in plant (*Arabidopsis*) to generate double-strand breaks and mutations at specific genomic sites. The DNA binding domain of a ZFN is made up of three to four zinc-finger arrays each recognizing 3-bp-long sequence. The amino acids that contribute to site-specific binding to the target DNA (−1, +2, +3, and +6 relative to the start of the zinc-finger α-helix) can be changed and modified to fit specific target sequences. Dimerization of the *Fok*I domain is crucial for its enzymatic activity. Thus, digestion of target DNA can be achieved when two ZFN monomers bind to their respective DNA target sequences and properly align with each other in reverse configuration. The two ZFN monomers are designed such that they will flank a 5- to 6-bp long sequence within the DNA target sequence, allowing the *Fok*I dimer to digest within that spacer sequence. ZFNs can be designed to cleave almost any long stretch of double-stranded DNA by modification of the zinc-finger DNA binding domain [14, 15]. More context-dependent effects between the repeat units are observed using ZFNs because of cross-talk between adjacent modules when assembled into a larger array [16]. ZFNs also face low targeting efficiency and specificity, and frequent off-target effects caused partly by nonspecific DNA binding [17].

The specificity of ZFN-mediated GE has been further refined by the development of zinc-finger nickases (ZFNickases) [18–22], which take advantage of the finding that induction of nicked DNA stimulates HR without activating the error-prone NHEJ repair pathway. Consequently, this approach leads to fewer off-target mutagenesis events than conventional DSB-induced methods for GE. Furthermore, there is evidence to suggest that four to six zinc-finger domains for each ZFN half enzyme significantly enhance activity and specificity [23–25].

12.3.3 Transcription Activator-Like Effector Nucleases

Though ZFNs helped in resolving some of the difficulties associated with MgNs such as limited number of target sites and easy engineering, there was a need to further improve GE. Christian *et al.*, [26] for the first time suggested that the zinc-finger arrays could be substituted with the DNA recognition domain of TALEs to create TALENs that recognize and cleave DNA targets. They first used two well-characterized TALEs, namely, AvrBs3 from the pepper pathogen *Xanthomonas campestris* pv. *vesicatoria* and PthXo1 from the rice pathogen *X. oryzae* pv. *oryzae*. Like ZFNs, TALENs consist of an engineered specific transcription activator-like effectors (TALEs) DNA binding domain and a nonspecific *Fok*I cleavage domain. TALE proteins obtained from genus *Xanthomonas* have central DNA binding domain, which consists 33-35 nearly identical long amino acid repeats, followed by the "half repeat," which contains only 20 amino acids. The amino acids of each monomer are highly conserved, except at the positions 12 and 13, which are hypervariable and called repeat-variable di-residues (RVDs). The first RVD residues are involved in forming a stabilizing contact with the backbone of the RVD loop and the second is involved in base-specific contact with DNA [27, 28]. The amino acid pairs NI, NG, HD, and NN of RVDs preferentially recognize adenine, thymine, cytosine, and guanine/adenine, respectively, and modulation of RVDs can recognize consecutive DNA bases. The recognition sites are always preceded by a thymine (T) before the first repeat in the array, and this is the only critical rule for TALE targeting [29]. The flanking region of the DNA binding domain in the TALEN protein affects TALEN activity in higher eukaryotes including plant species [30].

Though ZFN and TALENs are modular and have natural DNA binding property, they considerably differ in specificity efficiency and ease of designing. Single zinc-finger DNA binding domains recognize three base pairs, whereas TALE domains recognize only one base pair. This difference makes engineering of TALEN systems easy and convenient. TALENs also show less toxicity than ZFNs [31]. In spite of modular nature, TALE DNA binding monomers suffer from context-dependent specificity [32], which makes construction of novel TALE arrays labor intensive and costly. Another serious problem with TALEN is that the researchers cannot predict the activity at specific sites. Now a days, so many methods have been developed such as golden gate assay and platinum assay to overcome the above-mentioned limitations of novel TALEN engineering. Due to its several advantages, TALEN has been popular as a GE tool in various organisms including plants than ZFN [26, 33, 34].

12.3.4 CRISPR/Cas System: The Forerunner

CRISPRs are DNA loci containing short repetitions of base sequences and found in approximately 40% of sequenced bacterial genomes and 90% of sequenced archaea. CRISPRs are often associated with Cas genes that code for proteins related to CRISPRs. The CRISPR/Cas system is a prokaryotic immune system that confers resistance to foreign genetic elements such as plasmids and phages [35]. CRISPR provides a form of acquired immunity, and each repetition is followed by short segments of spacer DNA from previous exposures to a virus [36]. CRISPR spacers recognize and cut these exogenous genetic elements in a manner analogous to RNAi in eukaryotic organisms [35]. Eleven different CRISPR/Cas systems have been reported, which are grouped into three major types I to III. Type II CRISPR/Cas systems have been adapted as a GE tool. It differs with respect to the presence of protospacer adjacent motif (PAM) sequence and a second RNA, called trans-acting CRISPR RNA (tracrRNA). tracrRNA teams up with crRNA and helps crRNA in maturation and recruiting the Cas9 nuclease to DNA [37, 38]. The natural three-component system of type II CRISPR/Cas system was simplified by fusing together crRNA and tracrRNA, creating a single synthetic chimeric "guide" RNA (sgRNA or gRNA) (Figure 12.2d) [38, 39]. The ease of using a single RNA has led to the widespread adoption of gRNAs for genome engineering (Figure 12.3).

A significant advantage of the CRISPR/Cas9 system is that it utilizes RNA molecules that direct the nuclease to target a specific nucleic acid in the given genome. Designing and synthesizing RNA is easier and cheaper than the protein domains of ZFN and TALEN. CRISPR requires only a single construct for expression or RNA synthesis of sgRNA [40, 41]. Thus, the CRISPR/Cas9 system is simple, easy to design, and highly effective [42, 43]. CRISPR/Cas system provides greater efficiency over ZFNs and TALENs, and with this system mutations can be introduced in multiple genes at the same time by introducing multiple gRNAs (multiplexed mutations). No doubt the CRISPR/Cas system is more attractive than ZFNs and TALENs, but one of the important complications of these technologies is mutation introduced at nonspecific loci called as off-site effects. Off-site targeting further results in cell toxicity, and this is one of the hurdles to obtain transgenic plant through tissue culture. Moreover, availability of PAM sequence in adjacent to target site limits target site selection.

It has been identified that the 3′ end of the guide sequence presents target specificity, while mismatches at the 5′ end are tolerated [43, 44].

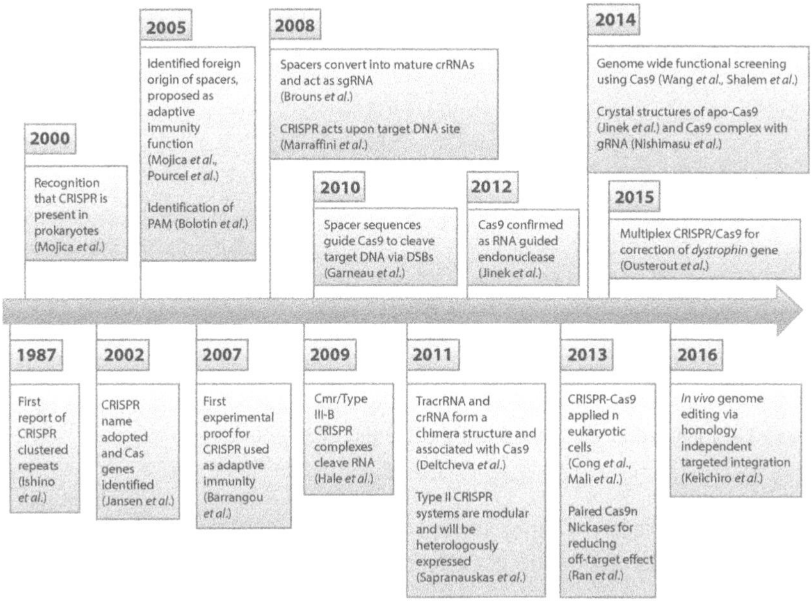

Figure 12.3 Historical timeline of the development of CRISPR/Cas technique [131].

A mutated version of Cas9 (D10A, Cas9 nickase; Cas9n) can be used to avoid potential off-target effects and to improve the specificity as it induces nicks (SSBs) in genome [39, 43, 45]. This strategy showed good promises in human cells [45]. Use of truncated gRNAs with a target sequence less than 20 nucleotides can also improve the specificity of CRISPR/Cas9 systems [46]. Finally the careful design of gRNAs [47, 48] and optimization of gRNA and Cas9 expression are the key factors to avoid off-target effects [44, 48]. The Cas9–guide RNA complex is more stable and showed long interaction with target sites that contain a PAM, but binding at nontarget sequences that do not have PAM is unstable and transient. This suggests that PAM plays an important role in stimulating Cas9 activity [49].

12.4 Targeted Mutations and Practical Considerations

12.4.1 Targeted Mutations

The GE can be employed in targeted mutagenesis, insertion, deletion, and replacement of selected gene of interest in the genome. Several genes have been modified using GE tools on the basis of the above applications (Figure 12.4).

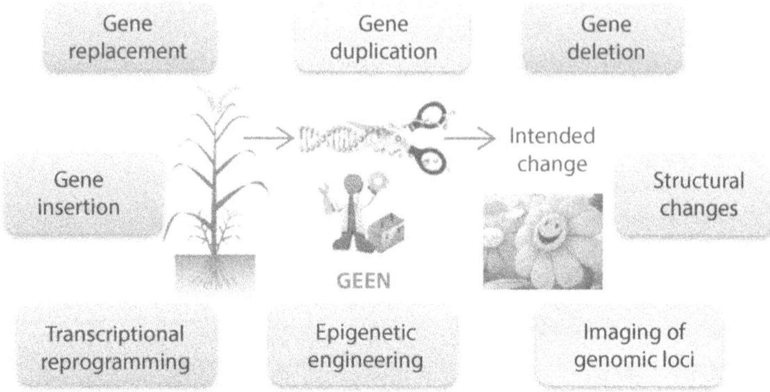

Figure 12.4 Intended modifications using GE in plants.

Mutagenesis: Targeted mutagenesis refers to deliberate change (addition/deletion/substitution) in the genetic structure directed at a specific site in the genome.

Gene insertion/replacement: Targeted gene insertion pertains to inserting the gene of interest at a specific site in genome. Gene replacement differs with targeted gene insertion, in which particular endogenous gene of an organism is replaced with new version of the same gene.

Gene excision: Involves removal or deletion of intended gene from the genome.

Targeted structural changes of genome: Targeted structural changes include large-scale addition, deletion, inversion, duplication, or translocation of DNA at intended site in the selected genome.

12.4.2 Steps Involved

On broad basis, we can divide experimental setup in four steps, *viz.*, identification of target sequence, vectors construction, transformation of vectors to plant, and finally screening of transgenic plants for intended GE. The basic flowchart of creating mutations using GE has been detailed in Figure 12.5.

Targets for GE

1. Structural genes
2. Regulatory genes
   a. Coding (Cis-elements)
   b. Noncoding (Regulatory)

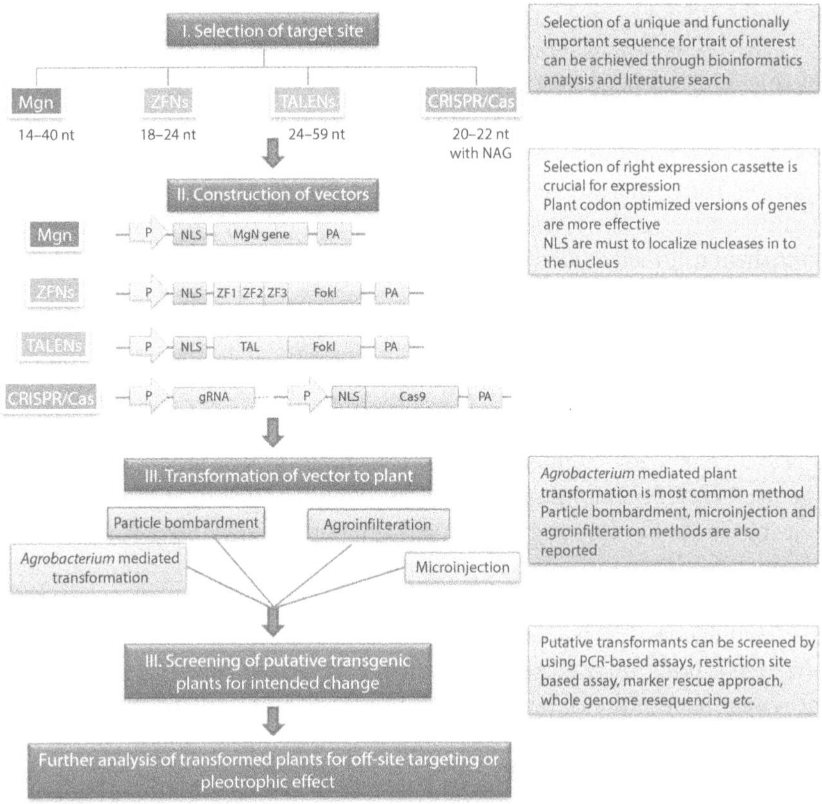

Figure 12.5 Basic strategies and important practical considerations for the GE in plants with the aid of different tools.

12.4.2.1 Selection of Target Sequence

Selection of target sequence is the first and most important step irrespective of any GE tools, as target site facilitates further binding and cleavage by designer nucleases. While selecting a target site, one should consider the following points: (1) For gene knockout experiment, identify such a sequence that is unique and important for functioning of gene and mutations at such site should destroy the function of a target gene. (2) The actual genomic sequence to be targeted must be 100% identical to the desired recognition sequence. (3) The target site must be unique in the genome of selected plant. BLAST search can be used to ensure that the identified target sequence is unique in the genome. (4) Multiple target sites in a single target gene can also be employed to maximize the chance of obtaining desired change. Considerations for selection of target site are applicable to all designer nucleases. But as CRISPR/Cas system has totally different components and mode of action, there is a need to consider extra points regarding selection of target sequence, *i.e.*, (1) choose a target sequence ranging from 19 to 20 nucleotides in length that is adjacent to an NGG (PAM) sequence on the 3′ end of the target sequence; (2) Select a target sequence coding the sense sequence of the target region or the antisense sequence. Thus, one can generate crRNA in two possible orientations, provided that it meets the PAM requirements on the 3′ end. Several web-based servers help us to select/identify probable target site through *in silico* analysis.

12.4.2.2 Designing Nucleases

Availability of engineered nuclease with desired specificity is the second important prerequisite. In case of MgNs, one can use naturally occurring MgNs or genetically engineered MgNs. If naturally occurring MgN needs to be used and if the homing site is not available in the target sequence, one must introduce the homing site to suit the particular natural MgN before utilizing it to cleave the DNA. In order to engineer novel MgNs, researchers modified recognition sites of existing MgNs such as *I-SecI, I-CreI, I-DMOI, etc.* by mutagenesis or by creating chimeric enzymes. There are several specialized methods of mutagenesis available to create novel MgN variants that recognize unique sequences with increased nuclease activity [50, 51]. Several methods are present to assess mutated endonucleases for their activity and altered specificities [52, 53]. One can also purchase pre-engineered MgNs as per demand from commercial suppliers such as Cellestic. For the construction of highly specific ZFNs, several methods are available such as modular assembly (MA), Oligomarizes pool engineering (OPEN), Context dependent assembly (CoDA), *etc.* [16, 54–56]. Some companies such as SangamoBioSciences and Sigma-Aldrich are involved in the development of novel ZFNs as per demand [55]. In spite of having many methods available that facilitate engineering of novel TALENs, the golden gate system, platinum TALEN, and platinum gate system are the important tools available to engineer custom TALENs [57] and, more conveniently, many web sources are available that help in engineering TALENs such as TALEN-NT [58], idTALE [59], and EENdb [60]. Pre-engineered TALENs can be made available by some commercial groups and companies as per demand.

Once designer nuclease coding DNA fragment with desired specificity is ready, then the next step is vector construction and plant transformation. Once we have MgNs with desired homing site, the next step is cloning of MgN coding DNA into suitable plant transformation vector with one or two nuclear localization signals (NLSs), suitable promoter, and other regulatory elements. There are two ways to clone ZFNs/TALENs; one can clone both

ZFNs/TALENs in the same vector or in separate vectors. Most of the researchers prefer the first one. In CRISPR/Cas system, designing of gRNA is the key step, which determines success of the experiment. Many online tools are available that help in gRNA designing. CRISPR DESIGN (http://www.rgenome.net/cas-designer/), CRISPR PLANT (http://www.genome.arizona.edu/CRISPR-), E-CRIS (http://www.e-crisp.org/E-CRISP/designcrispr.html), Cas-OFFinder (http://www.rgenome.net/cas-offinder/), Cas-Designer (http://www.rgenomc.net/cas-designer/), Cas9 Design (http://cas9.cbi.pku.edu.cn/database.jsp), CHOP CHOP (https://chopchop.rc.fas.harvard.edu/), CCTop (http://crispr.cos.uni-heidelberg.de/), CRISPR direct (http://crispr.dbcls.jp/), *etc.* are available for gRNA designing. These software packages are specifically designed for plants, in particular, *Arabidopsis*, rice, maize, tobacco, tomato, Medicago, *etc.*

CRISPR/Cas system includes three plasmids: Cas9 plasmid, target plasmid, and donor plasmid. The first two plasmids are of the obligate type, whereas donor plasmid is facultative, which means according to objectives, use of donor plasmid can be excluded. Cas9 plasmid: First clone artificially synthesized Cas9 gene version, which is codon optimized for plant. Attach one or more NLS to Cas9 gene in order to target it to nucleus. Use appropriate promoter such as CaMV35s for expression of Cas9 gene. There are some examples where scientists used different promoter other than CaMV35S such as AtUBQ, 35sPPDK, OsAct1, *etc.* [61, 62]. Based on objectives of the experiment, one can use one of the two forms of Cas9 gene, that is, nucleases or nikases. Nikases introduce single strand break and favor HR over NHEJ, which may reduce cellular toxicity. Target plasmid: Insert target sequence and sRNA sequence into a vector. For expression of target sequence along with sRNA, AtU6 and OsU6 promoters are widely used. Attach the target system to the 5′ end of the sRNA. Presence of 5′ G is necessary for transcription initiation with U6 promoter. One can also go for separate expression of crRNA and tracrRNA as per convenience. Donor plasmid: Construction of plasmid containing left and right homology arms flanking the sequence to be targeted is prerequisite for gene insertion, gene replacement, and other sequence integration. Choose homology sequence around 0.5 to 1 kb in length. One can use only one plasmid by combining all the three plasmids. One thing should be considered that all the components of CRISPR-Cas system (Cas gene, PAM sequence, crRNA, and tracrRNA) should be taken from only one organism. The second step is transformation of these plasmids to plant cell.

12.4.2.3 Transformation

The third step is transformation of construct in the plant. The success of GE experiment with any designer nuclease depends on the efficiency of transformation. Each plant responds differentially to different methods of transformation, so choose the appropriate transformation method for the plant under study. Protoplast transformation, *Agrobacterium* mediated transformation, biolistic method, and *in planta* transformation are commonly being employed [63].

12.4.2.4 Screening for Mutation

After development of genome edited plants, screen available plants for desired GE with any suitable method. Experimental results and transformation efficiency will depend on the objectives of study, plant species under study, and experimental setup. The transformed plants are screened for desired mutations using several methods like restriction digestion-based assays, *viz.*, cell-I endonuclease (mismatch cleavage detection assay), PCR-based assays, whole genome resequencing, surveyor assays [53], target gene sequencing, high-resolution melting analysis (HRMA), and Indel-based screening methods. Pre-engineered and customized ZFNs, TALENs, and CRISPR/Cas are also available from commercial suppliers such as Cellectis (http://www.cellectis.com), Sigma-Aldrich (https://www.sigmaaldrich.com/), Bioresearch and Life Technologies (https://www.biosearchtech.com), *etc.* as per the specifications of the experiment. For more information, readers are directed to latest reviews on practical considerations of efficient CRISPR/Cas experiment [64].

12.5 New Era: CRISPR/Cas9

The CRISPR/Cas9 system is a widely used tool among the four in current areas of science, and hence in this chapter, much has been discussed on the same. The CRISPR/Cas development has witnessed rapid advancements in the field of genome modification/engineering. Currently, the technology has seen various developments, including vector construction, delivery methods, screening assays, and in variants of Cas9. So in this section, we discussed technological developments of CRISPR/Cas.

12.5.1 Vector Construction

There are two types of vector systems available for CRISPR, namely, binary vectors and single vector system. Binary vector system is an old one, and it has its advantage for rapid initial testing of CRISPR/Cas system. A vector containing different gRNAs can be used for transformation of a plant, which is already expressing Cas9 protein. With the use of binary vector system different combination of Cas proteins can be tried for different gRNAs which further gives more flexibility and ease in designing experiment. Nowadays, a single vector containing Cas protein as well as gRNA is becoming more popular among researchers. In most of the single vector systems, RNA polymerase II based promoters such as CaMV 35S and ubiquitin are used for expression of the *Cas9* gene, whereas RNA polymerase III based promoters such as U6 and U3 are used for expression of gRNA. Such type of vector systems exploits mixed dual promoters. Nowadays, some modifications are available over this system such as dual polymerase II promoter and single polymerase II promoter. Dual polymerase II promoter based vectors use two different RNA polymerase II based promoters to drive expression of Cas gene and gRNA, whereas single polymerase II promoter based vectors use only one RNA polymerase II based promoters to drive expression of both *Cas* and gRNA. All these modifications help to reduce vector size, which ultimately results in increased transformation efficiencies and success rate [65]. Many companies supply CRISPR vectors based on objective of the experiment, and various variants are available as per user's choice (Table 12.2).

Table 12.2 Summary of various types of CRISPR plasmids for plants.

Available vectors*/Purpose	References
Plant expression/gene regulation pFGC-pcoCas9, pICH47742::2x35S-5'UTR-hCas9(STOP)-NOST, HBT-pcoCas9, pRGEB32, pKSE401, Cas9 MDC123, pHSE401, pBUN411, pRGEB31	[62, 95–98]
HR-based gene replacement or NHEJ-based insertion/deletion events (Cas9D10A nickase) pHSN501, pBUN501, pMonAID_nCas9-PmCDA_Hyg_ALS, pDicAID_nCas9-PmCDA-2A-NptII_ETR	[98]
Targeted knockdown of genes of interest (inactivated catalytic domain) pHSN6I01, pYPQ153, pBUN6I11, pEGB 35S:dCas9:Tnos (GB1191)	[98–100]
Targeted activation of the genes of interest pYPQ152, pHSN6A01, pBUN6A11	[66, 98–101]
Base editing pH-nCas9-PBE, pH-dCas9-PBE,	[102, 103]

*Sequences available at addgene plasmid repository (https://www.addgene.org/).

12.5.2 Delivery Methods

Particle bombardment, floral dip and in planta methods are commonly used for plant transformation in CRISPR guided genome editing. Virus mediated delivery, plasmid delivery, ribonucleo-protein complex delivery are some of the advanced plant transformation methods available now a days. Viral delivery is limited by low editing efficiency; nevertheless, many reports are available for virus mediated delivery of CRISPR components in plants [66–68]. Cas expression cassette can be transformed directly to plants using the PEG method and particle bombardment methods to protoplast. Cas9 protein and gRNA complex (ribonucleotide protein) and only mRNA of Cas9 and gRNA can be transformed to plants through gene gun and PEG mediated methods [69]. However, regeneration of protoplast remains a major challenge associated with such methods. Ribonucleoprotein complexes mediated transformation is quite interesting and is gaining importance. In this method, Cas9 and gRNA can be pre-assembled *in vitro* to make a ribonucleoprotein complex that is directly transformed into the target plant. The advantage of this method is that quick degradation of ribonucleoprotein by endogenous cellular processes resulted in absence of ribonucleoprotein to the next generation; rapid degradation also gives reduced off-target effects [70]. Some of the studies reported successful use of ribonucleoprotein in plants [64, 71].

12.5.3 CRISPR/Cas Variants

Streptococcus pyogenes's CRISPR/Cas systems are the most characterized and widely used [37]. Many Cas9 variants are available, which makes CRISPR/Cas system more flexible and easy (Table 12.3).

12.5.3.1 SpCas9 Nickases (nSpCas9)

This variant of Cas9 is developed by silencing each of the two nuclease domains of a key catalytic residue (D10A for HNH and H840A for RuvC), which only cleave a single strand of the DNA [45]. Single-strand break in the genome will be repaired with high fidelity, whereas DSB is processed by the NHEJ, which is error prone [72]. Many studies showed successful application of nickases to reduce offsite targeting [73].

12.5.3.2 Cas9 Variant without Endonuclease Activity

A catalytically inactive Cas9 variant has been engineered by introducing mutation in both endonuclease domains ($RuvC^-$ HNH^-). These variants will be useful in epigenome and transcriptional modifications of gene regulation.

Table 12.3 Summary of available Cas9 variants.

Variant of *Cas9*	**References**
Streptococcus pyogenes (*SpCas9*) D1135E variant (NGG)	[75, 77, 91, 92]
Streptococcus pyogenes *SpCas9* VRER variant (NGCG)	[75, 92]
Streptococcus pyogenes *SpCas9* EQR variant (NGAG)	
Streptococcus pyogenes *SpCas9* VQR variant (NGAN or NGNG)	
Staphylococcus aureus *SaCas9* (NNGRRT or NNGRR(N))	[45, 92]
Neisseria meningitidis *NmCas9* (NNNNGATT)	[92]
Streptococcus thermophilus *StCas9* (NNAGAAW)	
Treponema denticola *TdCas9* (NAAAAC)	[76]
Derivative of *Sp*Cas9-HF1 *SpCas9-HF2*	[75, 77]
Derivative of *Sp*Cas9-HF1 *SpCas9-HF3*	
Derivative of *Sp*Cas9-HF1 *SpCas9-HF4*	
Enhanced specificity *eSpCas9*	[75, 77, 92]
Cas9 variant without endonuclease activity	[93]
Cas9 nickases	[22, 42, 94]

12.5.3.3 FokI Fused Catalytically Inactive Cas9

Dimeric mode of action gives more on-target effect as compared to monomeric mode of action, and SpCas9 has been designed to adapt to a dimeric mode. This was achieved by mutating both nuclease domains of SpCas9 and creating "dead-Cas9" (dCas9) and further fusion with the FokI nuclease domain (fCas9) [74]. In some studies, Cas9 has also been fused to zinc finger, TALENs, and other DNA binding domains, which helped to increase its specificity.

12.5.3.4 Naturally Available and Engineered Cas9 Variants with Altered PAM

The requirement-specific PAM is quite stringent, which might hinder the selection of sequences that need to be targeted, particularly in repetitive regions of the genome. Engineering Cas9 with altered PAM sequence specificities is one possible solution. There are some reports where Cas9 was engineered with novel PAM sequences and their successful utilization of GE [75, 76]. Another solution is to use cas gene from different bacteria such as *Staphylococcus aureus, Neisseria meningitides, Streptococcus thermophilus, etc.*

12.5.3.5 Cas9 Variants for Increased On-Target Effect

The SpCas9–sgRNA complex might possess more energy than it required recognizing its target site, which might be the reason for off-site targeting. Fifteen SpCas9 variants have been constructed using substitutions in positions that form hydrogen bonds with the DNA backbone [75, 77].

12.5.3.6 CRISPR/Cpf1

CRISPR/Cpf1 is orthogonal to Cas9 from *Prevotella* and *Francisella* 1 and is based on type V CRISPR system. This system is found in some bacteria like *Primotella, Acidaminococcus, Francisella, Lachnospiraceae, etc.* Cpf1 has some properties different from Cas9 leveraging its usage (Table 12.4).

Table 12.4 Comparison between Cas9 and Cpf1 proteins.

Features	**Cpf1**	**Cas9**
Nature of cut ends	Generates sticky ends	Generates blunt ends
PAM sequence	Thymine rich	Guanine rich
gRNA	Smaller in length	Larger in length
Cut site/restriction site	Distal to PAM sequence	Near to PAM sequence
Type of CRISPR system	Type V	Type II
Protein size	Small	Large

Table 12.5 Brief differentiation between RNAi and CRISPR/Cas system.

RNAi mechanism	CRISPR/Cas system
Mechanism works at transcript level	System works at genomic level
Results in partial loss of targeted gene function	Results in complete loss of targeted gene function
Chance of getting targeted gene knock-out is high	Chance of getting targeted gene knock-out is less
It can help in disruption of gene function of host as well as parasites (viruses, insects, fungi, *etc.*)	It is mostly confined to host
Unable to disrupt a particular domain/motif or individual amino acid of gene	Possible to disrupt a particular domain/motif or individual amino acid of gene
Selectable marker genes are needed	Selectable marker genes are not needed
Effect (intended mutation) is heritable or nonheritable depending on the requirement	Resulted effect is always heritable
Can be suppressed by proteins of viruses or other parasites	Cannot be suppressed by proteins of viruses or other parasites
More versatile but less flexible technique	Less versatile but more flexible system
Deployed to study endogenous gene of host	Cannot be deployed to study housekeeping genes of host

Small protein size and small gRNA (nearly half of the gRNA of Cas9) facilitate easy and efficient delivery of ribonucleoproteins to plant cell. Many reports are coming up recently with application of Cpf1 in crop plants [78–83].

There are certain differences between RNA interference (RNAi) and CRISPR/Cas9 system and have been detailed in Table 12.5.

12.6 GE for Improving Economic Traits

The basic golden application of GE is exploring functional basis and functional characterization of genes. Functional characterization of genes is very much essential to understand the role of genes in a particular set of conditions, *e.g.*, in abiotic stress tolerance, disease resistance/plant–pathogen interaction, *etc.* GE has a practical application in improving economic traits like yield, tolerance or resistance to abiotic/biotic stresses, nutritional quality, nutrient use efficiency, *etc.* Development of climate-resilient SMART crops is a need of hour. The successful examples of GE in crops have been described in Table 12.6.

Recently, using the CRISPR/Cas9 system, the four yield-related genes, *viz.*, *Gn1a, DEP1, GS3*, and *IPA1*, were mutated. The edited plants of *gn1a, dep1*, and *gs3* mutants resulted in enhanced grain number, dense erect panicles, and larger grain size, respectively [84]. A more appealing application of GE has also been discussed below.

Table 12.6 Successful application of GE in model and important crop plants.

GE technique	Plant species	Reference
Meganucleases	Arabidopsis	[104]
	Tobacco	[105]
ZFNs	*Arabidopsis*	[22, 106, 107, 109, 110]
	Tobacco	[111–113]
	Rice	[114]
	Maize	[115, 116]
	Soybean	[23, 117]
TALENs	*Arabidopsis*	[26, 118]
	Tobacco	[35, 104]
	Rice	[36, 119, 120]
	Maize	[121, 122]
	Soybean	[76, 123]
CRISPR-Cas	*Arabidopsis*	[61, 85, 94, 97, 108, 124, 125]
	Tobacco	[62, 124, 125]
	Rice	[63, 84, 97, 108, 119, 125–128]
	Maize	[71, 97, 122]
	Soybean	[96]

12.6.1 Development of Next-Generation Smart Climate Resilient Crops

In order to cope with the current scenario of climate change, certain interventions are required in terms of modulating the plant stress signaling to respond to environment. The gene *OsNramp5* was knocked out using the CRISPR/Cas9 system, which resulted in the production of low Cd-accumulating *indica* rice without compromising yield [81]. Osakabe *et al.*, [85] showed mutation in a proton pump, *OST2*, using truncated gRNA and Cas combination resulted in altered stomatal closing in response to stress without any off-targeting in *Arabidopsis*. Similarly, genes involved in signaling of abiotic and biotic stresses like mitogen activated protein kinases (MAPK) or calcium signaling molecules also provide a good candidate for altering stress responses in crops.

12.6.2 Breaking Yield Incompatibility Barriers and Hybrid Breeding

Development of male sterile lines is critical for hybrid breeding. Zhou *et al.*, [86] developed new thermosensitive genic male sterility (TGMS) lines for potential hybrid breeding by targeting the gene tms5, which encodes the endonuclease *RNase Z* and controls the TGMS trait [86]. Overcoming the reproductive barrier, *i.e.*, genetic incompatibility in crossing the divergent populations, is very much essential for increasing yield levels in important food crops like rice. The allele/s conferring the hybrid male sterility can be identified and targeted for rescuing male fertility. An approach describing the same has been demonstrated in rice by knocking down one copy of *Sc-i* of the *Sc* locus that conferred japonica–indica hybrid male fertility [87]. The CRISPR/CAS9 system can also be potentially employed for exploiting heterosis by targeting genes involved in pollen development.

12.6.3 Creating New Variation through Engineered QTLs

Creation of new engineered quantitative trait loci (QTLs) or allelic variation in the gene pool is an exciting and unexplored application of GE. The regulatory sequences, *i.e.*, the cis-regulatory elements in promoters, can be mutated using CRISPR/Cas9 system and have an added advantage of fine-tuning the expression rather than deletion or insertion in the proteins that they encode. The best candidate genes for such editing are the promoters of yield genes wherein much of quantitative variation is generated for plant breeding. The plant CRISPR/Cas9 editing components are then crossed with the wild-type plant to create mutant plants with diverse phenotypic variability. This generates the much needed genetic variation for increasing the gene pool followed by selection of specific beneficial allele for crop improvement. This approach has been demonstrated in tomato for fruit architectural traits by Li *et al.*, [88] but needs to be applied in case of important cereal crops.

12.6.4 Transcriptional Regulation

Regulation of gene expression includes a wide range of mechanisms that are used by cells to increase/decrease/maintain production of specific gene products (protein or RNA). Targeting the regulatory element binding sites through GE, the regulation of endogenous gene can be altered. The regulatory transcription factors can be an ideal choice for such modification as they regulate major genes in case of stress and at various developmental stages in plants. Similarly, the small RNA, micro RNA, can be effectively targeted for downstream regulation of the gene it regulates. These applications are oriented more toward developing stress-resistant/tolerant crops for near future.

12.6.5 GE for Noncoding RNA, microRNA

MicroRNAs (miRNAs) are 20–24 nucleotide small regulatory, noncoding RNAs that are present ubiquitously and participate in most of the biological processes in plants at some or other point of life cycle. The plant miRNAs regulate genes encoding transcription factors, hormone receptor enzymes. These noncoding microRNAs can be edited precisely to regulate the genes that are under their control. For editing, it is very much essential that the function of microRNAs is fully known and has been characterized for its target at particular developmental stage or biological processes in plants. Thus, the functionally characterized miRNAs for a particular economic trait of interest across the plant species can be potential targets for GE. Several microRNAs have been known to regulate abiotic and biotic stresses, yield, and nutritional components in crops. Plant microRNA family databases are available and can be open-accessed for sequence, target sequences, and other characteristic features for fetching preliminary information on microRNAs for editing.

12.6.6 Epigenetic Modifications

Targeting the epigenetic genes involved in chromatin modifications and histone modifications can open a new arena of understanding the role and function of epigenetics in plant stress responses.

12.6.7 Gene Dosage Effect

The gene dosage effect on phenotypic character can also be modulated by knocking out the copies of duplicated genes. The probable targets of GE for improving various traits have been mentioned in Table 12.7 considering the example of rice.

12.7 Biosafety of GE Plants

In agriculture, improvements are need of the hour amidst challenges facing in today's formidable world. Water scarcity, whopping population, and land shrinkage envisage to embrace new technologies that enhance production or are capable of utilization of natural resources. To address these issues, rDNA technology-based GMOs have been considered as a potential technology in augmenting food production. For example, herbicide tolerance and insect resistance are the major crops produced by GMO technology, where the former ratio is around 90% [89]. In spite of wide adoption of GMO crops, social and political controversies coupled with intellectual property and regulation and plant-breeding issues have conciliated the significant impact of GMOs.

Unlike GMOs, emerging evidences indicate that the GE technique has showed promising results in maize, wheat, rice, soybean, tomato, and citrus. Interestingly, globally first meal with genome-edited foods (CRISPR) was prepared in Sweden by Stefan Jannson in the past summer [90]. The meal is known as "tagliatelle with CRISPRy fried vegetables" embellished with cabbage grown at Umea University's campus. A major impetus has been gained in CRISPR foods after seeing the interpretation of the Swedish Board of Agriculture. They have declared that the CRISPR/Cas genome-edited crops may not come under the premise of European Union's definition of a GMO because the DNA was removed from the system instead of new genes being inserted. They are also encouraging the researchers to adopt new technologies for the augmentation of food production.

Table 12.7 Probable target genes for GE in rice.

Sr. No.	Gene	Gene notation	Class of the trait	Trait
1	*Gibberellin 2-Oxidase3*	*GA2ox3*	Morphological trait	Dwarf
2	*Gibberellin 2-Oxidases9*	*GA2ox9*	Morphological trait	Dwarf
3	*Gibberellin 2-Oxidase6*	*GA2ox6*	Morphological trait	Dwarf
4	*SHORT GRAIN 1*	*SG1*	Morphological trait	Culm leaf
5	*SHORT GRAIN 1*	*SG1*	Morphological trait	Dwarf
6	*SHORT GRAIN 1*	*SG1*	Morphological trait	Panicle flower
7	*SHORT GRAIN 1*	*SG1*	Morphological trait	Seed
8	*Acetolactate synthase*	*ALS*	Resistance or tolerance	Other stress resistance
9	*Elicitor 5*	*EL5*	Morphological trait	Root
10	*Elicitor 5*	*EL5*	Morphological trait	Root

(Continued)

Table 12.7 Probable target genes for GE in rice. (*Continued*)

Sr. No.	Gene	Gene notation	Class of the trait	Trait
11	*Elicitor 5*	*EL5*	Morphological trait	Root
12	*Elicitor 5*	*EL5*	Morphological trait	Root
13	*Elicitor 5*	*EL5*	Morphological trait	Root
14	*Elicitor 5*	*EL5*	Morphological trait	Root
15	*Anthranilate synthase a-subunit 2*	*OASA2*	Physiological trait	Eating quality
16	*Glutamate decarboxylase 2*	*OsGAD2*	Others	Others
17	*OsIAA3*	*OsIAA3*	Morphological trait	Culm leaf
18	*OsIAA3*	*OsIAA3*	Morphological trait	Dwarf
19	*OsIAA3*	*OsIAA3*	Morphological trait	Root
20	Iron deficiency-responsive cis-acting element 1	*IDEF1*	Physiological trait	Germination dormancy

12.8 What's Next: Prospects

GE can be successfully employed in crops having economic importance in the present scenario of climate change and changing agricultural practices. A network or institutional collaborations should be developed for dealing with the genome-edited crops on a larger scale. This would encourage resource sharing and thus would require less time with trained human resources. The genome-edited crops for a particular economic trait should be used in various breeding programs for transferring the trait of interest for improving an elite local cultivar. This way the new breeding technology would enhance the efficiency of plant breeding, and the improved varieties can be released for cultivation in destined areas without any regulation. There is need to establish accurate phenotyping platforms that would be convenient for phenotyping the desirable trait on a large scale. Screening the genome-edited crops takes a longer time in development, which can be drastically cut down using phenotyping platforms like LemnaTech, Phene, *etc.* Such platforms can accelerate the process of phenotyping and further selection among the edited plants. Similarly, improvements in chemistries of high-throughput genotyping using next-generation sequencing platforms should be able to identify the edited base or region at a time. A combined strategy of high-throughput genotyping and phenotyping with networking will be a successful strategy for the advancement of GE.

References

1. Paszkowski, J., Baur, M., Bogucki *et al.*, Gene targeting in plants. *EMBO J.*, 7, 4021, 1988.
2. Puchta, H. and Fauser, F., Gene targeting in plants: 25 years later. *Int. J. Dev. Biol.*, 57, 629–637, 2013.
3. Bibikova, M., Beumer, K., Trautman, J.K. *et al.*, Enhancing gene targeting with designed zinc finger nucleases. *Science*, 300, 764, 2003.
4. Pardo, B., Gómez-González, B., Aguilera, A., DNA repair in mammalian cells. *Cell Mol. Life Sci.*, 66, 1039–1056, 2009.
5. Sonoda, E., Hochegger, H., Saberi, A., Taniguchi, Y., Takeda, S., Differential usage of non-homologous end-joining and homologous recombination in double strand break repair. *DNA Repair*, 5, 9, 1021–1029, 2006.
6. Arnould, S., Delenda, C., Grizot, S., Desseaux, C., Paques, F., Silva, G.H., Smith., J., The I-CreI meganuclease and its engineered derivatives: Applications from cell modification to gene therapy. *Protein Eng. Des. Sel.*, 24, 27–31, 2011.

7. Epinat, J.C., Arnould, S., Chames, P., Rochaix, P., Desfontaines, D., Puzin, C. *et al.*, A novel engineered meganuclease induces homologous recombination in yeast and mammalian cells. *Nucleic Acids Res.*, 31, 2952–2962, 2003.
8. Smith, J., Grizot, S., Arnould, S., Duclert, A., Epinat, J.C., Chames, P. *et al.*, A combinatorial approach to create artificial homing endonucleases cleaving chosen sequences. *Nucleic Acids Res.*, 34, e149, 2006.
9. Taylor, G.K., Petrucci, L.H., Lambert, A.R., Baxter, S.K., Jarjour, J., Stoddard, B.L., LAHEDES: The LAGLIDADG homing endonuclease database and engineering server. *Nucleic Acids Res.*, gks365, 2012.
10. Curtin, S.J., Voytas, D.F., Stupar, R.M., Genome engineering of crops with designer nucleases. *Plant Genome*, 5, 2, 42–50, 2012.
11. Boissel, S., Jarjour, J., Astrakhan, A., Adey, A., Gouble, A., Duchateau, P., Shendure, J., Stoddard, B.L., Certo, M.T., Baker, D., Scharenberg, A.M., megaTALs: A rare-cleaving nuclease architecture for therapeutic genome engineering. *Nucleic Acids Res.*, 42, 4, 2591–2601, 2013.
12. Kim, Y.G., Cha, J., Chandrasegaran, S., Hybrid restriction enzymes: Zinc finger fusions to Fok I cleavage domain. *PNAS*, 93, 1156–1160, 1996.
13. Lloyd, A., Plaisier, C.L., Carroll, D., Drews, G.N., Targeted mutagenesis using zinc-finger nucleases in *Arabidopsis*. *Proc. Natl. Acad. Sci. USA*, 102, 2232–2237, 2005.
14. Durai, S., Mani, M., Kandavelou, K., Wu, J., Porteus, M.H., Chandrasegaran, S., Zinc finger nucleases: Custom-designed molecular scissors for genome engineering of plant and mammalian cells. *Nucleic Acids Res.*, 18, 5978–5990, 2005.
15. Camenisch, T.D., Brilliant, M.H., Segal, D.J., Critical parameters for genome editing using zinc finger nucleases. *Mini Rev. Med. Chem.*, 8, 669–676, 2008.
16. Maeder, M.L., Thibodeau-Beganny, S., Osiak, A., Wright, D.A., Anthony, R.M., Eichtinger, M., Jiang, T., Foley, J.E., Winfrey, R.J., Townsend, J.A., Unger-Wallace, E., Rapid "open-source" engineering of customized zinc-finger nucleases for highly efficient gene modification. *Molec. Cell*, 31, 2, 294–301, 2008.
17. DeFrancesco, L., Erratum: Move over ZFNs. *Nat. Biotechnol.*, 30, 1, 112–112, 2011.
18. Kim, E., Kim, S., Kim, D.H., Choi, B.S., Choi, I.Y., Kim, J.S., Precision genome engineering with programmable DNA-nicking enzymes. *Genome Res.*, 22, 1327–1333, 2012.
19. Wang, J., Friedman, G., Doyon, Y., Wang, N.S., Li, C.J., Miller, J.C. *et al.*, Targeted gene addition to a predetermined site in the human genome using a ZFN-based nicking enzyme. *Genome Res.*, 22, 7, 1316–26, 2012.
20. Ramirez, C.L., Certo, M.T., Mussolino, C. *et al.*, Engineered zinc finger nickases induce homology-directed repair with reduced mutagenic effects. *Nucleic Acids Res.*, 40, 12, 5560–5568, 2012.

21. Gonzalez, B., Schwimmer, L.J., Fuller, R.P., Ye, Y., Asawapornmongkol, L., Barbas, C.F., Modular system for the construction of zinc-finger libraries and proteins. *Nat. Protoc.*, 5, 791–810, 2010.
22. Sander, J.D., Dahlborg, E.J., Goodwin, M.J., Cade, L., Zhang, F., Cifuentes, D., Pierick, C.J., Selection-free zinc-finger-nuclease engineering by context-dependent assembly (CoDA). *Nat. Methods*, 8, 1, 67, 2011.
23. McConnell, S.A., Takeuchi, R., Pellenz, S., Davis, L., Maizels, N., Monnat, R.J. Jr. *et al.*, Generation of a nicking enzyme that stimulates site-specific gene conversion from the I-AniI LAGLIDADG homing endonuclease. *Proc. Natl. Acad. Sci. USA*, 106, 5099–5104, 2009.
24. Bhakta, M.S. *et al.*, Highly active zinc-finger nucleases by extended modular assembly. *Genome Res.*, 23, 530–538, 2013.
25. Sood, R., Carrington, B., Bishop, K., Jones, M., Rissone, A., Candotti, F., Liu, P., Efficient methods for targeted mutagenesis in zebrafish using zinc-finger nucleases: Data from targeting of nine genes using CompoZr or CoDA ZFNs. *PloS One*, 8, 2, e57239, 2013.
26. Christian, M., Qi, Y., Zhang, Y., Voytas, D.F., Targeted mutagenesis of Arabidopsis thaliana using engineered TAL effector nucleases. *G3*, 3, 1697–1705, 2013.
27. Deng, D., Yan, C., Pan, X., Mahfouz, M., Wang, J., Zhu, J.K. *et al.*, Structural basis for sequence-specific recognition of DNA by TAL effectors. *Science*, 335, 720–723, 2012.
28. Mak, A.N., Bradley, P., Cernadas, R.A., Bogdanove, A.J., Stoddard, B.L., The crystal structure of TAL effector PthXo1 bound to its DNA target. *Science*, 335, 716–719, 2012.
29. Boch, J., Scholze, H., Schornack, S., Landgraf, A., Hahn, S., Kay, S., Lahaye, T., Nckstadt, A., Bonas, U., Breaking the code of DNA binding specificity of TAL-type III effectors. *Science*, 326, 1509–1512, 2009.
30. Voytas, D.F., Plant genome engineering with sequence-specific nucleases. *Annu. Rev. Plant Biol.*, 64, 327–350, 2013.
31. Mussolino, C., Morbitzer, R., Lütge, F., Dannemann, N., Lahaye, T., Cathomen, T., A novel TALE nuclease scaffold enables high genome editing activity in combination with low toxicity. *Nucleic Acids Res.*, 39, 9283–93, 2011.
32. Juillerat, A., Dubois, G., Valton, J., Thomas, S., Stella, S., Maréchal, A., Langevin, S., Benomari, N., Bertonati, C., Silva, G.H. *et al.*, Comprehensive analysis of the specificity of transcription activator-like effector nucleases. *Nucleic Acids Res.*, 42, 5390–5402, 2014.
33. Zhang, Y., Zhang, F., Li, X., Baller, J.A., Qi, Y., Starker, C.G., Voytas, D.F., Transcription activator-like effector nucleases enable efficient plant genome engineering. *Plant Physiol.*, 161, 20–27, 2013.
34. Li, T., Liu, B., Spalding, M.H., Weeks, D.P., Yang, B., High-efficiency TALEN-based gene editing produces disease-resistant rice. *Nat. Biotechnol.*, 30, 390–392, 2012b.

35. Marraffini, L.A. and Sontheimer, E.J., CRISPR interference limits horizontal gene transfer in staphylococci by targeting DNA. *Science*, 322, 1843–1845, 2008.
36. Grissa, I., Vergnaud, G., Pourcel, C., The CRISPRdb database and tools to display CRISPRs and to generate dictionaries of spacers and repeats. *BMC Bioinformatics*, 8, 172, 2007.
37. Deltcheva, E., Chylinski, K., Sharma, C.M., Gonzales, K., Chao, Y., Pirzada, Z.A., Eckert, M.R., Vogel, J., Charpentier, E., CRISPR RNA maturation by trans-encoded small RNA and host factor RNase III. *Nature*, 471, 602–607, 2011.
38. Jinek, M., Chylinski, K., Fonfara, I., Hauer, M., Doudna, J.A., Charpentier, E., A programmable dual-RNA-guided DNA endonuclease in adaptive bacterial immunity. *Science*, 337, 816–821, 2012.
39. Gasiunas, G., Barrangou, R., Horvath, P., Siksnys, V., Cas9–crRNA ribonucleoprotein complex mediates specific DNA cleavage for adaptive immunity in bacteria. *Proc. Natl. Acad. Sci.*, 109, E2579–E2586, 2012.
40. Jinek, M., East, A., Cheng, A., Lin, S., Ma, E., Doudna, J., RNA programmed genome editing in human cells. *eLife*, 2, e00471, 2013.
41. Jinek, M., Jiang, F., Taylor, D.W., Sternberg, S.H., Kaya, E., Ma, E., Anders, C. *et al.*, Structures of Cas9 endonucleases reveal RNA-mediated conformational activation. *Science*, 343, 1247997, 2014.
42. Mali, P., Yang, L., Esvelt, K.M., Aach, J., Guell, M., DiCarlo, J.E. *et al.*, RNA-guided human genome engineering via Cas9. *Science*, 339, 823–826, 2013a.
43. Cong, L., Ran, F.A., Cox, D., Lin, S., Barretto, R., Habib, N. *et al.*, Multiplex genome engineering using CRISPR/Cas systems. *Science*, 339, 819–823, 2013.
44. Fu, Y., Foden, J.A., Khayter., C., Maeder, M.L., Reyon, D., Joung, J.K. *et al.*, High-frequency off-target mutagenesis induced by CRISPR-Cas nucleases in human cells. *Nat. Biotechnol.*, 31, 822–826, 2013.
45. Ran, F.A., Hsu, P.D., Lin, C.Y., Gootenberg, J.S., Konermann, S., Trevino, A.E. *et al.*, Double nicking by RNA-guided CRISPR/Cas9 for enhanced genome editing specificity. *Cell*, 154, 1380–1389, 2013.
46. Fu, Y., Sander, J.D., Reyon, D., Cascio, V.M., Joung, J.K., Improving CRISPR-Cas nuclease specificity using truncated guide RNAs. *Nat. Biotechnol.*, 32, 279–284, 2014.
47. Mali, P., Aach, J., Stranges, P.B., Esvelt, K.M., Moosburner, M., Kosuri, S. *et al.*, CAS9 transcriptional activators for target specificity screening and paired nickases for cooperative genome engineering. *Nat. Biotechnol.*, 31, 833–838, 2013b.
48. Pattanayak, V., Lin, S., Guilinger, J.P., Ma, E., Doudna, J.A., Liu, D.R., High-throughput profiling of off-target DNA cleavage reveals RNA-programmed Cas9 nuclease specificity. *Nat. Biotechnol.*, 31, 839–843, 2013.
49. Sternberg, S.H., Redding, S., Jinek, M., Greene, E.C., Doudna, J.A., DNA interrogation by the CRISPR RNA-guided endonuclease Cas9. *Nature*, 507, 62–7, 2014.

50. Arnould, S., Perez, C., Cabaniols, J.P., Smith, J., Gouble, A., Grizot, S. *et al.*, Engineered I-Cre I derivatives cleaving sequences from the human XPC gene can induce highly efficient gene correction in mammalian cells. *J. Mol. Biol.*, 371, 49–65, 2007.
51. Grizot, S., Smith, J., Daboussi, F., Prieto, J., Redondo, P., Merino, N. *et al.*, Efficient targeting of a SCID gene by an engineered single-chain homing endonuclease. *Nucleic Acids Res.*, 37, 5405–5419, 2009.
52. Gao, H., Smith, J., Yang, M., Jones, S., Djukanovic, V., Nicholson, M.G., West, A., Bidney, D., Falco, S.C., Jantz, D., Lyznik., L.A., Heritable targeted mutagenesis in maize using a designed endonuclease. *Plant J.*, 61, 176–187, 2010.
53. Stoddard, B.L., Homing endonucleases: From microbial genetic invaders to reagents for targeted DNA modification. *Structure*, 19, 7–15, 2011.
54. Cathomen, T. and Joung, K.J., Zinc-finger nucleases: The next generation emerges. *Mol. Ther.*, 16, 1200–1207, 2008.
55. Scott, C.T., The zinc finger nuclease monopoly. *Nat. Biotechnol.*, 23, 915–918, 2005.
56. Alwin, S. *et al.*, Custom zinc-finger nucleases for use in human cells. *Mol. Ther.*, 12, 610–617, 2005.
57. Sakuma, T., Ochiai, H., Kaneko, T., Mashimo, T., Tokumasu, D., Sakane, Y. *et al.*, Repeating pattern of non-RVD variations in DNA-binding modules enhances TALEN activity. *Sci. Rep.*, 3, 3379, 2013.
58. Doyle, E.L., Booher, N.J., Standage, D.S., Voytas, D.F., Brendel, V.P., Vandyk, J.K., Bogdanove, A.J., TAL Effector-Nucleotide Targeter (TALE-NT) 2.0: Tools for TAL effector design and target prediction. *Nucleic Acids Res.*, 40, W117–W122, 2012.
59. Li, L., Piatek, M.J., Atef, A., Piatek, A., Wibowo, A., Fang, X., Sabir, J.S., Zhu, J.K., Mahfouz, M.M., Rapid and highly efficient construction of TALE-based transcriptional regulators and nucleases for genome modification. *Plant Mol. Biol.*, 78, 407–416, 2012.
60. Xiao, A., Wu, Y., Yang, Z., Hu, Y., Wang, W., Zhang, Y., Kong, L., Gao, G., Zhu, Z., Lin, S., Zhang, B., EENdb: A database and knowledge base of ZFNs and TALENs for endonuclease engineering. *Nucleic Acids Res.*, 41, D415–D422, 2013.
61. Mao, Y., Zhang, H., Xu, N. *et al.*, Application of the CRISPR–Cas system for efficient genome engineering in plants. *Mol. Plant*, 6, 2008, 2013.
62. Li, J.F., Norville, J.E., Aach, J., McCormack, M., Zhang, D., Bush, J., Church, G.M., Sheen, J., Multiplex and homologous recombination-mediated genome editing in *Arabidopsis* and *Nicotiana benthamiana* using guide RNA and Cas9. *Nat. Biotechnol.*, 31, 8, 688–691, 2013.
63. Xu, R., Li, H., Qin, R., Wang, L., Li, L., Wei, P., Yang, J., Gene targeting using the *Agrobacterium tumefaciens*-mediated CRISPR-Cas system in rice. *Rice*, 7, 1, 5, 2014.
64. Liang, Z., Chen, K., Li, T. *et al.*, Efficient DNA-free genome editing of bread wheat using CRISPR/Cas9 ribonucleoprotein complexes. *Nat. Commun.*, 8, 2017.

65. Lowder, L.G., Zhang, D., Baltes, N.J., Paul, J.W., Tang, X., Zheng, X., Voytas, D.F., Hsieh, T.F., Zhang, Y., Qi, Y., A CRISPR/Cas9 toolbox for multiplexed plant genome editing and transcriptional regulation. *Plant Physiol.*, 169, 2, 971–985, 2015.
66. Baltes, N.J., Gil-Humanes, J., Cermak, T., Atkins, P.A., Voytas, D.F., DNA replicons for plant genome engineering. *Plant Cell*, 26, 151–163, 2014.
67. Yin, C., Downey, S.I., Klages-Mundt, N.L., Ramachandran, S., Chen, X., Szabo, L.J. *et al.*, Identification of promising host-induced silencing targets among genes preferentially transcribed in haustoria of Puccinia. *BMC Genom.*, 16, 579, 2015.
68. Ali, Z., Abulfaraj, A., Idris, A., Ali, S., Tashkandi, M., Mahfouz, M.M., CRISPR/Cas9-mediated viral interference in plants. *Genome Biol.*, 16, 238, 2015.
69. Wolter, F. and Puchta, H., Knocking out consumer concerns and regulator's rules: Efficient use of CRISPR/Cas ribonucleoprotein complexes for genome editing in cereals. *Genome Biol.*, 18, 1, 43, 2017.
70. Kim, S., Kim, D., Cho, S.W., Kim, J., Kim, J.S., Highly efficient RNA-guided genome editing in human cells via delivery of purified Cas9 ribonucleoproteins. *Genome Res.*, 24, 2014, 1012–1019.
71. Svitashev, S., Schwartz, C., Lenderts, B., Young, J.K., Mark Cigan, A., Genome editing in Maize directed by CRISPR-Cas9 ribonucleoprotein complexes. *Nat. Commun.*, 7, 13274, 2016.
72. Dianov, G.L. and Hübscher, U., Mammalian base excision repair: The forgotten archangel. *Nucleic Acids Res.*, 41, 3483–3490, 2013.
73. Davis, L. and Maizels, N., Homology-directed repair of DNA nicks via pathways distinct from canonical double-strand break repair. *Proc. Natl. Acad. Sci. USA*, 111, 10, E924–E932, 2014.
74. Aouida, M., Eid, A., Ali, Z., Cradick, T., Lee, C., Deshmukh, H., Atef, A., AbuSamra, D., Gadhoum, S.Z., Merzaban, J., Bao, G., Mahfouz, M., Efficient fdCas9 synthetic endonuclease with improved specificity for precise genome engineering. *PLoS One*, 10, e0133373, 2015.
75. Kleinstiver, B.P., Pattanayak, V., Prew, M.S., Tsai, S.Q., Nguyen, N.T., Zheng, Z., Joung, J.K., High-fidelity CRISPR–Cas9 nucleases with no detectable genome-wide off-target effects. *Nature*, 529, 490–495, 2016.
76. Anderson, J.E., Kantar, M.B., Kono, T.Y. *et al.*, A roadmap for functional structural variants in the soybean genome.G3. *Genes Genomes Genet.*, 4, 7, 1307–1318, 2014.
77. Slaymaker, I.M., Gao, L., Zetsche, B., Scott, D.A., Yan, W.X., Zhang, F., Rationally engineered Cas9 nucleases with improved specificity. *Science*, 351, 84–88, 2016.
78. Begemann, M.B., Gray, B.N., January, E., Gordon, G.C., He, Y., Liu, H., Wu, X., Brutnell, T.P., Mockler, T.C., Oufattole, M., Precise insertion and guided editing of higher plant genomes using Cpf1 CRISPR nucleases. *Sci. Rep.*, 7, 1, 11606, 2017.
79. Kim, D., Alptekin, B., Budak, H., CRISPR/Cas9 genome editing in wheat. *Funct. Integr. Genom.*, 1–11, 2017.

80. Liu, L., Chen, P., Wang, M., Li, X., Wang, J., Yin, M., Wang, Y., C2c1-sgRNA complex structure reveals RNA-guided DNA cleavage mechanism. *Molec. Cell*, 65, 2, 310–322, 2017.
81. Tang, X., Lowder, L.G., Zhang, T., Malzahn, A.A., Zheng, X., Voytas, D.F., Zhong, Z., Chen, Y., Ren, Q., Li, Q., Kirkland, E.R., A CRISPR–Cpf1 system for efficient genome editing and transcriptional repression in plants. *Nat. Plants*, 3, 3, 17018, 2017.
82. Xu, R., Qin, R., Li, H., Li, D., Li, L., Wei, P., Yang, J., Generation of targeted mutant rice using a CRISPR-Cpf1 system. *Plant Biotechnol. J.*, 15, 6, 713–717, 2017.
83. Zaidi, S.S.E.A., Mahfouz, M.M., Mansoor, S., CRISPR-Cpf1: A new tool for plant genome editing. *Trends Plant Sci.*, 2017.
84. Li, M., Li, X., Zhou, Z. *et al.*, Reassessment of the four yield-related genes Gn1a, DEP1, GS3, and IPA1 in rice using a CRISPR/Cas9 system. *Frontiers Plant Sci.*, 7, 2016.
85. Osakabe, Y., Watanabe, T., Sugano, S.S., Ueta, R., Ishihara, R., Shinozaki, K., Osakabe, K., Optimization of CRISPR/Cas9 genome editing to modify abiotic stress responses in plants. *Sci. Rep.*, 6, 26685, 2016.
86. Zhou, H., He, M., Li, J., Chen, L., Huang, Z., Zheng, S., Zhu, L., Ni, E., Jiang, D., Zhao, B. and Zhuang, C., Development of commercial thermo-sensitive genic male sterile rice accelerates hybrid rice breeding using the CRISPR/Cas9-mediated TMS5 editing system. *Scientific reports*, 6: 37395, 2016.
87. Shen, R., Wang, L., Liu, X., Wu, J., Jin, W., Zhao, X., Xie, X., Zhu, Q., Tang, H., Li, Q., Chen, L., Genomic structural variation-mediated allelic suppression causes hybrid male sterility in rice. *Nat. Commun.*, 8, 1, 1310, 2017.
88. Li, X., Wang, Y., Chen, S., Tian, H., Fu, D., Zhu, B., Luo, Y., Zhu, H., Lycopene is enriched in tomato fruit by CRISPR/Cas9-mediated multiplex genome editing. *Frontiers Plant Sci.*, 9, 2018.
89. James, C., Global Status of Commercialized Biotech/GM Crops. ISAAA Brief No. 46, ISAAA, Ithaca, NY, USA, 2013.
90. Zhang, D., Li, Z., Li, J.F., Targeted gene manipulation in plants using the CRISPR/Cas technology. *J. Genet. Genom.*, 43, 5, 251–62, 2016.
91. Leenay, R.T., Maksimchuk, K.R., Slotkowski, R.A., Agrawal, R.N., Gomaa, A.A., Briner, A.E., Barrangou, R., Beisel, C.L., Identifying and visualizing functional PAM diversity across CRISPR-Cas systems. *Mol. Cell*, 62, 1, 137–147, 2016.
92. Jaskula-Ranga, V. and Zack, D.J., grID: A CRISPR-Cas9 guide RNA database and resource for genome-editing. bioRxiv., 2016.
93. Zentner, G.E. and Henikoff, S., Epigenomic editing made easy. *Nat. Biotech.*, 33, 606–607, 2015.
94. Fauser, F., Schiml, S., Puchta, H., Both CRISPR/Cas-based nucleases and nickases can be used efficiently for genome engineering in *Arabidopsis thaliana*. *Plant J.*, 79, 348–359, 2014.

95. Nekrasov, V., Staskawicz, B., Weigel, D. *et al.*, Targeted mutagenesis in the model plant *Nicotiana benthamiana* using Cas9 RNA-guided endonuclease. *Nat. Biotechnol.*, 31, 691–693, 2013.
96. Belhaj, K., Chaparro-Garcia, A., Kamoun, S. *et al.*, Plant genome editing made easy: Targeted mutagenesis in model and crop plants using the CRISPR/Cas system. *Plant Methods*, 9, 39, 2013.
97. Xie, K., Zhang, J., Yang, Y., Genome-wide prediction of highly specific guide RNA spacers for CRISPR–Cas9-mediated genome editing in model plants and major crops. *Mol. Plant*, 7, 5, 923–926, 2014.
98. Xing, H.L., Dong, L., Wang, Z.P. *et al.*, A CRISPR/Cas9 toolkit for multiplex genome editing in plants. *BMC Plant Biol.*, 14, 1, 327, 2014.
99. Qi, L.S., Larson, M.H., Gilbert, L.A., Doudna, J.A., Weissman, J.S., Arkin, A.P., Lim, W.A., Repurposing CRISPR as an RNA-guided platform for sequence-specific control of gene expression. *Cell*, 152, 5, 1173–1183, 2013.
100. Sarrion-Perdigones, A., Vazquez-Vilar, M., Palací, J. *et al.*, GoldenBraid 2.0: A comprehensive DNA assembly framework for plant synthetic biology. *Plant Physiol.*, 162, 3, 1618–1631, 2013.
101. Vazquez-Vilar, M., Bernabé-Orts, J.M., Fernandez-del-Carmen, A., Ziarsolo, P., Blanca, J., Granell, A., Orzaez, D., A modular toolbox for gRNA–Cas9 genome engineering in plants based on the GoldenBraid standard. *Plant Methods*, 12, 1, 10, 2016.
102. Zong, Y., Wang, Y., Li, C., Zhang, R., Chen, K., Ran, Y., Qiu, J.L., Wang, D., Gao, C., Precise base editing in rice, wheat and maize with a Cas9-cytidine deaminase fusion. *Nat. Biotechnol.*, 35, 5, 438, 2017.
103. Shimatani, Z., Kashojiya, S., Takayama, M., Terada, R., Arazoe, T., Ishii, H., Teramura, H., Yamamoto, T., Komatsu, H., Miura, K., Ezura, H., Targeted base editing in rice and tomato using a CRISPR-Cas9 cytidine deaminase fusion. *Nat. Biotechnol.*, 35, 5, 441, 2017.
104. Mahfouz, M.M., Li, L., Shamimuzzaman, M. *et al.*, De novo-engineered transcription activator-like effector (TALE) hybrid nuclease with novel DNA binding specificity creates double-strand breaks. *PNAS*, 108, 2623–2628, 2011.
105. Antunes, M.S., Smith, J.J., Jantz, D., Medford, J.I., Targeted DNA excision in *Arabidopsis* by a re-engineered homing endonuclease. *BMC Biotechnol.*, 12, 1, 86, 2012.
106. Osakabe, K., Osakabe, Y., Toki, S., Site-directed mutagenesis in *Arabidopsis* using custom-designed zinc finger nucleases. *PNAS*, 107, 12034–12039, 2010.
107. Zhang, F., Maeder, M.L., Unger-Wallace, E. *et al.*, High frequency targeted mutagenesis in *Arabidopsis thaliana* using zinc finger nucleases. *PNAS*, 107, 26, 12028–12033, 2010.
108. Zhang, D., Wang, Z., Wang, N. *et al.*, Tissue culture-induced heritable genomic variation in rice, and their phenotypic implications. *PloS One*, 9, 5, 2014.
109. Pater, S., Pinas, J.E., Hooykaas, P.J. *et al.*, ZFN-mediated gene targeting of the *Arabidopsis* protoporphyrinogen oxidase gene through Agrobacterium-mediated floral dip transformation. *Plant Biotechnol. J.*, 11, 510–515, 2013.

110. Qi, Y., Starker, C.J., Zhang, F., Baltes, N.J., Voytas, D.F., Tailor-made mutations in *Arabidopsis* using zinc finger nucleases. *Methods Mol. Biol.*, 1062, 193–209, 2014.
111. Wright, D.A., Thibodeau-Beganny, S., Sander, J.D., Winfrey, R.J., Hirsh, A.S., Eichtinger, M., Joung, J.K., Standardized reagents and protocols for engineering zinc finger nucleases by modular assembly. *Nat. Protocols*, 1, 3, 1637, 2006.
112. Townsend, J.A., Wright, D.A., Winfrey, R.J., Fu, F., Maeder, M.L., Joung, J.K., Voytas, D.F., High-frequency modification of plant genes using engineered zinc-finger nucleases. *Nature*, 459, 442–445, 2009.
113. Petolino, J.F., Worden, A., Curlee, K., Connell, J., Moynahan, T.L.S., Larsen, C., Russell, S., Zinc finger nuclease-mediated transgene deletion. *Plant Mol. Biol.*, 73, 6, 617–628, 2010.
114. Cantos, C., Francisco, P., Trijatmiko, K.R., Slamet-Loedin, I., Chadha-Mohanty, P.K., Identification of "safe harbor" loci in indica rice genome by harnessing the property of zinc-finger nucleases to induce DNA damage and repair. *Frontiers Plant Sci.*, 5, 302, 2014.
115. Shukla, V.K., Doyon, Y., Miller, J.C., DeKelver, R.C., Moehle, E.A., Worden, S.E., Mitchell, J.C., Arnold, N.L., Gopalan, S., Meng, X., Choi, V.M., Precise genome modification in the crop species *Zea mays* using zinc-finger nucleases. *Nature*, 459, 7245, 437, 2009.
116. Ainley, W.M., Sastry-Dent, L., Welter, M.E. *et al.*, Trait stacking via targeted genome editing. *Plant Biotechnol. J.*, 11, 9, 1126–1134, 2013.
117. Curtin, S.J., Anderson, J.E., Starker, C.G., Baltes, N.J., Mani, D., Voytas, D.F., Stupar, R.M., Targeted mutagenesis for functional analysis of gene duplication in legumes. *Methods Mol. Biol.*, 1069, 25–42, 2013.
118. Cermak, T., Doyle, E.L., Christian, M., Wang, L., Zhang, Y., Schmidt, C. *et al.*, Efficient design and assembly of custom TALEN and other TAL effector-based constructs for DNA targeting. *Nucleic Acids Res.*, 39, e82, 2011.
119. Shan, Q., Wang, Y., Chen, K., Liang, Z., Li, J., Zhang, Y., Zhang, K., Liu, J., Voytas, D.F., Zheng, X., Zhang, Y., Gao, C., Rapid and efficient gene modification in rice and Brachypodium using TALENs. *Mol. Plant*, 2013.
120. Shan, Q., Zhang, Y., Chen, K. *et al.*, Creation of fragrant rice by targeted knockout of the *OsBADH2* gene using TALEN technology. *Plant Biotechnol. J.*, 13, 791–800, 2015.
121. Feng, Z., Mao, Y., Xu, N., Zhang, B., Wei, P., Yang, D.L., Wang, Z., Zhang, Z., Zheng, R., Yang, L., Multigeneration analysis reveals the inheritance, specificity, and patterns of CRISPR/Cas-induced gene modifications in *Arabidopsis*. *Proc. Natl. Acad. Sci. USA*, 111, 4632–7, 2014. PMID:24550464; http://dx.doi.org/10.1073/pnas.1400822111.
122. Liang, Z., Zhang, K., Chen, K. *et al.*, Targeted mutagenesis in *Zea mays* using TALENs and the CRISPR/Cas system. *J. Genet. Genom.*, 41, 63–68, 2014.
123. Haun, W., Coffman, A., Clasen, B.M., Demorest, Z.L., Lowy, A., Ray, E., Retterath, A., Stoddard, T., Juillerat, A., Cedrone, F., Mathis, L., Improved

soybean oil quality by targeted mutagenesis of the fatty acid desaturase 2 gene family. *Plant Biotechnol. J.*, 12, 7, 934–940, 2014.
124. Jiang, W., Zhou, H., Bi, H., Fromm, M., Yang, B., Weeks, D.P., Demonstration of CRISPR/Cas9/sgRNA-mediated targeted gene modification in Arabidopsis, tobacco, sorghum and rice. *Nucleic Acids Res.*, 41, 20, e188–e188, 2013.
125. Woo, J.W., Kim, J., Kwon, S.I., Corvalán, C., Cho, S.W., Kim, H., Kim, S.G., Kim, S.T., Choe, S., Kim, J.S., DNA-free genome editing in plants with pre-assembled CRISPR-Cas9 ribonucleoproteins. *Nat. Biotechnol.*, 33, 11, 1162, 2015.
126. Miao, J., Guo, D., Zhang, J., Huang, Q., Qin, G., Zhang, X., Wan, J., Gu, H., Qu, L.J., Targeted mutagenesis in rice using CRISPR-Cas system. *Cell Res.*, 23, 10, 1233, 2013.
127. Zhou, H., Liu, B., Weeks, D.P., Spalding, M.H., Yang, B., Large chromosomal deletions and heritable small genetic changes induced by CRISPR/Cas9 in rice. *Nucleic Acids Res.*, 42, 17, 10903–10914, 2014.
128. Wang, F., Wang, C., Liu, P., Lei, C., Hao, W., Gao, Y., Liu, Y.G., Zhao, K., Enhanced rice blast resistance by CRISPR/Cas9-targeted mutagenesis of the ERF transcription factor gene OsERF922. *PloS One*, 11, 4, e0154027, 2016.
129. Malzahn, A., Lowder, L., Qi, Y., Plant genome editing with TALEN and CRISPR. *Cell Biosci.*, 7, 1, 21, 2017.
130. Osakabe, Y. and Osakabe, K., Genome editing with engineered nucleases in plants. *Plant Cell Physiol.*, 56, 3, 389–400, 2014.
131. Singh, V., Gohil, N., Ramírez García, R., Braddick, D., Fofié, C.K., Recent advances in CRISPR-Cas9 genome editing technology for biological and biomedical investigations. *J. Cell Biochem.*, 119, 1, 81–94, 2018.

13

Regulation of Gene Expression by Global Methylation Pattern in Plants Development

Vrijesh Kumar Yadav[1]*, Krishan Mohan Rai[2], Nishant Kumar[1] and Vikash Kumar Yadav[3]

[1]State Forensic Science Laboratory, Hotwar, Ranchi, Jharkhand
[2]Fiber and Biopolymer Research Institute, Department of Plant and Soil Science, Texas Tech University, Lubbock, TX
[3]Department of Plant Biology, Uppsala BioCenter, Swedish University of Agricultural Sciences, Uppsala, Sweden

Abstract

Regulation of the copy number of genes through epigenetics is maintained by methylation of nucleic acid, modifications of histones, and change in the structure of chromosome into a high level of packaging. Chromatin modifications are associated with the changes in chromatin states that result from alterations in the histones and modifications in the specific proteins and small RNAs that associate with a genomic region. Methylation of nucleic acid and modifications of histone are considered as the traits concerned mainly with epigenetics. In recent years, regulation of the copy number of genes through methylation of nucleic acid is emerging as a new area of interest in the field of research in plants, which has significant impact on the developmental process and transcriptional regulation. DNA methylation is involved in various aspects in plants like evolution of plant species, defenses, resistance, shoot regeneration, sex determination, and developmental pathways. In genomes, DNA methylation is partitioned into CG and non-CG methylation pattern, where non-CG methylation is known as CHG and CHH. Methylation pattern has been classified into maintenance and *de novo* DNA methylation, which occur by DNA methyl transferase (DNMtase). DNMtase reported in plants is classified mainly into two categories, that is, maintenance methyl transferase and *de novo* methyl transferase. Genome-level study of

**Corresponding author*: yadav.vrijesh@gmail.com

Rintu Banerjee, Garlapati Vijay Kumar, and S.P. Jeevan Kumar (eds.) OMICS-Based Approaches in Plant Biotechnology, (287–302) © 2019 Scrivener Publishing LLC

nucleic acid methylation in plants like *Arabidopsis*, maize, and rice has shown that H3K9me2-dependent pathway, ribonucleic acid directed nucleic acid methylation pathway, and mobile siRNAs are involved in the regulation of the copy number of genes. Recent technological advances have sped up the research in the area of omics-based genomic approaches and high-throughput analysis for the involvement of nucleic acid methylation in plants.

Keywords: Chromatin modifications, nucleic acid methylation, DNA methyl transferase (DNMtase), DNA methylation, *de novo* methylation

13.1 Introduction

Chromatin-like structure of nucleic acid occurs due to tight packaging and nucleosome occupancy at the nucleosome-depleted region (NDR), which determines the pattern of gene expression, which in turn envisages high nucleosome occupancy linked with variable gene expression and low nucleosome occupancy associated with stable gene expression [1–3] (Figure 13.1). Change in chromatin structure as an epigenetic phenomenon is influenced by DNA modifications and histone core modulation.

The term "epigenetics" was first coined by C.H. Waddington (1942). The concept of epigenetics as beyond genetics has changed over time. Now it is believed that epigenetics is evolving by DNA methylation, chromatin

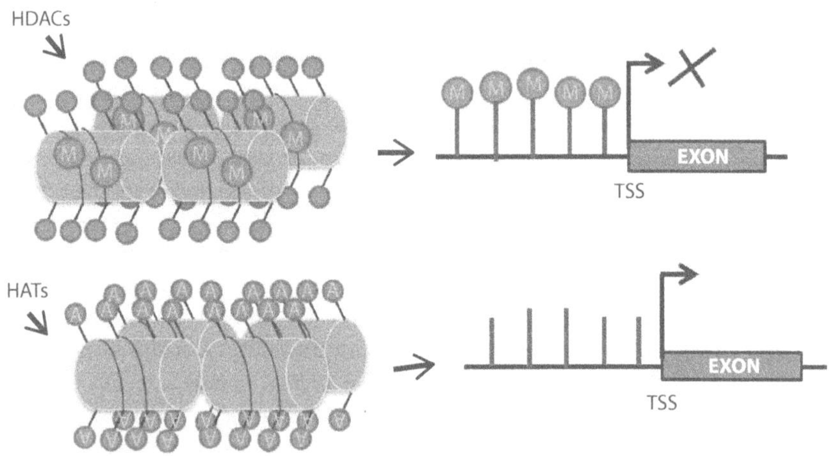

Figure 13.1 Schematic of packaging of eukaryotic genome and histone modification associated with DNA methylation. HDACs enzymes cause deacetylation of histone of methylated DNA (M), which leads to transcriptional silencing. HATs enzymes cause acetylation of histone tails (A) of demethylated DNA, which promotes transcriptional activation.

modifications, and noncoding RNA [4]. Epigenetics describes the stable changes in expression of genes at the time of development and cell division [5]. In recent years, epigenetics has emerged as the area that has significant impact on the eukaryotic cellular development and gene regulation. These alterations in the accessibility of genome are responsible for the differential transcriptional state of different cell types [6].

DNA methylation and histone modifications are major epigenetic marks. Mechanisms of epigenetic modification include the following processes:

1. Chromatin remodeling
2. DNA methylation
3. Histone modification
4. Noncoding RNAs and
5. Deposition of histone variants

The methylation pattern of nucleic acid is a crucial factor in the developmental progress of plants, which is affected by environmental stimuli. DNA methylation, which may persist at the time of replication of nucleic acid during meiosis and mitosis, is relatively stable as compared with other epigenetic phenomena. DNA methylation is not only involved in control of expression patterns of genes but also responsible for the stability of genomes [7–9]. Genome stability is important in various aspects in plant development like defenses, resistance, shoot regeneration, and sex determination. However, transposable DNA elements (mobile DNA) can induce mutations that may lead to genome instability. Methylation of DNA is very crucial both in plant and animal developmental process. DNA methylation is involved in the developmental process of reproductive tissues (asexual or sexual), developmental pathways, the evolution of plant species, plant development and physiology, floral development, endosperm, and growth of tissue in flowering plants and fungi like *Neurospora* [10, 11].

13.2 Nucleic Acid Methylation Targets in the Genome

In nucleic acid methylation, the addition of a methyl group on cytosine sites in the DNA forms 5-methylcytosine, which is conserved throughout in plant and animal kingdoms [12] (Figure 13.2).

In plants, nucleic acid methylation pattern on nucleotide stretch CG, CHG, and CHH is represented as mCG, mCCG, and mCGG [13]. Cytosine methylation imparts various roles in genome regulation in plants as well as in animals. In the genomes of both plants and animals, methylation is

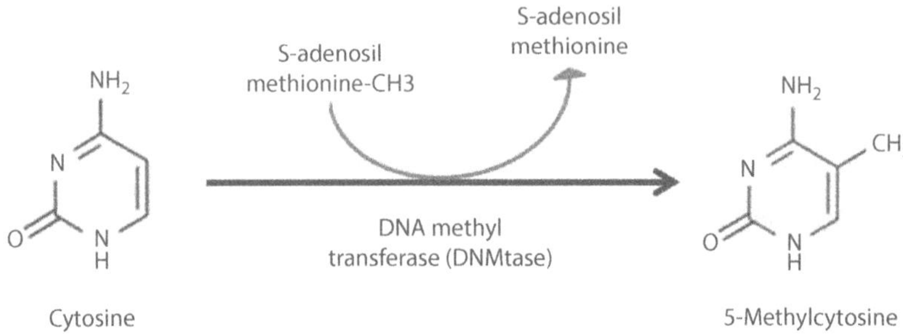

Figure 13.2 Methylation of nucleic acid on cytosine nucleotide site.

partitioned into symmetric nucleotides as CG and nonsymmetric nucleotides as CH (non-CG), where H is any base other than G [14–16]. The non-CG methylation, which is best characterized in plants, is known as CHG and CHH [8, 14]. However, methylation of nucleic acid also occurred beyond the CG, CHG, and CHH nucleotides stretch revealed by whole genome methylation profiling in *Arabidopsis thaliana* [17, 18].

13.3 Nucleic Acid Methyl Transferase (DNMtase)

Methylation of nucleic acid DNA is a robust process that is maintained by DNMtases as well as demethylases of DNA [19]. The methyltransferases are classified into maintenance methyltransferase and *de novo* methyltransferase (Figure 13.3). Methyltransferase such as methyltransferase 1 and chromomethylase 3 comes under the maintenance methyltransferase category. Domain rearrangement methyltransferase has been categorized as *de novo* DNMtase in plants [8, 20–23].

Methyltransferase 1 (MET1) is a homolog of animal DNMtases DNMT1, which is involved in the maintenance of 5mC on symmetric CG methylation [8, 24]. The methylation of nucleic acid at CHG sites is maintained by plant-specific DNMtase CMT3 [12]. However, CHH methylation is maintained by *de novo* DNMtase DRM2 [8, 23]. However, another class of DNMtases, DNMT2, is found in all kingdoms which is involved in the methylation of tRNAAsp [25–26]. However, methylated bases in DNA in plants are controlled through DNA demethylation process by Demeter (DME), Demeter-like 2 (DML2), and Demeter-like 3 (DML3). On the genome level, DNA demethylation in various tissues of plants like central cells and endosperms is maintained by DME [27–30]. Repressor of Silencing1 (ROS1) is involved in the demethylation of genomic regions of genes [31–33].

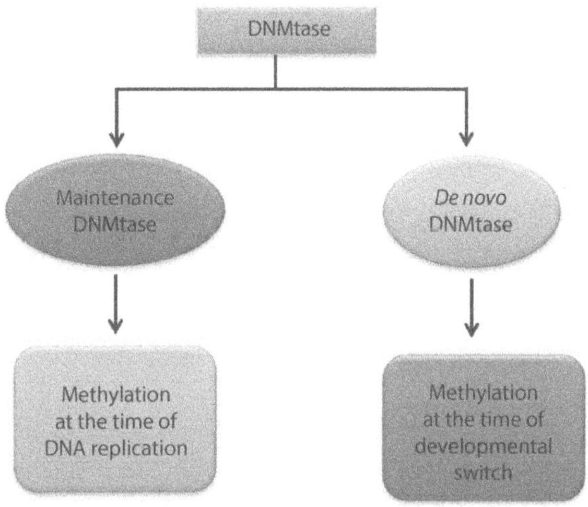

Figure 13.3 Schematic classification of DNA methyltransferase in plants. DNMtases are classified into maintenance DNMtase, which is involved in methylation at the time of nuclei duplication, and *de novo* DNMtase, which is involved in methylation during developmental switch.

Under epigenetic regulation, the most extensively studied methylation pattern of nucleic acid is DNA methylation, which is involved in the following processes:

1. Transcription
2. Genomic imprinting
3. Chromatin structure
4. Embryonic development
5. Chromosomal inactivation

13.4 Genomic DNA Methylation and Expression Pattern

The design of DNA methylation and correlation with the expression patterns in plant developmental stages is yet not well known. However, genome-wide DNA methylation analysis in upstream regulatory region (promoters) in *Arabidopsis* shows the relationship with transcriptional silencing of genes, actively transcribed genes, and transposable (mobile DNA) and upstream regulatory regions [34–38] (Figure 13.4).

DNA methylation design in nature can vary among individual organisms of the same species, which is affected by environmental stimuli.

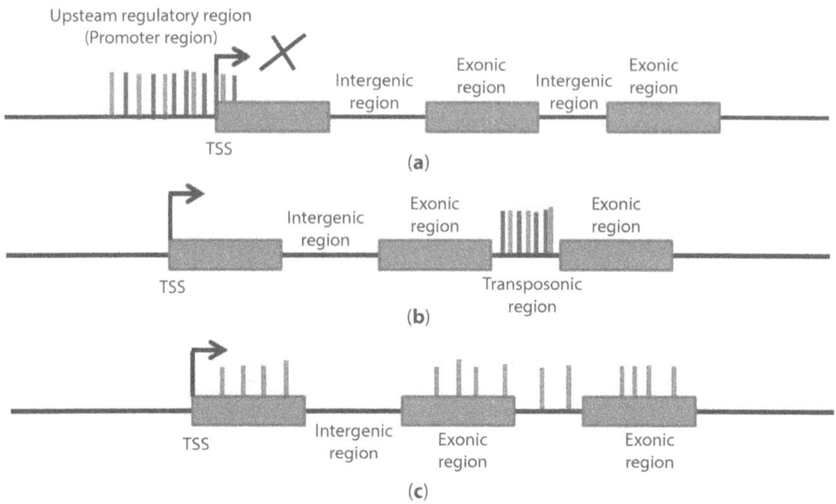

Figure 13.4 Schematic diagram of DNA methylation pattern. Expression of genes in *Arabidopsis* at different position in the genome, i.e., promoter region, intergenic region, and exonic region. Red bar represents DNA methylation on CG region, and blue bar represents DNA methylation on non-CG regions. (a) DNA methylation on promoter suppresses the gene. (b) DNA methylation on intergenic region at transposons results in expression of genes, while loss of methylation results in immature transcript. (c) In the coding regions, CG DNA methylation patterns are responsible for transcription of active genes.

Although it is important for the process of developmental control and responses to different environmental stimuli [39, 40]. Recent studies on *Arabidopsis* mutants have shown that DNA methylation is involved in the process, which promotes out-crossing and prevents self-mating, determination of sexual trait, organogenesis of plants, and genomic imprinting [41–43].

13.5 Pattern of DNA Methylation in Early Plant Life

A recent study has demonstrated that genomic regions rich in CG and CHG methylation pattern remain unchanged in early embryos and mature embryos. On the other hand, genomic regions rich in CHH methylation vary at the early embryo stage compared with mature embryo, and it has been noted that CHH methylation was higher during mature embryos [44].

In the mature embryo, hyper methylation on the CHH region of genome is controlled by the H3K9me2-dependent pathway, RNA directed DNA methylation (RdDM) pathway, and mobile siRNAs mediated process

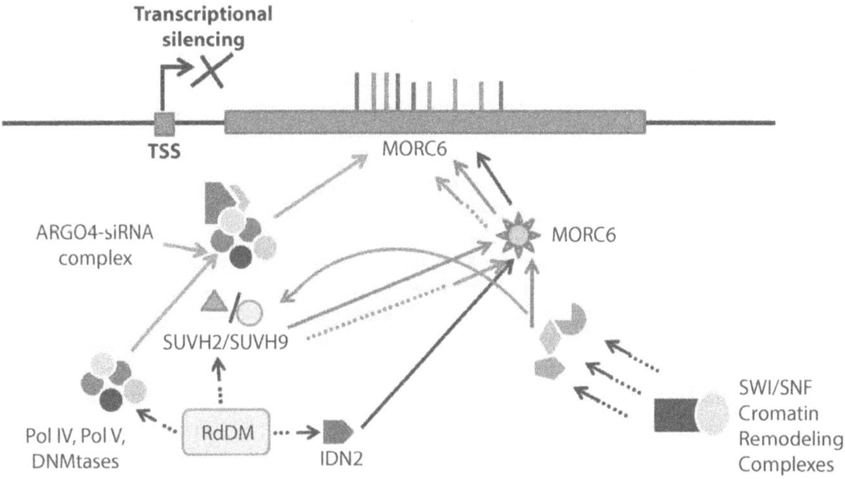

Figure 13.5 Schematic diagram of ribonucleic acid mediated methylation. SUVH2/9 component shows direct interaction (dotted red arrow) or with SWI/SNF components to direct transcriptional silencing (blue arrow). In the next cycle, SUVH2/9, MORC6 complex, IDN2, SWI/SNF complex components SWI3B/SWI3C/SWI3D, and Pol IV/V/DNMtases mediate methylation in nucleic acid. In another cycle, ARGO4–siRNA complex mediates silencing of nucleic acid by methylation.

(Figure 13.5). Genome-level study on DNA methylation has shown that Microrchidia 6 (MORC6) functions in transcriptional silencing at loci targeted by the components of the RdDM pathway (Figure 13.5).

MORC6 is involved in transcriptional silencing of their target loci by heterochromatin condensation or H3K9 dimethylation [45–47]. Hypomethylation of the genomic region originates because of DNA demethylation on the CC methylated region and the low expression level of DNMtases in the endosperm of *Arabidopsis*, maize, and rice [48–51]. Hypomethylation of the genomic region in endosperm leads to CHH hypermethylation in the embryo through siRNAs mediated process. In columella cells and meristematic tissues, CHH hypermethylation persists because of maintenance of DNA methylation as in the mature embryo stage [52].

13.6 DNA Methylation Pattern in Mushroom

Knowledge of the role and pattern of DNA methylation in fungi-like mushroom is scanty. Recent work on mushrooms such as Basidiomycotina and Ascomycotina has revealed a new class of DNMtases, Rad8class, which is fungi-specific and belongs to the SNF2 family. Other DNMtases reported in the fungi *Pleurotus ostreatus* are Dnmt1a, Dnmt1b, Dnmt1c, and Rad8. These DNMtases have organ- and stage-specific expression patterns and show higher expression levels in dikaryotic fungi than in monokaryotic fungi during growth. Dnmt1c is involved in cap maturation, Dnmt1a and Dnmt1b are involved in veil development, whereas Rad8 is involved in early fruiting body formation [53].

13.7 Methylation Pattern in Tumor

Genome-wide methylation analysis in *A. thaliana* has shown that T-DNA of crown galls is highly methylated. However, the promoter region shows a low level of methylation, which is mediated by the siRNA pathway [54] (Figure 13.5). Endo-reduplication is a process that increases the ploidy level in many plant species including *A. thaliana*, which causes methylation changes [55, 56]. It has been observed that the increased level of phytohormones like auxin and cytokinin in the crown gall of *A. thaliana* causes these methylation changes [57, 58] as T-DNA contains gene (IaaM, IaaH) for these phytohormones [59, 60]. Increased level of other phytohormones like salicylic acid (SA), ethylene, and abscisic acid (ABA) has been found in the crown galls of stems [61, 62]. In addition, it has also been reported that in the genome of *A. thaliana*, hypermethylation is rich in heterochromatic regions, which are highly dominated with the mobile DNA (transposable elements) and are involved in transcriptional gene silencing as shown in Ref. [63].

13.8 DNA Methylation Analysis Approaches

In the field of methylation, earlier studies were focused to find out the methylation pattern and extent of methylation of the genes of interest [64]. Later on, during the advent of techniques in the field of research, DNA methylation study was shifted from locus specific to genomic scale and global scale. Construction of genomic maps and single-cell methylome is being aimed in the area of DNA methylation study [65].

1. Locus-specific DNA methylation
2. Genome-wide DNA methylation
3. Global DNA methylation

13.8.1 Locus-Specific DNA Methylation

Numerous methods have been reported for locus-specific DNA methylation analysis, which applied techniques such as

i. Restriction enzymes approaches
ii. Bisulfite conversion approaches
iii. Next-generation sequencing platform (pyrosequencing platform)

These techniques were employed in the estimation of methylation in single-nucleotide polymorphisms, on a particular locus and on CpG sites within the genome or gene of interest. The extent of DNA methylation pattern is determined either qualitatively or quantitatively using the techniques.

13.8.2 Genome-Wide and Global DNA Methylation

Technological advances have geared up research in the fields of genomics and molecular biology entering into an era of rapid discovery, omics-scale approaches, and high-throughput analysis.

- Matrix-assisted laser desorption ionization—time of flight (MALDI-TOF) Sequenome
- Methylation-specific PCR (MSP)
- Bisulfate-Sanger sequencing (BS sequencing)
- Pyrosequencing (next-generation sequencing platform)
- Chromatin immunoprecipitation (CHIP)—CHIP on CHIP assay
- Differential methylation hybridization—tilling array
- Restriction landmark genomic scanning
- Electron spray ionization (ESI)—mass spectrometry
- Luminometric methylation assay (LUMA)
- Whole genome sequence analysis by bioinformatics analysis

Site-specific digestion characteristics of restriction enzymes are employed in DNA methylation studies, which cleave the unmethylated regions of DNA and leave the methylated DNA, and different DNA restriction enzymes show different sensitivity toward methylation pattern in the genome [66]. Genomic-level DNA methylation sensitive restriction enzymes (MREs) cleavage trait is coupled with sequencing technologies and being utilized

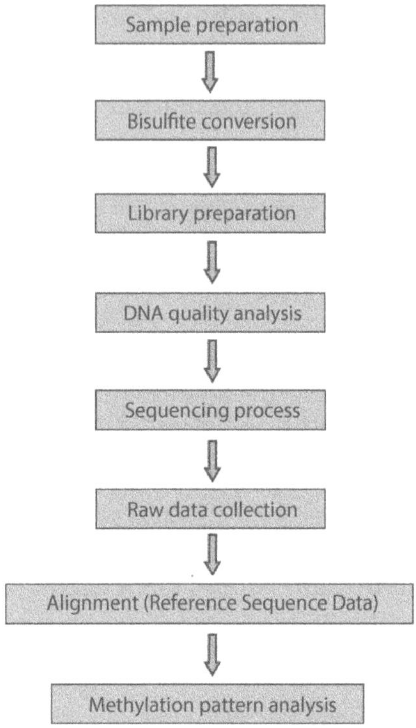

Figure 13.6 Flowchart for the analysis of global DNA methylation in genomes.

in the prediction of methylation status genome wide. Thus, sequencing strategy on the genome level reveals the unmethylated sites within the genome [67, 68] (Figure 13.6).

13.8.3 Whole Genome Sequence Analysis by Bioinformatics Analysis

The biostatistical analysis of DNA methylation on the genome scale includes processing of genome data available, the extent of DNA methylation, profiling of genome data, identification of extent of DNA methylated regions (DMRs), and the status of methylome [69] (Figure 13.7). The biostatistical analysis of DNA methylation study includes the following points:

- Analysis based on relative abundance of methylated and unmethylated loci.
- Analysis of fragments by comparing relative abundance.
- Biostatistical analysis for differential methylation.

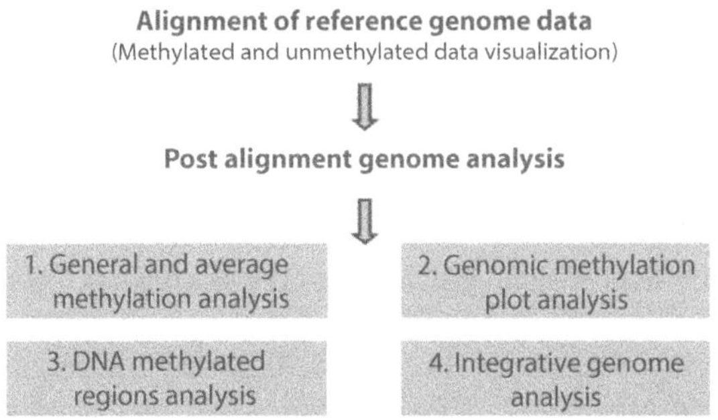

Figure 13.7 Flowchart for bioinformatics analysis of whole genome methylation.

References

1. Tirosh, I. and Barkai, N., Two strategies for gene regulation by promoter nucleosomes. *Genom. Res.*, 18, 7, 1084–1091, 2008.
2. Field, Y., Kaplan, N., Fondufe-Mittendorf, Y., Moore, I.K., Sharon, E., Lubling, Y., Widom, J., Segal, E., Distinct modes of regulation by chromatin encoded through nucleosome positioning signals. *PLoS Comput. Boil.*, 4, 11, 1000216, 2008.
3. Choi, J.K. and Kim, Y.J., Intrinsic variability of gene expression encoded in nucleosome positioning sequences. *Nat. Genet.*, 41, 4, 498, 2009.
4. Choudhuri, S., Epigenetic landscape to small noncoding RNA—Some important milestones in the history of epigenetics research. *Tox. Mech. Meth.*, 21, 4, 252–274, 2011.
5. Waddington, C.H., *The epigenetics of birds*, Cambridge University Press, 2014.
6. Marstrand, T.T. and Storey, J.D., Identifying and mapping cell-type-specific chromatin programming of gene expression. *Proc. Natl. Acad. Sci.*, 111, 6, 645–654, 2014.
7. Chan, S.W.L., Henderson, I.R., Jacobsen, S.E., Gardening the genome: DNA methylation in *Arabidopsis thaliana*. *Nat. Rev. Genet.*, 6, 5, 351, 2005.
8. Law, J.A. and Jacobsen, S.E., Establishing, maintaining and modifying DNA methylation patterns in plants and animals. *Nat. Rev. Genet.*, 11, 204–220, 2010.
9. Gan, H., Wen, L., Liao, S., Lin, X., Ma, T., Liu, J., Song, C.X., Wang, M., He, C., Han, C., Tang, F., Dynamics of 5-hydroxymethylcytosine during mouse spermatogenesis. *Nat. Commun.*, 4, 1995, 2013.
10. Finnegan, E.J., Peacock, W.J., Dennis, E.S., DNA methylation, a key regulator of plant development and other processes. *Curr. Opin. Genet. Dev.*, 10, 217–223, 2000.
11. Filippovich, S., Bachurina, G.P., Kritskii, M.S., Effect of 5-azacytidine on the light-sensitive formation of sexual and asexual reproductive structures in wc-1 and wc-2 mutants of *Neurospora crassa*. *Prikl. Biokhim. Mikrobiol.*, 40, 466–471, 2004.

12. Zhu, Y., Stevens, R.G., Hoffman, A.E., Tjonneland, A., Vogel, U.B., Zheng, T., Hansen, J., Epigenetic impact of long-term shiftwork: Pilot evidence from circadian genes and whole-genome methylation analysis. *Chronobiol. Int.*, 28, 10, 852–861, 2011.
13. Gouil, Q. and Baulcombe, D.C., DNA methylation signatures of the plant chromomethyltransferases. *PLoS Genet.*, 12, 12, 1006526, 2016.
14. Feng, S., Cokus, S.J., Zhang, X., Chen, P.Y., Bostick, M., Goll, M.G. *et al.*, Conservation and divergence of methylation patterning in plants and animals. *Proc. Natl. Acad. Sci. USA*, 107, 19, 8689–8694, 2010.
15. Lister, R. and Mukamel, E., Turning over DNA methylation in the mind. *Front. Neurosci.*, 9, 252, 2015.
16. Schultz, M.D., He, Y., Whitaker, J.W., Hariharan, M., Mukamel, E.A., Leung, D. *et al.*, Human body epigenome maps reveal noncanonical DNA methylation variation. *Nature*, 523, 7559, 212–216, 2015.
17. Cokus, S.J., Feng, S., Zhang, X., Chen, Z., Merriman, B., Haudenschild, C.D. *et al.*, Shotgun bisulphite sequencing of the Arabidopsis genome reveals DNA methylation patterning. *Nature*, 452, 7184, 215–219, 2008(2010).
18. Lister, R., O'Malley, R.C., Tonti-Filippini, J., Gregory, B.D., Berry, C.C., Millar, A.H. *et al.*, Highly integrated single-base resolution maps of the epigenome in *Arabidopsis*. *Cell*, 133, 3, 523–536, 2008.
19. Furner, I.J. and Matzke, M., Methylation and demethylation of the *Arabidopsis* genome. *Curr. Opin. Plant Biol.*, 14, 137–141, 2011.
20. Ronemus, M.J., Galbiati, M., Ticknor, C., Chen, J., Dellaporta, S.L., Demethylation-induced developmental pleiotropy in *Arabidopsis*. *Science*, 273, 654–657, 1996.
21. Lindroth, A.M., Cao, X., Jackson, J.P. *et al.*, Requirement of chromomethylase3 for maintenance of CpXpG methylation. *Science*, 292, 2077–2080, 2001.
22. Cao, X. and Jacobsen, S.E., Locus-specific control of asymmetric and CpNpG methylation by the DRM and CMT3 methyltransferase genes. *Proc. Natl. Acad. Sci. USA*, 99, 16491–16498, 2002.
23. Cao, X., Aufsatz, W., Zilberman, D., Mette, M.F., Huang, M.S., Matzke, M. *et al.*, Role of the DRM and CMT3 methyltransferases in RNA-directed DNA methylation. *Curr. Biol.*, 13, 2212–2217, 2003.
24. Jones, L., Ratcliff, F., Baulcombe, D.C., RNA-directed transcriptional gene silencing in plants can be inherited independently of the RNA trigger and requires Met1 for maintenance. *Curr. Biol.*, 11, 747–757, 2001.
25. Jeltsch, A., Nellen, W., Lyko, F., Two substrates are better than one: Dual specificities for Dnmt2 methyltransferases. *Trends. Biochem. Sci.*, 31, 306–308, 2006.
26. Zemach, A. and Zilberman, D., Evolution of eukaryotic DNA methylation and the pursuit of safer sex. *Curr. Biol.*, 20, 17, 780–785, 2010.
27. Gehring, M., Huh, J.H., Hsieh, T.F. *et al.*, DEMETER DNA glycosylase establishes MEDEA polycomb gene self-imprinting by allele-specific demethylation. *Cell*, 124, 495–506, 2006.

28. Choi, Y., Gehring, M., Johnson, L. *et al.*, DEMETER, a DNA glycosylase domain protein, is required for endosperm gene imprinting and seed viability in *Arabidopsis*. *Cell*, 110, 33–42, 2002.
29. Gehring, M., Bubb, K.L., Henikoff, S., Extensive demethylation of repetitive elements during seed development underlies gene imprinting. *Science*, 324, 1447–1451, 2009.
30. Hsieh, T.F., Ibarra, C.A., Silva, P. *et al.*, Genome-wide demethylation of *Arabidopsis* endosperm. *Science*, 324, 1451–1454, 2009.
31. Agius, F., Kapoor, A., Zhu, J.K., Role of the *Arabidopsis* DNA glycosylase/lyase ROS1 in active DNA demethylation. *Proc. Natl. Acad. Sci. USA*, 103, 11796–11801, 2006.
32. Gong, Z., Morales-Ruiz, T., Ariza, R.R. *et al.*, ROS1, a repressor of transcriptional gene silencing in *Arabidopsis*, encodes a DNA glycosylase/lyase. *Cell*, 111, 803–814, 2002.
33. Zhu, J., Kapoor, A., Sridhar, V.V. *et al.*, The DNA glycosylase/lyase ROS1 functions in pruning DNA methylation patterns in *Arabidopsis*. *Curr. Biol.*, 17, 54–59, 2007.
34. Park, Y.D., Papp, I., Moscone, E.A., Iglesias, V.A., Vaucheret, H., Matzke, A.J.M., Matzke, M.A., Gene silencing mediated by promoter homology occurs at the level of transcription and results in meiotically heritable alterations in methylation and gene activity. *Plant J.*, 9, 2, 183–194, 1996.
35. Stam, M., Viterbo, A., Mol, J.N., Kooter, J.M., Position-dependent methylation and transcriptional silencing of transgenes in inverted T-DNA repeats: Implications for posttranscriptional silencing of homologous host genes in plants. *Molec. Cell. Biol.*, 18, 11, 6165–6177, 1998.
36. Jones, L., Hamilton, A.J., Voinnet, O., Thomas, C.L., Maule, A.J., Baulcombe, D.C., RNA–DNA interactions and DNA methylation in post-transcriptional gene silencing. *Plant Cell*, 11, 12, 2291–2301, 1999.
37. Zhang, X., Yazaki, J., Sundaresan, A., Cokus, S., Chan, S.W.L. *et al.*, Genomewide high-resolution mapping and functional analysis of DNA methylation in *Arabidopsis*. *Cell*, 126, 1189–1201, 2006.
38. Zilberman, D., Gehring, M., Tran, R.K., Ballinger, T., Henikoff, S., Genome-wide analysis of *Arabidopsis thaliana* DNA methylation uncovers an interdependence between methylation and transcription. *Nat. Genet.*, 39, 1, 61, 2007.
39. He, Y.F., Li, B.Z., Li, Z., Liu, P., Wang, Y., Tang, Q., Ding, J., Jia, Y., Chen, Z., Li, L., Sun, Y., Tet-mediated formation of 5-carboxylcytosine and its excision by TDG in mammalian DNA. *Science*, 333, 6047, 1303–1307, 2011.
40. Song, S.J., Poliseno, L., Song, M.S., Ala, U., Webster, K., Ng, C., Beringer, G., Brikbak, N.J., Yuan, X., Cantley, L.C., Richardson, A.L., MicroRNA-antagonism regulates breast cancer stemness and metastasis via TET-family-dependent chromatin remodeling. *Cell*, 154, 2, 311–324, 2013.
41. Shiba, H., Kakizaki, T., Iwano, M., Tarutani, Y., Watanabe, M., Isogai, A., Takayama, S., Dominance relationships between self-incompatibility alleles controlled by DNA methylation. *Nat. Genet.*, 38, 3, 297, 2006.

42. Martin, A., Troadec, C., Boualem, A., Rajab, M., Fernandez, R., Morin, H., Pitrat, M., Dogimont, C., Bendahmane, A., A transposon-induced epigenetic change leads to sex determination in melon. *Nature*, 461, 7267, 1135, 2009.
43. Li, Y. and Sasaki, H., Genomic imprinting in mammals: Its life cycle, molecular mechanisms and reprogramming. *Cell Res.*, 21, 3, 466, 2011.
44. Bouyer, D., Kramdi, A., Kassam, M., Heese, M., Schnittger, A., Roudier, F., Colot, V., DNA methylation dynamics during early plant life. *Gen. Boil.*, 18, 1, 179, 2017.
45. Moissiard, G., Cokus, S.J., Cary, J., Feng, S., Billi, A.C. *et al.*, MORC family ATPases required for heterochromatin condensation and gene silencing. *Science*, 336, 1448–1451, 2012.
46. Lorkovic, Z.J., Naumann, U., Matzke, A.J., Matzke, M., Involvement of a GHKL ATPase in RNA-directed DNA methylation in *Arabidopsis thaliana*. *Curr. Biol.*, 22, 933–938, 2012.
47. Liu, X.S., Wu, H., Ji, X., Stelzer, Y., Wu, X., Czauderna, S., Shu, J., Dadon, D., Young, R.A., Jaenisch, R., Editing DNA methylation in the mammalian genome. *Cell*, 167, 1, 233–247, 2016.
48. Ibarra, C.A., Feng, X., Schoft, V.K., Hsieh, T.F., Uzawa, R., Rodrigues, J.A. *et al.*, Active DNA demethylation in plant companion cells reinforces transposons methylation in gametes. *Science*, 337, 1360–4, 2012.
49. Jullien, P.E., Susaki, D., Yelagandula, R., Higashiyama, T., Berger, F., DNA methylation dynamics during sexual reproduction in *Arabidopsis thaliana*. *Curr. Biol.*, 22, 1825–30, 2012.
50. Rodrigues, J.A., Ruan, R., Nishimura, T., Sharma, M.K., Sharma, R., Ronald, P.C. *et al.*, Imprinted expression of genes and small RNA is associated with localized hypomethylation of the maternal genome in rice endosperm. *Proc. Natl. Acad. Sci. USA*, 110, 7934–7939, 2013.
51. Park, K., Kim, M.Y., Vickers, M., Park, J.S., Hyun, Y., Okamoto, T. *et al.*, DNA demethylation is initiated in the central cells of *Arabidopsis* and rice. *Proc. Natl. Acad. Sci. USA*, 113, 15138, 2016.
52. Kawakatsu, T., Stuart, T., Valdes, M., Breakfield, N., Schmitz, R.J., Nery, J.R. *et al.*, Unique cell-type-specific patterns of DNA methylation in the root meristem. *Nat. Plants.*, 2, 16058, 2016.
53. Huang, R., Ding, Q., Xiang, Y., Gu, T., Li, Y., Comparative analysis of DNA methyltransferase gene family in fungi: A focus on Basidiomycota. *Front. Plant Sci.*, 7, 1556, 2016.
54. Dunoyer, P., Himber, C., Voinnet, O., Induction, suppression and requirement of RNA silencing pathways in virulent *Agrobacterium tumefaciens* infections. *Nat. Genet.*, 38, 258–263, 2006.
55. Lee, H.S. and Chen, Z.J., Protein-coding genes are epigenetically regulated in *Arabidopsis* polyploids. *Proc. Natl. Acad. Sci. USA*, 98, 6753–6758, 2001.
56. Lee, H.O., Davidson, J.M., Duronio, R.J., Endoreplication: Polyploidy with purpose. *Gen. Dev.*, 23, 2461–2477, 2009.

57. Galbraith, D.W., Harkins, K.R., Knapp, S., Systemic endopolyploidy in *Arabidopsis thaliana*. *Plant Physiol.*, 96, 985–989, 1991.
58. Ishida, T., Adachi, S., Yoshimura, M., Shimizu, K., Umeda, M. *et al.*, Auxin modulates the transition from the mitotic cycle to the endocycle in *Arabidopsis*. *Development*, 137, 63–71, 2010.
59. Thomashow, L.S., Reeves, S., Thomashow, M.F., Crown gall oncogenesis: Evidence that a T-DNA gene from the *Agrobacterium* Ti plasmid pTiA6 encodes an enzyme that catalyzes synthesis of indoleacetic acid. *Proc. Natl. Acad. Sci. USA*, 81, 5071–5075, 1984.
60. Akiyoshi, D.E., Klee, H., Amasino, R.M., Nester, E.W., Gordon, M.P., T-DNA of *Agrobacterium tumefaciens* encodes an enzyme of cytokinin biosynthesis. *Proc. Natl. Acad. Sci. USA*, 81, 5994–5998, 1984.
61. Aloni, R., Wolf, A., Feigenbaum, P., Avni, A., Klee, H.J., The never ripe mutant provides evidence that tumor-induced ethylene controls the morphogenesis of *Agrobacterium tumefaciens* induced crown galls on tomato stems. *Plant Physiol.*, 117, 841–849, 1998.
62. Hansen, H. and Grossmann, K., Auxin-induced ethylene triggers abscisic acid biosynthesis and growth inhibition. *Plant Physiol.*, 124, 1437–1448, 2000.
63. Veselov, D., Langhans, M., Hartung, W., Aloni, R., Feussner, I. *et al.*, Development of *Agrobacterium tumefaciens* C58-induced plant tumors and impact on host shoots are controlled by a cascade of jasmonic acid, auxin, cytokinin, ethylene and abscisic acid. *Planta*, 216, 512–522, 2003.
64. Lisanti, S., Omar, W.A., Tomaszewski, B., De Prins, S., Jacobs, G., Koppen, G. *et al.*, Comparison of methods for quantification of global DNA methylation in human cells and tissues. *PLoS One*, 8, e79044, 2013.
65. Laird, P.W., Principles and challenges of genomewide DNA methylation analysis. *Nat. Rev. Genet.*, 11, 191–203, 2010.
66. New England BioLabs. https://www.neb.com/tools-and-resources/selection-charts/dam-dcm-and-cpg-methylation.
67. Maunakea, A.K., Nagarajan, R.P., Bilenky, M., Ballinger, T.J., D'Souza, C., Fouse, S.D. *et al.*, Conserved role of intragenic DNA methylation in regulating alternative promoters. *Nature*, 466, 253–257, 2010.
68. Li, D., Zhang, B., Xing, X., Wang, T., Combining MeDIP-seq and MRE-seq to investigate genome-wide CpG methylation. *Methods*, 72, 29–40, 2015.
69. Bock, C., Analysing and interpreting DNA methylation data. *Nat. Rev. Genet.*, 13, 705–719, 2012.

14

High-Throughput Phenotyping: Potential Tool for Genomics

Kalyani M. Barbadikar[1]*, Divya Balakrishnan[1], C. Gireesh[1], Hemant Kardile[2], Tejas C. Bosamia[3] and Ankita Mishra[4]

[1]Biotechnology Section, ICAR-Indian Institute of Rice Research (IIRR), Rajendranagar, Hyderabad, Telangana, India
[2]Biotechnology Section, ICAR-Central Potato Research Institute (CPRI), Shimla, Himachal Pradesh, India
[3]Biotechnology Section, ICAR-Directorate of Groundnut Research (DGR), Junagadh, Gujarat, India
[4]Institute of Biotechnology, Professor Jayashankar Telangana State Agricultural University (PJTSAU), Hyderabad, India

Abstract
The progress and achievements of genomics need to be complemented with precise phenotyping for establishing accurate relationship between genotype, phenotype, and environment. High-throughput plant phenotyping is the comprehensive assessment of plant complex traits based on physiological parameters using mechanization, sensors, cameras, and robotics for genetic gains. Physiological traits like growth dynamics, photosynthetic status, water content, shoot biomass, yield traits, panicle traits, root architecture, imbibitions and germination rates, height, flowering time, etc. are determined in field as well as in the controlled conditions. Various high-throughput phenotyping (HTP) platforms and pipelines determine plant traits nondestructively on the basis of imaging using a combination of different wavelengths or cameras at different stages of plant life cycle. HTP can be employed in determining the quantitative trait loci, marker-assisted selection, association mapping, forward and reverse genetics including functional annotation of genes, and establishing relationship between the gene and its function. In this chapter, we discuss HTP including imaging techniques, platforms in field and controlled conditions, and potential applications in crop improvement.

**Corresponding author*: kalyaniaau@gmail.com

Rintu Banerjee, Garlapati Vijay Kumar, and S.P. Jeevan Kumar (eds.) OMICS-Based Approaches in Plant Biotechnology, (303–322) © 2019 Scrivener Publishing LLC

Keywords: Phenome, high-throughput phenotyping, genotype, image processing, crop improvement

14.1 Introduction

The growing concern regarding food security demands crop improvement for sustainable agriculture. Tremendous progress has been observed in the past decade in the field of genomics, especially high-throughput genotyping and next-generation sequencing technologies. This progress is nevertheless not complemented by high-throughput plant phenotyping, and it still remains a bottleneck in the crop improvement. Precision farming with reduced input costs, higher productivity, and higher income for farmers is a need of an hour to meet the growing demand for food grain production. A way forward to achieve this is the conglomeration of traditional methods, mechanization, robotics, sensors, bioinformatics, geoinformatics, space technology, automated data recordings, and statistics. Plant phenotyping is the comprehensive assessment of plant complex traits based on physiological parameters. It comprehensively assesses plant growth, performance, and composition, involving the effects of the physiological trait to biochemical or biophysical processes. It involves mechanized noninvasive phenotyping on a large scale, which provides far better resolution, precision, and handling with reduced cost and labor than conventional phenotyping and hence termed high throughput. High-throughput plant phenotyping is essential for assessing the effect of genomic variants to decipher complex traits and to hasten the advancements in understanding gene function and environmental responses. In today's scenario, such studies need much recognition, development, and adoption of precise basic and applied studies.

14.2 Relation of Phenotype, Genotype and Environment

The major aim of genetics is to understand phenotypic characteristics and their variation developed through a complex network of interactions between genetic and environmental factors. The phenotype is the expression of a genotype as a result of its interaction with its environment in the ecosystem. The characterization of phenotype in multiple levels considering various environmental and external factors affecting the phenotype collectively results in phenomics [1, 2]. Phenomics is the translation of genes or the whole genome into the phenotype of plants through recent advances in genomics, and analysis of large datasets relating to the traits under consideration. It has broad applications in applied and basic biology and crop improvement to dissect the complex genetic mechanisms. The phenotype is the outcome of genotype, gene interactions, along with environmental influence. Plant phenotyping involves various levels of dimensions and resolution of a genotype from the molecular to the whole plant, and the environments, from controlled conditions, field, and natural ecosystems [3]. Integration of this information is essential to understand a genotype and design the breeding and genetics studies for crop improvement (Figure 14.1). However, characterization of phenomes lags much behind the developments in the area to characterize genomes.

As the phenotype is the outcome of genotype and its interaction with the environment, to obtain precise phenotypic variations, it is very much essential to replicate the environmental conditions for screening and cataloguing every phenotypic variation that occurs in the genotype. Multidimensional, high-resolution data on agronomical, physiological, and morphological traits describing the phenotype in optimal, biotic, and abiotic stress conditions would enable mapping of genetic elements to biological function at the desired level of accuracy. Detection of large numbers of proteins and metabolites is becoming popular and helps in deciphering the genotypic variation. However, there are several challenges for their accessibility and expenditure incurred. Physiological measurements of processes such as photosynthesis, nutrient uptake, and transport are common, but there is a lack of high throughput in large-scale measurements. Therefore, imaging and spatial technologies are increasingly gaining acceptance in large-scale phenotyping, but accuracy is the lacuna. A combination of technologies with the proper analysis is very much needed for precise estimation of phenomes. High-throughput phenotyping (HTP) platforms and computer algorithms based on image analysis are available for laboratory and greenhouse settings (http://www.lemnatec.com/; http://www.plantaccelerator.org.au/) and are not used in a widespread manner. Remote sensing technologies can help in identifying unique phenotypic variation in transitional plant breeding if they are cost-effective [4]. Phenomics studies are resource intensive due to their high throughput and will support several studies due to large-scale data generation, and the quality of the data depends on the existing genetic diversity, growth conditions employed, phenotypic assays, and further collection, storage, and interpretation of data.

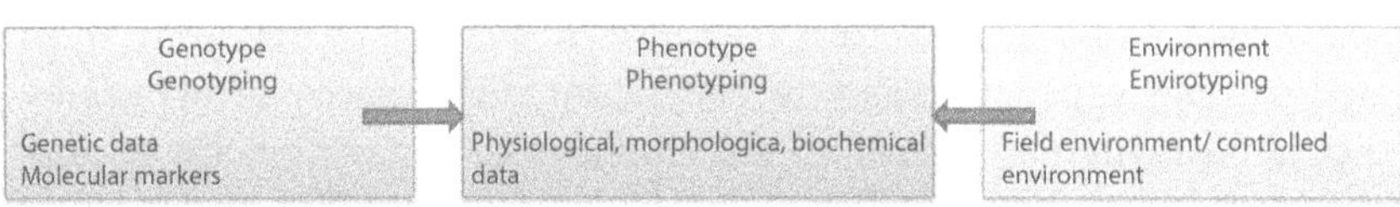

Figure 14.1 Relationship between genotype, phenotype, and environment.

Even though genomic studies have been done extensively in many of the major and minor food, horticultural, and industrial crops, there exists a gap in achieving required targets in crop improvement mainly due to the limitations in appropriate phenotyping [5, 6]. The availability of HTP, analytical and bioinformatic approaches to handling large-scale data, and the development of biological models to link phenome with genome will help to explore complex biological phenomena in the future. In the applied research view, large-scale, multi-environmental, and multipopulation studies are gaining momentum due to the precise information they are generating. General and specific adaptation of genotypes to different ecologies is due to interaction effects of genotype and genotype × environment [7]. Selection of the suitable environmental condition to maximize the grain yield requires knowledge on the different components contributing to phenotypic variations [8]. Studies on genotype × environment interaction on grain yield have been reported for different ecosystems. Many of these studies have reported that genotype × environment interaction plays a major role as compared to genotypic effects in yield performance [9]. HTP with large-scale mechanized methods results in a higher resolution, precision, and reduced labor costs than conventional phenotyping. This will enable the accurate identification of the specific role of genotype and environment and provides the information on the linkage of plant genomes to phenomes [6].

14.3 Features of HTP

Traditional phenotyping is majorly labor- intensive, time consuming, low throughput, costly, and frequently destructive to plants with low phenotyping efficiency [10]. Phenomics utilizes the large-scale collection of phenotypic data and analysis compared to traditional phenotyping and has wider applications in crop improvement [11]. Comprehensive assessment of plant growth and its physiological and biochemical changes in response to environment helps in understanding gene function and environmental responses in an advanced level [12]. HTP is based on automated plant handling systems in smart houses and greenhouses with anticipatory environmental controls for trait interpretation to exploit phenotypic variation, closing the "phenotyping gap." Automation enables flexible growth conditions to elicit and measure stress responses in plants in a relatively short period of time to generate growth curves, measure relative growth rates, and obtain the time of maximum growth [13]. The imaging helps acquire an image in particular wavelength at regular intervals. The summary of various image-based phenotyping is given in Table 14.1.

Table 14.1 Summary of image-based phenotyping.

Imaging	Tools	Parameters	Reference
Visible light imaging	Cameras sensitive in the visible spectral range	Projected area, growth dynamics, shoot biomass, yield traits, panicle traits, root architecture, imbibition and germination rates, early embryonic axis growth, height, size morphology, flowering time	[22, 23]
Fluorescence imaging	Fluorescence cameras and setups	Photosynthetic status (variable fluorescence), quantum yield, nonphotochemical quenching, leaf health status, shoot architecture	[24, 25]
Thermal imaging	Near-infrared cameras	Canopy or leaf temperature, insect infestation of grain	[26, 27]
Near-infrared imaging	Near-infrared cameras, multispectral line scanning cameras, active thermography	Water content composition parameters for seeds, leaf area index	[28, 29]
Hyperspectral imaging	Near-infrared instruments, spectrometers, hyperspectral cameras, thermal cameras	Leaf and canopy water status, leaf and canopy health status, panicle health status, leaf growth, coverage density	[24, 30, 31]
3D imaging	Stereo camera systems, time-of-flight cameras	Shoot structure, leaf angle distributions, canopy structure, root architecture, height	[32, 33]

(Continued)

Table 14.1 Summary of image-based phenotyping. (*Continued*)

Imaging	**Tools**	**Parameters**	**Reference**
Laser imaging	Laser scanning instruments with widely different ranges	Shoot biomass and structure, leaf angle distributions, canopy structure, root architecture, height, stem	[34, 35]
MRI	Magnetic resonance imagers	Morphometric parameters in 3D, water content	[36–38]
PET	Positron emission detectors for short-lived isotopes	Transport partitioning, sectorality, flow velocity	[37]
CT	X-ray computed tomography and X-ray digital radiography	Tillers, morphometric parameters in 3D, grain quality	[39, 40]

Image-based phenotyping approaches provide the great promise to monitor the general plant growth and the effects of biotic and abiotic stress conditions. Every platform has its own specialties in terms of measuring the growth parameters; same platforms can be further exploited to monitor the plant health under the biotic and abiotic stress conditions. It is evident from the many studies on the abiotic stresses, where plant growth has often become limited. The effect of this kind of stress on growth parameter generally remained unnoticeable at the early stage and often remains undetected. Recording the observation on these types of parameters with the human eye is difficult and generally ends up with the error-prone phenotyping. Phenotyping of a large number of germplasms becomes very laborious if we want to analyze several phenotypes. Effect of various stresses sustains for a long time; measurement of this becomes really difficult. Continuous monitoring of plant phenotype is required to analyze the effect of stress conditions. As we know, the phenotype is the result of interaction between a genetic component of plant and its environment. Measurement of the environmental component is also very important because phenotypes respond to the environment and also shape their environment. Life history can influence the expression and intensity of a specific phenotype at certain developmental stages, and its exact estimation often becomes difficult. In order to get rid of these major limitations, we require digital vision. Digital eye in terms of image-based platforms can efficiently monitor the plant growth by recoding the observations on the various growth parameters; for instance, RGB imaging platforms have been successfully used for the screening of the water stress and salt-tolerant wheat and barley genotypes [14]. Chlorophyll content can be the indicator of the plant health; under the stress conditions, its availability becomes limited. Fluorescence imagining platform can be efficiently utilized for measuring the chlorophyll fluorescence. The same platform has been utilized for the chlorophyll fluorescence measurement for quantitative assessment of drought survival in Arabidopsis [15].

Similar is the case under biotic stress conditions where characterizing plant–pathogen interactions with such methods also has many challenges, in part due to the diverse range of symptoms and multiple scales at which the disease occurs. Image-based phenotyping methods enable quantification of spatial and temporal dynamics for plant–pathogen interactions. It allows measurement of symptom area with increased time resolution and automated imaging, which, in turn, facilitated to observe an enhanced rate of early disease accumulation for the pathogens. It has demonstrated the potential of imagining platform for accessing the pathogen's ability to establish infection for overall virulence [16]. Apart from this, plant–pathogen interactions have been successfully studied in many areas using image-based platforms [17–21].

Plants are characterized on the basis of images taken time to time for structure and composition and processed further for analysis. These images correspond to the function of plants like biomass accumulation (yield) and chlorophyll content in leaves in different stress environments. The data are stored and can be retrieved at will along with report integration.

14.4 HTP Pipeline and Platforms

HTP platforms are designed for nondestructive analysis of different crop plants, which utilize image hardware for monitoring plant growth by taking snapshots at multiple data points and at different wavelengths (Figure 14.2). These images are analyzed by the image analysis software, and the data are transformed into a readable, reproducible format being stored in databases. Different pipelines are designed and platforms are available for HTP. The major HTP platforms include LemnaTec, Plant Accelerator, TraitMill platform, LeafAnalyser, LAMINA, GROWSCREEN 3D, PHENOPSIS platform, PHENODYN platform, GERMINATOR platform, and HYPOTrace. Automation and use of calibrated image analysis should provide more accurate, objective, and faster analyses than visual assessments. In India, Phenospex high-throughput field screening facility (FieldScan) is available at the International Crops Research Institute for the Semi-Arid Tropics, ICRSAT, Hyderabad. LemnaTec HTP platform

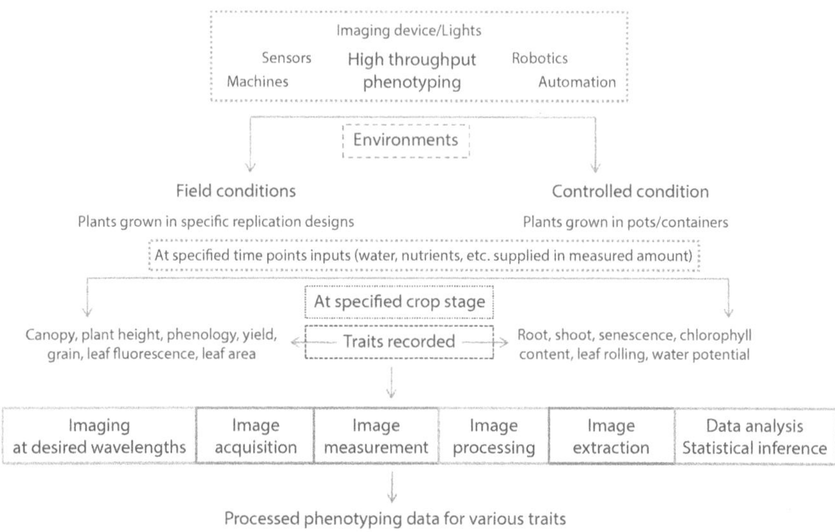

Figure 14.2 Flowchart for HTP.

is available at ICAR–Central Research Institute for Dryland Agriculture (CRIDA), Hyderabad, ICAR–Indian Agricultural Research Institute, Delhi, and ICAR–Indian Institute of Horticultural Research (IIHR), Bangalore.

High-throughput plant phenotyping platforms [41] and open-source image analysis pipelines are being utilized to measure phenotypic traits in the segregating population in different plant species [42–44].

14.5 Controlled Environment-Based Phenotyping

HTP under controlled environment is especially essential for evaluating and assessing stress situations faced by the plants in the lab or in greenhouse conditions. There is always a need to develop varieties having multiple traits of economic importance like disease resistance, tolerance to salinity, drought, cold, heat stresses for coping up with challenges of climate change, and changing dynamics of diseases/pests. Hence, there is a need to screen new germplasm, landraces, cultivars, or varieties for tolerance or resistance to particular stress. Stress imposition for phenotyping forms an integral part of such experiments, which cannot be taken care in the field conditions due to limitations in stress imposition and danger of disease/pest/insect's outbreak or spread. Moreover, specified nutrient composition, photoperiod, and temperature can be controlled as per the experimental requirements [45, 46]. For nutrient use efficiency, the plants can be phenotyped for various physiological and morphological traits like the root system, fluorescence (Fv/Fm ratio), etc. in the absence of a particular macro- and micronutrient like nitrogen or phosphorous. The agar-based media have been used for phenotyping of root traits in various crops to assess the potential of roots in overcoming the drought stress [47]. Automated disease phenotyping platforms are available wherein the disease severity can be recorded and the plant–pathogen interaction can be studied (https://www.lemnatec.com). The controlled phenotyping can be employed in studying developmental phases in plant growth in varying environments.

14.6 Field-Based High-Throughput Plant Phenotyping (Fb-HTPP)

Precise and high-throughput phenotyping under field condition is one of the major bottlenecks in accurate estimation of phenotypic effects of quantitative trait loci (QTL) controlling economically important traits, yield, biotic tolerance, and abiotic tolerance. In the beginning, high-throughput phenotypic facilities were developed to generate phenotypic data from the controlled environment such as growth chamber and greenhouses. As phenotypic data from the controlled condition are poorly correlated to phenotypic data in field condition, therefore, development of platforms for phenotyping in field condition for different crops is imperative for accurate estimation of phenotypic values. Several techniques of HTP like sensors, field robotic, and navigation system need to be integrated into new architect based on crop specificity, crop dimension, crop production, and phenology. HTP of breeding populations in field condition has a very high potential to accelerate breeding program and improve selection efficiency by precise estimation of phenotypic values. Field-based HTPP has been successfully employed in several crops for accurate estimation of phenotypic values, namely in wheat [48], cotton [49], sorghum [50], and soybean [51]. The phenomics data generated from field-based HTPP have also been used for identification of QTLs associated with various economically important traits, viz., cotton [49], maize [44], and rice [52].

Barker *et al.*, [63] developed a mobile platform to carry various sensors for high-throughput field phenotyping. The hardware and software designs were modular, allowing easy sensor addition and removal and flexible system expansion. Fernandez *et al.*, [50] developed and tested Phenobot 1.0, an auto-steered and self-propelled field-based HTP platform for tall dense canopy crops, such as sorghum (*Sorghum bicolor*). It was equipped with laterally positioned and vertically stacked stereo RGB cameras, which collected from 307 diverse sorghum lines, and images were reconstructed in 3D for feature extraction. A HTP method for scoring plant stand density in field condition was developed using spectral reflectance in wheat by Kipp *et al.*, [48].

Vehicle-based multispectral active sensor involving two sensors (GreenSeeker and CropCircle) was employed in wheat for HTP of early seedling vigor in field condition to accelerate the breeding process [48]. A total of 200 plots, 12 m in length, were measured within 75 min. A novel spectral plant vigor index (EPVI) was developed to precisely estimate the plant vigor at tillering. Digital images based on EPVI and RAGP (relative amount of green pixels) were significantly related with $r^2 = 0.98$ to each other in 2 years indicating the precision of estimated values. The findings showed that EPVI was an accurate scoring method for the high-throughput screening of large field trials. The rapidity and accuracy of this novel method may contribute to enhanced selection at early growth stages. A high-clearance tractor with a set of sensors was architected to estimate data on canopy properties including canopy temperature, reflectance, and height mapping population of 95 recombinant inbred lines in cotton (*Gossypium hirsutum* L.) under contrasting irrigation regimes in field condition to identify QTL associated with canopy traits under high temperature and water deficit condition [49]. The HTPP system precisely estimated the canopy temperature, normalized difference vegetation index (NDVI), height, and leaf area index (LAI) under well-watered and water-limited field conditions at different times on multiple days across 2010–2012. The results showed that the genomic position of some QTL controlling HTPP canopy traits was similar with those of QTL identified for agronomic and physiological traits. The novel use of a field-based HTPP system to study the genetic basis of stress-adaptive traits in cotton and the potential to accelerate development of stress-resilient cotton cultivars was demonstrated [49].

Constraints in field condition limit the precise estimation of phenotypes of quantitative traits in crop plants. Thus, development of effective field-based high-throughput technology is indeed imperative to reduce the gap between genomic and phenotypic data for accurately estimation of QTL effect and to acceleration of breeding programs.

14.7 Applications of HTP

Multidisciplinary holistic characterization of a plant in terms of integrating it with genotypic and phenotyping information results in rapid progress in breeding programs (Figure 14.3). HTP is an essential component in studying the genotype–phenotype map, identifying genetic basis of complex

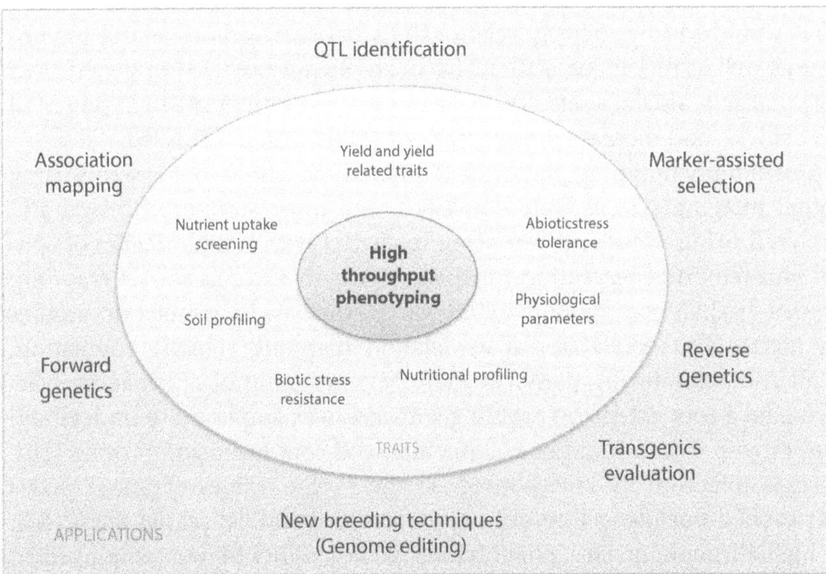

Figure 14.3 Applications of HTP.

traits, evaluation of transgene in different conditions, forward genetic studies and screening of mapping, association and mutant populations, or segregating mapping populations. There was a tremendous advancement in genotyping observed in the last few decades and supported several studies in identifying major genes and QTLs of agronomic importance. Linking traits to the associated genes and gene variants thus requires a rigorous phenotyping in every experiment either in the field or controlled conditions. Phenotyping the traits at multiple data points at various stages thus is a priori for getting tangible outcomes from the experiment.

14.7.1 Marker-Assisted Selection and QTL Detection

Mapping QTL or genes for agronomic traits and employing marker-assisted selection for desirable phenotypic variation are extremely important for modern crop breeding. As mapping and MAS linking of genotype with the phenotype, precise and large-scale phenotyping has a major role to play. Potential applications of HTP are mainly envisaged for detection of quantitative trait loci and crop improvement through marker-assisted selection. Large populations of genetic variants are being employed with the goal of sampling variation in many or all genes using well-tested and high-throughput standard operating procedures with maximum accuracy. When key features of the growth conditions are well defined, analyzed in detail, and closely monitored, it supports precise linkage mapping. Thus, HTP would ideally identify relationships between genotype and phenotype as well as reveal correlations between phenotypes [53] or genetic loci [54]. Large-scale data sets also help in increasing the statistical power of the analysis and increase the accuracy of QTLs or genes studied.

Custom phenotyping platforms to phenotype whole root systems using digital imaging system were employed, and images were processed and analyzed using novel software RootReader3D with a high degree of spatial and temporal resolution for diverse growth systems for several crop species [55]. It was demonstrated for genome-wide association studies on maize (*Zea mays*) nested association mapping (NAM) population. Plant root monitoring platform PlaRoM, consisting of an imaging platform and a root extension profiling software, was employed to understand root growth dynamics, root architecture, and root extension profiles [56]. Disease infection in leaves based on the Fv/Fm values of image pixels was carried out using threshold approach to delimit diseased areas [57]. A high-throughput rice phenotyping facility (HRPF) was employed to monitor different traits throughout growth period and after harvest in rice by combining HTP and genome-wide association studies [30]. Siddiqui *et al.*,

[57] and Hairmansis *et al.*, [58] employed infrared imaging to phenotype physiological parameters in rice under salt stress. These studies emphasized the potential of HTP over traditional phenotyping techniques and can provide valuable information on the genetic architecture of important traits.

14.7.2 Forward and Reverse Genetics

Forward and reverse genetics refers to the approaches of determining the genetic basis of phenotype and determining the function of the gene by assessing the phenotypic effects due to the genotype. HTP plays a major role in the studies to evaluate the genotypes' more rapid and precise ways and support forward phenomics and reverse phenomics [59]. Mutagenesis is carried out to the creation of allelic variations, and HTP is very much essential to phenotype thousands of such mutants in the field or under controlled conditions. RNA interference or gene knockout, transgenics, and site-directed mutagenesis create many phenotypes, which need to be selected for a particular associated trait. For functional analysis and appropriate annotation of genes, it is very much essential to understand its role in organ development or stress tolerance. Such studies would involve HTP for understanding the relationship between genotype and phenotype.

14.7.3 New Breeding Techniques

New breeding technologies include genome editing using sequence specific nucleases or engineered nucleases, viz., meganucleases, Zinc finger nucleases (ZFNs), TALENs, and CRISPR-Cas9. The intended change or mutation is precise and ranges from a single nucleotide to a few bases creating a large number of phenotypes for the desired trait of interest. Thus, it is very much essential to screen the generated plants for the intended change relating to the trait of interest [60]. Moreover, it is very much essential to co-relate the sequencing data with the phenotype to infer the linkage between intended change and the deleted/inserted/replaced site. HTP can be employed to screen such a large number of mutations for precision in results [61].

14.7.3.1 Envirotyping

Envirotyping is an upcoming area that refers to the assessment of environmental impacts on plants over time, i.e., developmental stage [62]. Multiple environmental data in terms of crop growth as influenced by soil, climate, biotic/abiotic factors, and cropping system are very much essential for future breeding and crop modeling. For envirotyping over multiple environments, it is very much essential to phenotype on a high-throughput level to co-relate the genotype, phenotype, environment, and time. HTP is the heart of envirotyping, and phenotyping data over multiple environments can be very well used for prediction of crop performance in the target environment.

14.8 Conclusion and Future Thrust

HTP is becoming an area of high priority research in the era of genomics especially to utilize the massive genomics generated by next sequencing technologies. The precision, accuracy, sensitivity, and reproducibility make HTP central to crop improvement and a significant tool for genomics. The HTP platforms can be well utilized for detecting the quantitative trait loci, fine mapping, linkage analysis, and studying various traits functional annotation of genes. Still, there are scopes for improving the controlled and field-based phenotyping, but with the advent of newer technologies, algorithms, software, mechanization, and drones, the gap is expected to bridge soon. Especially, the plants generated through genome editing, which is very much in vogue nowadays, can be phenotyped on a large scale. Networks have already been developed to efficiently use such facilities, and researchers should take benefit of it by phenotyping the plant germplasm.

References

1. Araus, J.L., Kefauver, S.C., Zaman-Allah, M., Olsen, M.S., Cairns, J.E., Translating high-throughput phenotyping into genetic gain. *Trends in Plant Sci.*, 23, 5, 451–466, 2018.
2. Dhondt, S., Wuyts, N., Inze, D., Cell to whole-plant phenotyping: The best is yet to come. *Trends Plant Sci.*, 18, 428–439, 2013.
3. Yang, N., Lu, Y., Yang, X., Huang, J., Zhou, Y., Ali, F., Wen, W., Liu, J., Li, J., Yan, J., Genome wide association studies using a new nonparametric model reveal the genetic architecture of 17 agronomic traits in an enlarged maize association panel. *PLoS Genet.*, 10, 9, e1004573, 2014.
4. Gutierrez, M., Reynolds, M.P., Klatt, A.R., Association of water spectral indices with plant and soil water relations in contrasting wheat genotypes. *J. Exp. Bot.*, 61, 3291–3303, 2010.

5. Grobkinsky, D.K., Svensgaard, J., Christensen, S., Roitsch, T., Plant phenomics and the need for physiological phenotyping across scales to narrow the genotype-to-phenotype knowledge gap. *J. Exp. Bot.*, 66, 18, 5429–5440, 2015.
6. Houle, D., Govindaraju, D.R., Omholt, S., Phenomics: The next challenge. *Nat. Rev. Genet.*, 11, 855–866, 2010.
7. Falconer, D.S. and Mackay, T.F.C., *Introduction to Quantitative Genetics*, 4th ed., Longmans Green, Harlow, Essex, UK, 1996.
8. Gauch, H.G. and Zobel, R.W., AMMI analysis of yield trials, in: *Genotype-by-Environment Interaction*, M.S. Kang and H.G. Gauch (Eds.), pp. 85–122, CRC Press, Boca Raton, 1996.
9. Balakrishnan, D., Subrahmanyam, D., Badri, J., Raju, A.K., Rao, Y.V., Beerelli, K., Mesapogu, S., Surapaneni, M., Ponnuswamy, R., Padmavathi, G., Babu, V.R., Neelamraju, S., Genotype × environment interactions of yield traits in back-cross introgression lines derived from *Oryza sativa* cv. Swarna/*Oryza nivara*. *Front. Plant Sci.*, 7, 1530, 2016.
10. Yang, W., Duan, L., Chen, G., Xiong, L., Liu, Q., Plant phenomics and high-throughput phenotyping: Accelerating rice functional genomics using multidisciplinary technologies. *Curr. Opin. Plant Biol.*, 16, 180–187, 2013.
11. Heffner, E.L., Jannink, J.L., Sorrells, M.E., Genomic selection accuracy using multifamily prediction models in a wheat breeding program. *The Plant Genome*, 4, 1, 65–75, 2011.
12. Masuka, B., Araus, J.L., Das, B., Sonder, K., Cairns, J.E., Phenotyping for abiotic stress tolerance in maize. *J. Integr. Plant Biol.*, 54, 4, 238–249, 2012.
13. Sozzani, R. and Benfey, P.N., High-throughput phenotyping of multicellular organisms: Finding the link between genotype and phenotype. *Genome Biol.*, 12, 219, 2011.
14. Munns, R., James, R.A., Sirault, X.R., Furbank, R.T., Jones, H.G., New phenotyping methods for screening wheat and barley for beneficial responses to water deficit. *J. Exp. Bot.*, 61, 3499–3507, 2010.
15. Woo, N.S., Badger, M.R., Pogson, B.J., A rapid, non-invasive procedure for quantitative assessment of drought survival using chlorophyll fluorescence. *Plant Methods*, 4, 27, 2008.
16. Mutka, A.M., Fentress, S.J., Sher, J.W., Berry, J.C., Pretz, C., Nusinow, D.A., Bart, R., Quantitative, image-based phenotyping methods provide insight into spatial and temporal dimensions of plant disease. *Plant Physiol.*, 172, 650–660, 2016.
17. Mahlein, A.K., Steiner, U., Hillnhütter, C., Dehne, H.W., Oerke, E.C., Hyperspectral imaging for small-scale analysis of symptoms caused by different sugar beet diseases. *Plant Methods*, 8, 3, 2012.
18. Rousseau, C., Belin, E., Bove, E., Rousseau, D., Fabre, F., Berruyer, R., Guillaumès, J., Manceau, C., Jacques, M.A., Boureau, T., High throughput quantitative phenotyping of plant resistance using chlorophyll fluorescence image analysis. *Plant Methods*, 9, 17, 2013.

19. Bauriegel, E. and Herppich, W., Hyperspectral and chlorophyll fluorescence imaging for early detection of plant diseases, with special reference to Fusarium spec. infections on wheat. *Agric.*, 4, 32–57, 2014.
20. Baranowski, P., Jedryczka, M., Mazurek, W., Babula-Skowronska, D., Siedliska, A., Kaczmarek, J., Hyperspectral and thermal imaging of oilseed rape (*Brassica napus*) response to fungal species of the genus Alternaria. *PLoS One*, 10, e0122913, 2015.
21. Raza, S.E., Prince, G., Clarkson, J.P., Rajpoot, N.M., Automatic detection of diseased tomato plants using thermal and stereo visible light images. *PLoS One*, 10, e0123262, 2015.
22. Clark, R.T., MacCurdy, R.B., Jung, J.K., Shaff, J.E., McCouch, S.R., Aneshansley, D.J., Kochian, L.V., Three-dimensional root phenotyping with a novel imaging and software platform. *Plant Physiol.*, 156, 455–465, 2011.
23. Ikeda, M., Hirose, Y., Takashi, T., Shibata, Y., Yamamura, T., Komura, T., Doi, K., Ashikari, M., Matsuoka, M., Kitano, H., Analysis of rice panicle traits and detection of QTLs using an image analyzing method. *Breed. Sci.*, 60, 55–64, 2010.
24. Moshou, D., Bravo, C., Oberti, R., West, J., Bodria, L., McCartney, A., Ramon, H., Plant disease detection based on data fusion of hyper-spectral and multi-spectral fluorescence imaging using Kohonen maps. *Real-Time Imaging*, 11, 75–83, 2005.
25. Bürling, K., Hunsche, M., Noga, G., Quantum yield of non-regulated energy dissipation in psii (y (no)) for early detection of leaf rust (*Puccinia triticina*) infection in susceptible and resistant wheat (*Triticum aestivum* L.) cultivars. *Prec. Agric.*, 11, 703–716, 2010.
26. Manickavasagan, A., Jayas, D., White, N., Thermal imaging to detect infestation by *Cryptolestes ferrugineus* inside wheat kernels. *J. Stored Prod. Res.*, 44, 186–192, 2008.
27. Jones, H.G., Serraj, R., Loveys, B.R., Xiong, L., Wheaton, A., Price, A.H., Thermal infrared imaging of crop canopies for the remote diagnosis and quantification of plant responses to water stress in the field. *Funct. Plant Biol.*, 36, 978–989, 2009.
28. Bolon, Y.T., Haun, W.J., Xu, W.W., Grant, D., Stacey, M.G., Nelson, R.T., Gerhardt, D.J., Jeddeloh, J.A., Stacey, G., Muehlbauer, G.J., Phenotypic and genomic analyses of a fast neutron mutant population resource in soybean. *Plant Physiol.*, 156, 240–253, 2011.
29. Cook, J.P., McMullen, M.D., Holland, J.B., Tian, F., Bradbury, P., Ross-Ibarra, J., Buckler, E.S., Flint-Garcia, S.A., Genetic architecture of maize kernel composition in the nested association mapping and inbred association panels. *Plant Physiol.*, 158, 824–834, 2012.
30. Yang, W., Guo, Z., Huang, C., Duan, L., Chen, G., Jiang, N., Fang, W., Feng, H., Xie, W., Lian, X., Wang, G., Luo, Q., Zhang, Q., Liu, Q., Xiong, L., Combining high-throughput phenotyping and genome-wide association

studies to reveal natural genetic variation in rice. *Nat. Commun.*, 5, 5087, 2014.
31. Busemeyer, L., Mentrup, D., Möller, K., Wunder, E., Alheit, K., Hahn, V., Maurer, H.P., Reif, J.C., Würschum, T., Müller, J., Breedvision—A multi-sensor platform for non-destructive field-based phenotyping in plant breeding. *Sensors*, 13, 2830–2847, 2013.
32. Klose, R., Penlington, J., Ruckelshausen, A., Usability study of 3D time-of-flight cameras for automatic plant phenotyping. *Bornimer Agrartechnische Berichte*, 69, 93–105, 2009.
33. Biskup, B., Scharr, H., Schurr, U., Rascher, U., A stereo imaging system for measuring structural parameters of plant canopies. *Plant Cell Environ.*, 30, 1299–1308, 2007.
34. Paulus, S., Behmann, J., Mahlein, A.K., Plümer, L., Kuhlmann, H., Low-cost 3D systems: Suitable tools for plant phenotyping. *Sensors*, 14, 3001–3018, 2014.
35. Fang, S., Yan, X., Liao, H., 3D reconstruction and dynamic modeling of root architecture *in situ* and its application to crop phosphorus research. *Plant J.*, 60, 1096–1108, 2009.
36. Hillnhütter, C., Sikora, R., Oerke, E.C., van Dusschoten, D., Nuclear magnetic resonance: A tool for imaging belowground damage caused by *Heterodera schachtii* and *Rhizoctonia solani* on sugar beet. *J. Exp. Bot.*, 63, 319–327, 2012.
37. Poorter, H., Bühler, J., van Dusschoten, D., Climent, J., Postma, J.A., Pot size matters: A meta-analysis of the effects of rooting volume on plant growth. *Funct. Plant Biol.*, 39, 839–850, 2012.
38. Rascher, U., Blossfeld, S., Fiorani, F., Jahnke, S., Jansen, M., Kuhn, A.J., Matsubara, S., Märtin, L.L., Merchant, A., Metzner, R., Non-invasive approaches for phenotyping of enhanced performance traits in bean. *Funct. Plant Biol.*, 38, 968–983, 2011.
39. Karunakaran, C., Jayas, D., White, N., Detection of internal wheat seed infestation by *Rhyzopertha dominica* using x-ray imaging. *J. Stored Prod. Res.*, 40, 507–516, 2004.
40. Garbout, A., Munkholm, L.J., Hansen, S.B., Petersen, B.M., Munk, O.L., Pajor, R., The use of pet/ct scanning technique for 3D visualization and quantification of real-time soil/plant interactions. *Plant Soil*, 352, 113–127, 2012.
41. Honsdorf, N., March, T.J., Berger, B., Tester, M., Pillen, K., High-throughput phenotyping to detect drought tolerance QTL in wild barley introgression lines. *PLoS ONE*, 9, 5, e97047, 2014.
42. Klukas, C., Chen, D., Pape, J.M., Integrated analysis platform: An open-source information system for high-throughput plant phenotyping. *Plant Physiol.*, 65, 2, 506–518, 2014.
43. Muraya, M.M., Chu, J., Zhao, Y., Junker, A., Klukas, C., Reif, J.C., Altmann, T., Genetic variation of growth dynamics in maize (*Zea mays* L.) revealed

through automated non-invasive phenotyping. *Plant J.*, 2017, https://doi.org/10.1111/tpj.13390.
44. Zhang, X., Huang, C., Wu, D., Qiao, F., Li, W., Duan, L., Wang, K., Xiao, Y., Chen, G., Liu, Q., Xiong, L., Yang, W., Yan, J., High-throughput phenotyping and QTL mapping reveals the genetic architecture of maize plant growth. *Plant Physiol.*, 173, 1554–1564, 2017.
45. Kuijken, R.C., van Eeuwijk, F.A., Marcelis, L.F., Bouwmeester, H.J., Root phenotyping: From component trait in the lab to breeding. *J. Exp. Bot.*, 66, 18, 5389–5401, 2015.
46. Barbadikar, K.M., Phule, A.S., Madhav, M.S., Senguttuvel, P., Subrahmanyam, D., Balachandran, S.M., Kumar, A.P., Water scarcity: Driving force for root studies in rice. *ICAR-IIRR Newsletter*, 14, 1, 7, 2016.
47. Singh, A., Ganapathysubramanian, B., Singh, A.K., Sarkar, S., Machine learning for high-throughput stress phenotyping in plants. *Trends in Plant Sci.*, 21, 2, 2016.
48. Kipp, S., Mistele, B., Baresel, P., Schmidhalter, U., High-throughput phenotyping early plant vigour of winter wheat. *Eur. J. Agron.*, 52, 271–278, 2014.
49. Pauli, D., Andrade-Sanchez, P., Carmo-Silva, A.E., Gazave, E., French, A.N., Heun, J., Hunsaker, D.J., Lipka, A.E., Setter, T.L., Strand, R.J., Thorp, K.R., Wang, S., White, J.W., Gore, M.A., Field-based high-throughput plant phenotyping reveals the temporal patterns of quantitative trait loci associated with stress-responsive traits in cotton. *G3 (Bethesda)*, 6, 865–879, 2016.
50. Fernandez, M.G.S., Bao, Y., Tang, L., Schnablea, P.S., A High-throughput, field-based phenotyping technology for tall biomass crops. *Plant Physiol.*, 174, 2008–2022, 2017.
51. Bai, G., Ge, Y., Hussain, W., Baenziger, P.S., Graef, G., A multi-sensor system for high throughput field phenotyping in soybean and wheat breeding. *Comput. Electron. Agric.*, 128, 181–192, 2016.
52. Tanger, P., Klassen, S., Mojica, J.P., Lovell, J.T., Moyers, B.T., Baraoidan, Naredo, M.E.B., McNally, K.L., Poland, J., Bush, D.R., Leung, H., Leach, J.E., McKay, J.K., Field-based high throughput phenotyping rapidly identifies genomic regions controlling yield components in rice. *Sci. Reports*, 7, 42839, 2017.
53. Schauer, N., Semel, Y., Roessner, U., Gur, A., Balbo, I., Carrari, F., Pleban, T., Perez-Melis, A., Bruedigam, C., Kopka, J., Willmitzer, L., Zamir, D., Fernie, A.R., Comprehensive metabolic profiling and phenotyping of interspecific introgression lines for tomato improvement. *Nat. Biotechnol.*, 24, 4, 447–454, 2006.
54. Gerke, J., Lorenz, K., Ramnarine, S., Cohen, B., Gene–environment interactions at nucleotide resolution. *PLoS Genet.*, 6, 9, e1001144, 2010.
55. Clark, R.T., Famoso, A.N., Zhao, K., Shaff, J.E., Craft, E.J., Bustamante, C.D., McCouch, S.R., Aneshansley, D.J., Kochian, L.V., High-throughput two-dimensional root system phenotyping platform facilitates genetic analysis of root growth and development. *Plant Cell Environ.*, 36, 2, 454–66, 2012.

56. Yazdanbakhsh, N. and Fisahn, J., High throughput phenotyping of root growth dynamics, lateral root formation, root architecture and root hair development enabled by PlaRoM. Funct. *Plant Biol.*, 36, 938–946, 2009.
57. Siddiqui, Z.S., Cho, J.I., Park, S.H., Kwon, T.R., Ahn, B.O., Lee, G.S., Jeong, M.J., Kim, K.W., Lee, S., Park, S.C., Phenotyping of rice in salt stress environment using high-throughput infrared imaging. *Acta. Bot. Croat.*, 73, 1, 149–158, 2014.
58. Hairmansis, A., Berger, B., Tester, M., Roy, S.J., Image-based phenotyping for non-destructive screening of different salinity tolerance traits in rice. *Rice*, 7, 16, 2014.
59. Camargo, A.V. and Lobos, G.A., Latin America: A development pole for phenomics. *Front. Plant Sci.*, 7, 1729, 2016.
60. Schaart, J.G., van de Wiel, C.C.M., Lotz, L.A.P., Smulders, M.J.M., Opportunities for products of new plant breeding techniques. *Trends Plant Sci.*, 21, 5, 438–449, 2016.
61. Rani, R., Yadav, P., Barbadikar, K.M., Baliyan, N., Malhotra, E.V., Singh, B.K., Kumar, A., Singh, D., CRISPR/Cas9: A promising way to exploit genetic variation in plants. *Biotechnol. Lett.*, 38, 12, 1991–2006, 2016.
62. Xu, Y., Envirotyping for deciphering environmental impacts on crop plants. *Theor. Appl. Genet.*, 129, 4, 653–673, 2016.
63. Barker, J. III., Zhang, N., Sharon, J., Steeves, R., Wang, X., Wei, Y., Poland, J., Development of a field-based high-throughput mobile phenotyping platform. *Comput. Electron. Agric.*, 122, 74–85, 2016, https://doi.org/10.1016/j.compag.2016.01.017

Index

Lightning Source UK Ltd.
Milton Keynes UK
UKHW020310230219
337627UK00001B/2/P

9 781119 509936